PENGUIN BOOKS

THE DIVERSITY OF LIFE

'I was moved by Ed Wilson's *The Diversity of Life*. No other book in this or recent years so fulfilled my yearning for a thoughtful and readable text on natural history' James Lovelock, *The Times Literary Supplement*

'An important book ... Wilson celebrates the complexity of evolution precisely to emphasize the extent of the catastrophe we face' Christopher Lehmann-Haupt, *The New York Times*

'What is the biodiversity of which Professor Wilson writes with such conviction and authority? What is the scope of it? How did it originate? Where is it found? How are humans diminishing it? And why should it matter that humans are diminishing it? Professor Wilson offers answers to all these questions in engaging and non-technical prose ... [full of] original and fascinating insights' John Terborgh, *New York Review of Books*

'A passionate defence of life's variety written, at its best, in the dispassionate terms of a master of scientific ecology' Steve Jones, *London Review of Books*

'His prose is assured, calm and classical; in addition to theoretical explanation, he is capable of writing evocatively about his own experiences in the field. The result is canonical authority with a human face' Marek Kohn, *New Statesman*

'Wilson's passion for the beauty and mystery of nature, coupled with his adherence to scientific methods and his unsurpassed professional standing, give the work the possibility of being the most important book since Rachel Carson's *Silent Spring*' Charles A. Radin, *Boston Globe*

ABOUT THE AUTHOR

Edward O. Wilson is a University Research Professor and Honorary Curator in Entomology at the Museum of Comparative Zoology, Harvard University. A native of Alabama, he has been a member of the Harvard faculty since 1956. His field research has taken him to countries all over the world. His many contributions to our understanding of the biological world include the books *Sociobiology*; *The Insect Societies*; *Biophilia*; *On Human Nature*, winner of the Pulitzer Prize; *Naturalist*, winner of the *Los Angeles Times* Book Prize; and *In Search of Nature*. *On Human Nature*, *Naturalist* and *In Search of Nature* are also published in Penguin. Edward O. Wilson is also the co-author, with Bert Hölldobler, of the Pulitzer Prize-winning book, *The Ants*. His many scientific awards include the National Medal of Science, the International Prize for Biology from Japan, the Crafoord Prize of the Royal Swedish Academy of Sciences and the King Faisal International Prize for Science from Saudi Arabia.

The Diversity of Life was awarded the Sir Peter Kent Conservation Book Prize for the best book published on environmental issues in 1993.

EDWARD O. WILSON

The
Diversity
of Life

PENGUIN BOOKS

To my mother
Inez Linnette Huddleston
in love and gratitude

PENGUIN BOOKS

Published by the Penguin Group
Penguin Books Ltd, 27 Wrights Lane, London W8 5TZ, England
Penguin Putnam Inc., 375 Hudson Street, New York, New York 10014, USA
Penguin Books Australia Ltd, Ringwood, Victoria, Australia
Penguin Books Canada Ltd, 10 Alcorn Avenue, Toronto, Ontario, Canada M4V 3B2
Penguin Books India (P) Ltd, 11, Community Centre, Panchsheel Park, New Delhi – 110 017, India
Penguin Books (NZ) Ltd, Private Bag 102902, NSMC, Auckland, New Zealand
Penguin Books (South Africa) (Pty) Ltd, 5 Watkins Street,Denver Ext 4, Johannesburg 2094, South Africa

Penguin Books Ltd, Registered Offices: Harmondsworth, Middlesex, England

First published in the USA by The Belknap Press of Harvard University Press 1992
First published in Great Britain by Allen Lane The Penguin Press 1993
Published in Penguin Books 1994
Reprinted with a new Foreword 2001
1

Contents

Biodiversity at the Start of the New Century

T WELVE YEARS have swept by since the multiple-authored book *BioDiversity* introduced the title word into the English language, and eight years since the appearance of the work before you, *The Diversity of Life*. In that decade conservation biology, the new discipline explained by these works, has grown at an explosive rate. By 1992 biodiversity was well enough known to be a central concern of the Earth Summit held in Rio de Janeiro. The global Convention on Biodiversity drafted at that conference has been ratified by almost 170 countries. Biodiversity courses are now taught at innumerable colleges and universities in the United States and elsewhere. Natural history museums have rewritten their agendas to focus on the study and conservation of ecosystems and biotas. And conservation organizations routinely base their programs on conservation biology; in response to the scientific content of the subject, they count their successes not just by the salvaging of tigers and eagles but also the protection of entire ecosystems that harbor such star species.

Humanity Versus the Natual World. What then is the status of biodiversity as this new century begins? Without doubt, worldwide awareness of its problems and promise has grown dramatically. Many countries, including tropical "megadiversity" nations such as Brazil, Colombia, and Indonesia, have added reserves and adjudicated—in principle at least—the practice of biological conservation as part of national policy. There is some justification for optimism, guarded by realism and shadowed by sensible apprehension, that the world is turning the corner in at least its attitude toward the rest of life.

Still, those who monitor the diversity of life are especially apprehensive because with all the good intentions of many scientists and policy makers, the growth of human population and the depletion of natural resources continues unabated. The prospects for biodiversity can be summarized by the following imagery of the bottleneck. The world's human population will increase by about a third before peaking within a century or so, then commence a slow decline. If the number at maximum is not much greater than 8 billion, everyone can, in theory at least, be housed and fed. However, the already intense pressures on the last remnants of wild biodiversity might easily grow fatal for a majority of the remaining ecosystems and their distressed species of plants and animals. The only way to carry biodiversity safely through the bottleneck of this critical period is by a combination of scientific and technological innovation, abatement of population growth, and environmental education, guided by a redirection of moral purpose.

Consider some of the basic facts and projections that define the bottleneck. The world population is now passing 6 billion. The average number of children produced per woman is about 2.6 and falling. To achieve zero population growth the average number of children per woman must drop to 2.1. (The extra tenth of a child, or one child per ten women, compensates for child mortality.) If the 2.1 level is reached very soon and held, there will be 7.7 billion people on Earth by 2050, leveling off at 8.5 billion in 2150. If a slight negative growth is reached, at 2.0 children per woman, the world population will peak at 7.8 billion and drop to 5.6 billion, a bit below the current level, also by 2150. To these estimates, unfortunately, must be added an even more worrisome estimate: if the replacement level is held even slightly above 2.1, the world population will continue to soar. Thus, only 2.2 children per woman will yield 12.5 billion by 2050, a potentially catastrophic doubling of present levels. In general, whatever the magnitude of increase, most of it will occur in the developing world, which is already economically impoverished.

Why would a doubling of the global population be so bad? After all, average per capita production is rising worldwide, a statistic often cited cheerfully by bottom-line economists. The reason for concern is that the increase is being achieved by an accelerating consumption of Earth's nonrenewable capital. Natural-resources experts generally agree that we are running out of arable land and water as a result of rising total demand, and that the finiteness of land and water represents a wall beyond which humanity cannot pass. There are no new continents to colonize, and displacement of other people by conquest and genocide is now generally frowned upon. As a species we are at long last approaching the Malthusian limit. The amount of cropland available per person, having dropped from 0.23 hectares in 1950 to 0.12 hectares in the mid-1990s, is continuing downward. The productivity per hectare has been elevated by fertilizer, Green-Revolution genetic improvement of crop strains, and other technological innovations, but the rate of improvement is diminishing, and its maintenance also depends on water, the other principal declining resource. Meanwhile, the population is shooting upward in most countries.

To the strain on the food and water resources can be added the aftershock of personal economic striving that follows the population explosion. Billions of people are not only producing children in excess of zero population growth but also taking every means available to improve their own and their family's personal security and welfare. The true measure of the human impact on Earth's environment is therefore not in pure population numbers but in the "ecological footprint," the amount of productive land needed to supply each person with food, water, energy, living space, commerce, transportation, and waste management. The footprint is far more than just land that each person uses directly. It is also composed of small pieces obtained from many places, some even from other continents—where food is grown, petroleum drawn, clothing manufactured, roads built, waste dumped, and public buildings raised. The ecological footprint of the average person in the developing countries is half a hectare; that of one citizen of the United States is 5 hectares. For the entire current world population to pull itself up to U.S. levels with existing technology would require two more planet Earths.

That amount of bootstrapping, of course, is not going to happen. The greatest challenge of the twenty-first century is to settle humanity down and accommodate 8 to 10 billion people with a decent standard of living before they wreck the planet. Meanwhile, the last remaining natural environments—the world's forests, grasslands, semideserts, and wetlands—are shrinking before the onslaught. Humanity's responsibility to the rest of life and to future generations is clear: bring with us as much of the environment and biodiversity through the bottleneck as possible.

Viewed this way, the study of biodiversity in all its aspects, from its origin, distribution, and extinction to its use, management, and preservation, is of paramount importance to human affairs. Here then is a brief review of some subjects in this field in which significant progress has been made since the 1992 publication of *The Diversity of Life*.

The Evolution of the Species Concept. Biologists still find it useful to divide living diversity into a hierarchy of three levels: ecosystem, species, and gene. An ecosystem is a local community of species (plants, animals, and microorganisms) plus their physical environment. Familiar examples include a pond, an estuary, and an alpine grassland atop a forested mountain. Because ecosystems are so often hard to delimit, and because genes are difficult to identify and count, the unit of choice in biodiversity studies remains the species, which is relatively easy to diagnose and has moreover been the central object of research for over two hundred years.

Throughout most of the twentieth century, the prevailing definition of the species has been embodied in the "biological species concept," which holds that a species is most usefully defined as a population or series of populations of organisms capable of freely interbreeding with one another under natural conditions. In short, a species is a closed gene pool that perpetuates itself *in nature*. The production of hybrids in gardens and zoos does not justify combining two species in classification if they stay apart in nature. Unfortunately, as concrete as the biological species concept appears in abstract, it often fails in practice. For example, it cannot be applied to the large number of organisms that reproduce asexually. During the 1990s, dissatisfaction with the concept resulted in the promotion of an alternative category, the phylogenetic species. In this perspective, the most meaningful species is a genetically distinct population with a monophyletic lineage; that is, all members are derived from a single ancestral species. It is of little concern if the populations thus delimited indiscriminately breed with one another in nature. So long as the delimited population comprises individuals of the same coherent lineage that are distinguishable to a mutually agreed-upon amount from those of other populations, it can be ranked as a species by the purely phylogenetic definition.

The contest between the two opposing yet internally consistent species concepts creates a dilemma. Both criteria they respectively use are desirable. The advantage of the biological species concept is its recognition that closed gene pools are entities which for the most part have been irreversibly launched on an independent course of evolution. Given enough time, closed gene pools at first barely distinguishable from other closed pools are destined to diverge genetically and eventually to be very distinguishable. In contrast, the advantage of the phylogenetic species concept is that it reflects rigorously the past history of the lineages of groups of related species without reference to their reproductive status or hypothetical future history.

The cross-cutting criteria of interbreeding posed by the two concepts are actually not contradictory. They can be joined to create a synthetic species concept as follows: *a sexual species is a population that is both reproductively isolated in nature and monophyletic*. Suppose a given monophyletic sexual population is geographically—physically, not genetically—isolated from other, similar populations (as it would be, for example, if it were on an island) so that its ability to hybridize or to avoid hybridization is indecipherable. Its status is thus subjective. The population can be called a species if it is markedly different, or it can be called a subspecies (that is, a geographical race) if it is only slightly different. The same subjective criteria can be applied for populations that reproduce asexually. In short, the test of reproductive isolation is desirable but not crucial.

The new emphasis on monophyly provides greater precision and reliability in reconstructing the ancestry of species but does not constitute a fundamental shift away from the species concept already used by most practicing taxonomists. As a rule, specialists on the classification of organisms have embraced both reproductive isolation and monophyly for most of this century while conceding that the assignment of reproductive relationships among closely related but geographically isolated populations is little more than educated guesswork.

Why does all this semantic juggling matter? Because the current trend of the theory of classification is toward a higher degree of objectivity and consensus than existed in the past. Thanks to new techniques in molecular biology and statistical inference, a truly synthetic biological species concept, providing considerable information about each genetically distinguishable population, seems attainable. That goal is in turn pivotally important in ecology and conservation biology. How species are classified determines the number recognized, hence our sense of the magnitude of both local and global biodiversity. It affects the evaluation of the status of individual populations in conservation planning, in particular whether populations are ranked as species, subspecies, or unnamed entities. And finally, the refined species concept conforms more closely to the emerging picture of how biodiversity is created by evolution.

How many species? The estimates I made in 1986 and cited in *The Diversity of Life* (1992) put the number of recognized living species in the world—in other words, those formally described and bearing two-part scientific names—at approximately 1.4 million. About 13,000 additional "new" species are recognized each year. Thus in the decade since 1986 more than 100,000 species have been added, bringing the total as I counted it to 1.5 million. Meanwhile, the numbers in some important groups have been revised upward—in particular in the insects, the largest of all groups, from 751,000 in the 1980s to 865,000 in 1998. A similar elevation has been made for the fungi, from 47,000 to 69,000 species. A commonly cited total world figure in the late 1990s, suggested for example in the *Global Biodiversity Assessment* of the United Nations Environment Programme (1995), is 1.75 million species. But this does not take into account the number of formal species names that have been erroneously applied to species named by earlier investigators, requiring an eventual reduction of the global number by 10 or even 20 percent.

The bottom line is that only a very rough estimate can be made of the number of recognized species in the world. It is reasonable on the basis of existing evidence to suppose that the figure lies somewhere between 1.5 and 1.7 million. Until a central register is established, however, with a census of the biodiversity of all of the groups of plants, animals, and microorganisms entered and constantly updated, the exact figure will remain elusive.

Even more distant, with a margin of uncertainty a hundred times greater, is the true number of species on Earth, both those described and the far larger number undescribed. As cited by the *Global Biodiversity Assessment* in 1995, estimates made during the past two decades range from 3,635,000, which is the sum of the lowest estimates for each group in turn by various authors, to 111,655,000, the sum of the highest estimates of each group by the same authors. A large part of the latter figure, 100 million, comes from the insects. Most entomologists now agree that this maximum number for insects is much too high, perhaps tenfold in excess of the actual number. "Working figures" for all groups, including insects, have tended to fall close to 10 million; the *Global Biodiversity Assessment* number, which constitutes no more than an educated guess and leans to conservative sentiment, is 13,620,000.

At first glance there may be a temptation to regard such counting and extrapolation as just numerology, playing with numbers for their own sake. But the ongoing attempt has an important practical purpose. It helps to identify the blank spaces on the biodiversity map, and it points the way to original and productive research. For example, as I write, about 98 percent of the living bird species of the world are believed to be known. In sharp contrast, one authority has reckoned that only 1.5 percent (15,000 out of 1 million) chromophyte algal species have been described. The past ten years have witnessed the discovery of large pools of species diversity in previously obscure or unknown groups, as illustrated by the following two examples:

• One third of all recognized species of organisms are parasites. Most higher organisms with larger bodies, including insects, have species that live in or on them. Further, a majority of the parasites are highly specialized for one or several host species. Because they are typically hidden as well, and often do not inflict obvious harm on the hosts, parasite species are seldom recognized until the host species themselves have been well studied. An extreme example, of many that can be cited, is provided by the mycoplasmas. In addition to being among the smallest of the bacteria, mycoplasmas lack cell walls, making them exceptionally hard to detect. Yet even a modest amount of research has yielded a wide array of species living in the tissues of insects and plants. Suppose that each free-living species had on average one parasite of such an obscure group limited to it, a modest assumption. Then the global count must be increased by many millions of undiscovered species.

• Bacteria continue to be the "black hole" of biodiversity, their depths unplumbed. That ignorance will not last indefinitely. The subject has been opened dramtically in the past decade by new molecular technology. Through most of the history of microbiology, perhaps fewer than 1 percent of bacterial species could be successfully cultured in large enough numbers to allow analysis by anatomical study and standard chemical tests. Now systematists are able to bypass the petri dish in favor of direct genetic analysis. A favored current method is to sequence the 165 rRNA segment, comprising about 1,500 base pairs whose variation allows not only the separation of the species but an estimation of their evolutionary relationships.

As a consequence of these advances, discoveries concerning the diversity of bacteria, as well as of archaea, the other major group of microbes, are now expanding swiftly. It turns out, for example, that while some bacterial species are widespread, especially those in the open sea, others are quite local in distribution in the manner typical of terrestrial animals and plants. Since bacteria and archaea are so small, with a million or more representing thousands of species to be found in a single gram of soil or of marine silt, their potential variety is mindboggling. Their prior obscurity was such that even common and widespread species are being recognized for the first time. For example, the extremely small members of the chlorophyll-bearing genus *Prochlorococcus*, discovered only in the late 1980s, are now recognized as one of the most abundant elements of oceanic plankton. Some newly found varieties of bacteria, called lithoautotrophs, survive on inorganic materials inside solid rock over 3 kilometers below the Earth's surface.

To summarize the prospect for biological exploration, Earth is biologically still a mostly unexplored planet. The opportunities for basic research, which promise us great scientific and practical rewards, seem limitless. Their pursuit will help make the next hundred years the Century of the Environment.

Extinction and Hot Spots. The extinction process as a biological phenomenon has assumed major importance in present-day studies of biodiversity. In 1995 the National Biological Service of the U.S. Department of the Interior (an agency subsequently dismantled to satisfy antienvironmental forces in the House of Representatives) released the first analysis of the status of ecosystems in the United States as a booklet, *Endangered Ecosystems of the United States: A Preliminary Assessment of Loss and Degradation.* Virtually all types of ecosystems, it reported, have been diminished by human activity. The greatest losses have occurred in wetlands and forests. Many of America's grasslands, savannas, and barrens communities have been cut to 2 percent or less of their original cover, with a corresponding threat to the native species living in them. Other ecosystem types reduced to an extreme degree are old-growth stands in the eastern deciduous forests, longleaf pine forests of the southeastern coastal plain, Atlantic white-cedar stands, and freshwater streams in the Mississippi alluvial plain. Surveys in other countries report comparable degrees of loss. By the late 1980s, for example, about three quarters of the world's original forest had been destroyed, including 50 percent of rain forests in both tropical and temperate regions.

Conservation biologists have continued to refine measures of endangerment in both ecosystems and species. The system for species introduced in 1994 by the International Union for Conservation of Nature and Natural Resources (IUCN), the source of the Red Data Books and Red Lists, is worth summarizing here for its authority and degree of documentation:

Extinct: There is no reasonable doubt that the last individual of the species (or subspecies) has died.

Extinct in the wild: The species survives only in cultivation, captivity, or in environments away from the species' native range.

Critically endangered: The species is at extremely high risk of extinction in the wild in the immediate future, with continued survival very unlikely.

Endangered: The species is at very high risk of extinction in the near future; in many cases, however, it is salvageable by conservation efforts.

Vulnerable: The species comprises shrinking populations, and extinction is a good possibility in the near or medium-range future.

Lower risk: The species is evidently safe for the time being, even if dependent on current conservation measures.

Around the world, biologists, like doctors in a critical-care ward, are monitoring the descent of selected individual species through the above categories on their road to extinction. Special interest is focused on the last survivors of vertebrate and plant species and how to save them. Among the many extremely rare, critically endangered species in the world can be counted the Catalina mountain mahogany, down to 6 adult trees in a gully on the southwest side of Santa Catalina Island, California; the Seychelles magpie robin, increasing slowly and at last report comprising 61 individuals on two islands; and four species of honeycreeper, among the most distinctive species of Hawaiian birds, limited to a scrap of forest on the Hawaiian island of Maui and represented by no more than several individuals, which by the time you read this have likely perished.

Experts generally agree that on a worldwide basis the causes of extinction, which are virtually all due to human activity, can be rank-ordered from the top down as follows: habitat destruction or degradation, the spread of exotic (nonnative) species, pollution, overharvesting, and disease. The data that measure the factors endangering U.S. species, as compiled by David S. Wilcove and his co-workers in 1998 (*BioScience*, 48(8): 607-615), are habitat loss, 88 percent; exotics, 46 percent; pollution, 20 percent; overharvesting, 14 percent; and disease, 2 percent. These numbers add up to more than 100 percent because the destructive factors seldom work alone. As habitat shrinks, for example, alien species and human hunters penetrate the interiors more easily, pollution reaches closer to the core populations, and disease strikes more easily to decimate the survivors.

What then is the rate of extinction? For local populations extinction can be total or nearly so. Virtually all the songbirds of Guam have been wiped out by the voracious brown tree snake, an exotic introduced from somewhere within its native range in the Solomon Islands, Papua New Guinea, and northern Australia shortly after the Second World War. Half of the 300 or more endemic cichlid fish species of Lake Victoria have been terminated by pollution and, especially, the introduction of the predatory Nile perch in midcentury and surface-choking water hyacinth decades earlier. On a global scale, the rates of extinction cannot yet be measured in absolute numbers except for exceptionally well-studied groups such as birds, mammals, and flowering plants. But the means exist to make rough indirect estimates of the *percentage* of ongoing extinctions. For example, studies conducted since 1988 by independent researchers using different techniques have variously estimated the rate of species extinction per decade to lie between 1 and 10 percent; the average of all the estimates combined is 6 percent.

Most local and global estimates, other than those taken from detailed direct observations of the fate of particular species, are based on the relation between the area of a habitat and the sustainable number of species inhabiting it. When the habitat is reduced in size, as when a forest is progressively clear-cut, the number of species declines in a predictable fashion. This reduction varies according to the kind of organism and habitat from about the sixth root to the third root of the reduction of area. A central and very commonly encountered value is a decline equal to the fourth root of area reduction. Its magnitude can be stated most clearly by the following rule of thumb: a loss of 90 percent in area of the habitat—that is, to 10 percent of the original cover—results in a reduction of 50 percent in the number of species that the habitat can indefinitely support.

Substantial reductions in habitat area, many beyond 90 percent, have already occurred in numerous places around the world. Recent studies have shown that they unfortunately tend to be concentrated in those regions where biodiversity is richest. The relation between shrinkage of habitat area and the loss of species, either through reduction of populations to critically endangered and endangered levels (hence "commitment" to early extinction) or outright extinction, has been well substantiated in many studies on the effects of deforestation and other habitat conversion beyond those described in *The Diversity of Life* in 1992. They include birds in the forests of the eastern United States, Indonesia, and the Philippines, and mammals in the national parks of western North America.

This area effect is not due solely to the loss of living space. Fragmentation alone—breaking available space into pieces—has its own impact, by increasing the length of the edge of the habitat patches relative to their area. Recent studies on the Amazonian rain forest, which is now being both clear-cut and fragmented at an alarming rate, have determined that the physical environment of the remaining forest is altered as much as 100 meters in from the edge, by drying and greater exposure to wind damage. Trees become more fragile, their biomass declines, and deep-forest plant and animal species disappear. Parallel monitoring in forests in the midwestern United States has revealed that populations of native birds decline as predators and parasites such as cowbirds penetrate more efficiently to their nesting places.

A major development in conservation biology in the 1990s was the growing awareness and closer examination of the destructive effects of exotic species. A freshet of articles, together with appositely titled books such as *Strangers In Paradise* (Daniel Simberloff et al., eds., Island Press, 1997), *Alien Invasion* (Robert Devine, National Geographic Society, 1998), and *America's Least Wanted* (Bruce A. Stein and Stephanie R. Flack, eds., The Nature Conservancy, 1998), have brought the problem to the attention of a broad readership of environmental scientists and the general reading public. The problems caused by exotics for both native species and humanity are indeed severe, as these works suggest. The world's biodiversity is being homogenized at an ascending rate as more and more alien species invade even relatively natural environments. The United States, for example, is now home to red imported fire ants, ripsaw catfish, starlings, Asian tiger mosquitoes, green crabs, leafy spurge, Chinese tallow, Brazilian pepper, and hundreds of others, some beautiful and interesting but all pests. As they take hold over broad areas, they push native plants and animals back, some toward extinction, and as a consequence global biodiversity steadily declines. By the late 1990s, 11 percent of free-living plant species in the United States were exotics. The figure for Ontario was 28 percent, for the British Isles 43 percent, and for Hawaii 44 percent.

Those who consider such species potential assets should take another look. Here are a few of the environmental changes wrought by alien species in the American Great Lakes:

Sea lamprey: decline of the native lake trout

Purple loosestrife: loss of waterfowl habitat and decline of native wetland plant species

Common carp: destruction of fish and waterfowl habitat

Eurasian watermilfoil: reduction of aquatic recreational resources

Zebra mussel: fouling of hard substrate, including water pipes, and the reduction of populations of native mollusks.

On the positive side, chinook and coho salmon, both introduced into the Great Lakes, have become valuable sports fishes. But even that advantage is bought at the price of reduction of the native species on which they prey. Overall, the costs of immigrants far outweigh the rather narrow and meager benefits they bring.

A sinister but poorly understood enhancement of the extinction of species is caused by inbreeding. The process is basically simple: when populations of breeding individuals drop to low levels—say, below 100—individuals are far more likely to mate with close relatives. The genetic similarity between kin makes it more probable that their offspring will be homozygous for genes (that is, possessing two of each kind instead of just one) that are deleterious or lethal. As a result of the increased mortality and reduced fertility, the entire population is more likely than otherwise to suffer a final descent to extinction.

Recent studies of song sparrows and white-footed mice in North America have demonstrated that inbred individuals do indeed fare less well in wild populations. However, mathematical models also suggest that in very small populations, random changes in the environment and in birth and death schedules are likely to cause extinction before inbreeding comes into play as an important factor. Empirical data are not yet adequate to settle the matter one way or the other. Still, in one field study of wild Finnish populations of a butterfly, the Glanville fritillary, more inbred populations vanished than outbred but otherwise comparable populations (*Nature*, 392: 442, 1998).

The principle that endangerment and extinction are rarely caused by a single factor is illustrated by the decline of frog populations, a trend that stirred worldwide publicity in the late 1990s. For a decade precipitous drops in populations of many species had been observed in Africa, Australia, the United States, and Central and South America. At least two species, the northern gastric brooding frog of Australia and the golden toad of Costa Rica (and probably many others yet to be verified), spiraled to extinction. The catastrophe raised alarm among environmentalists and the public. The question was asked: Are frogs our canary in the mine, whose death when carried there warns of silent but deadly changes that threaten our own existence?

A flurry of activity to find the reason for the frog die-off ensued, and multiple causes were quickly discovered. In the Midwest and Great Lakes region of the United States, a rise in multiple and missing legs and other abnormalities turned out to be due to water pollutants, as expected. In at least some other parts of the world, reduction of natural habitat has played a role. The increase in ultraviolet radiation by the thinning of the protective ozone layer, leading to aborted embryonic development, has also been suggested. Another, potentially decisive factor is the spread of a newly discovered species of chytrid fungus that causes a fatal skin disease. It appears to be especially important in frog populations of Australia and Central America.

All of these known and suspected sources of rising frog mortality are enhanced if not caused outright by human activity. Even the chytrid fungus is transported by humans from one continent to another, turning up in imported captive frogs, for example, in the United States.

Why should frogs be more vulnerable than other animal groups? The answer is believed to lie in their damp, porous skin, which allows easy access to pollutants and parasites. It may also be found in their amphibious existence. Most frog species live in the water as larvae, but on land as adults. Each habitat is subject to area loss, and each contains different pollutants and disease organisms, putting frogs in double jeopardy.

Hot Spots. In the 1990s systematists and biogeographers focused more intently on biodiversity mapping and preservation. Data were mined from the "focal" groups, those already well enough known, including birds, mammals, reptiles, amphibians, fishes, flowering plants, butterflies, tiger beetles, mollusks, crayfish, and a few other smaller groups, to allow sound analysis. The result has been an increase in the number recognized and a clearer definition of the hot spots. Some geographic areas contain ecosystems that are both threatened and the home of relatively larger numbers of species found nowhere else. In addition to those cited in 1992 in *The Diversity of Life*, others highlighted are the *cerrado* or savanna of central Brazil, the Great Lakes of East Africa, and the pantanal wetlands of the South American south-central heartland. Overdue attention is now also being paid to marine ecosystems, especially coral reefs, the widely threatened "rain forests of the sea."

Some important generalizations have emerged from these collective studies. One is the frequent discordance of hot spots for different groups of organisms. In a tropical forest wilderness it is a safe bet that local areas with the maximum diversity of trees also harbor a maximum diversity or at least very high numbers of species of shrubs, herbaceous plants, birds, ants, frogs, and most other principal groups of organisms, as well as the most biodiverse river systems and lakes. But in many other regions, such as the wildlands of Madagascar and northeastern Australia, which contain a mosaic of savanna, desert, and forest patches, hot spots differ according to groups of organisms. Bees and wasps, for example, are likely to be concentrated in the arid habitats and frogs and butterflies in the woodland. Nonetheless, when the biogeographical information is put together, it is still possible to delimit hot spots, and to do so with greater precision than in previous years.

Hot-spot mapping is a practical necessity. Those who wish to restrict conservation measures in deference to pure economic expansion are prone to claim that the mapping of endangered species imperils the "human" use of large tracts of land worldwide. Farmers are encouraged to believe that the discovery of a rare species on their woodlots will turn the land into a nature reserve.

This opposition is based on a gross misapprehension. In fact, hot spots cover only a very small fraction of Earth's surface. A 1997 survey by the staff of Conservation International, for example, revealed that the 17 hottest spots of the world occupy only 1.3 percent of the land surface yet contain a fourth of the planet's terrestrial vertebrate species and 40 percent of its plants. This study and others yielding comparable conclusions have resulted in a more precise and compelling policy guideline among conservation organizations: concentrate first on saving the hot spots and the remaining wilderness areas.

With these facts in hand, added to knowledge of the many benefits that biodiversity and natural environments bring, it should be possible to enlist a strong commitment to conservation from the countries most affected. After all, it is their own biological riches that they are being asked to protect.

Why care? It has been my experience that people tend to react to the evidences of species extinction with three stages of denial.

The *first stage* is cheerful resignation: "Why worry? Extinction is natural, isn't it?" And yes, species—as is often pointed out—have been dying through the more than 3.5 billion years of biological history without permanent harm to the biosphere. It has been estimated that 98 percent of the species that ever existed have vanished. The response then continues, "Evolution has always replaced extinct species with new ones, has it not?"

But these statements, while true, conceal a terrible twist. After the Mesozoic extinction spasm 65 million years ago, which ended the Age of Reptiles, and after each of the other four greatest previous spasms spaced over 400 million years at very roughly 100-million-year intervals, evolution required about 10 million years to restore the predisaster levels of diversity. That is an extremely long time for future generations to wait because of the damage we are inflicting on the environment within a few decades. Equally serious, evolution cannot perform as in previous ages if natural environments have been crowded out by artificial ones, the phenomenon known by biologists as "the death of birth." For example, Hawaii and most of the Polynesian islands have been either entirely deforested or largely covered by introduced plants.

In the *second stage* of denial, people ask, "Why do we need so many species anyway? Why care, especially since the vast majority are bugs, weeds, and fungi?" It is too easy to dismiss the creepy crawlies of the world, forgetting that less than a century ago and before the rise of the modern conservation movement native birds and mammals were dismissed with the same callous indifference. Now the value of the little things in the natural world has become compellingly clear. Recent experimental studies on whole ecosystems support what was long suspected: in most cases, the more species living in an ecosystem, the higher its productivity and the greater its ability to withstand drought and other kinds of environmental stress. Since we depend on an abundance of functioning ecosystems to cleanse our water, enrich our soil, and manufacture the very air we breathe, biodiversity is clearly not an inheritance to be discarded carelessly.

In addition to creating a habitable environment for human beings, wild species are the source of new pharmaceuticals, crops, fibers, and other products that help sustain our lives. The varied examples of such discoveries cited in *The Diversity of Life* continue to be multiplied. A new sense of opportunity is especially prominent in the search for novel antibiotics and other pharmaceuticals.

The clinching argument for the protection of species, however, may in the end prove to be moral. *Homo sapiens* is a brilliant and proud citizen of the biosphere, but Earth is where we originated and will stay. Who are we to destroy the planet's Creation? Each species around us is a masterpiece of evolution, exquisitely adapted to its environment. Species existing today are thousands to millions of years old. Their genes, having been tested by adversity over countless generations, engineer a staggeringly complex mix of biochemical devices that promote the survival and reproduction of the organisms carrying them.

Even if all these arguments are granted, the *third stage* of denial still emerges: "Why rush to save all the species now? We have more important things to do. Why not keep live specimens in zoos and botanical gardens and return them to the wild later?" The grim answer is that all the zoos in the world today can sustain tiny populations of at most 2,000 species of mammals, birds, reptiles, and amphibians out of a total of about 24,000 known to exist. The world's botanical gardens would be even more overwhelmed by the 250,000 plant species still alive on Earth. These refuges are invaluable in helping to save a few endangered species. So is freezing embryos in liquid nitrogen. But such stopgap measures cannot come close to solving the problem as a whole. To add to the difficulty, no one has even conceived a plan that might save the millions of kinds of insects, fungi, and other ecologically vital small organisms. Even if that prodigious feat could be accomplished, and scientists were poised to return species to independence, the forests, savannas, and other ecosystems in which many lived would no longer exist. To reassemble them would be like unscrambling an egg.

It is the universal consensus of conservation scientists and other environmental professionals that to protect the biosphere it will be necessary to maintain the natural environments in which wild species live. But given how rapidly these refuges are being destroyed, even that straightforward solution seems a daunting task. To keep the Creation, we need all the science, technology, and moral commitment that can be mustered in the service of ecology. Thus is the word ecology derived: *oikos*, our home; *logos*, its study and understanding.

E. O. W.

HARVARD UNIVERSITY

22 JUNE 2000

Violent Nature,
Resilient Life

Storm over the Amazon

IN THE AMAZON BASIN the greatest violence sometimes begins as a flicker of light beyond the horizon. There in the perfect bowl of the night sky, untouched by light from any human source, a thunderstorm sends its premonitory signal and begins a slow journey to the observer, who thinks: the world is about to change. And so it was one night at the edge of rain forest north of Manaus, where I sat in the dark, working my mind through the labyrinths of field biology and ambition, tired, bored, and ready for any chance distraction.

Each evening after dinner I carried a chair to a nearby clearing to escape the noise and stink of the camp I shared with Brazilian forest workers, a place called Fazenda Dimona. To the south most of the forest had been cut and burned to create pastures. In the daytime cattle browsed in remorseless heat bouncing off the yellow clay and at night animals and spirits edged out onto the ruined land. To the north the virgin rain forest began, one of the great surviving wildernesses of the world, stretching 500 kilometers before it broke apart and dwindled into gallery woodland among the savannas of Roraima.

Enclosed in darkness so complete I could not see beyond my outstretched hand, I was forced to think of the rain forest as though I were seated in my library at home, with the lights turned low. The forest at night is an experience in sensory deprivation most of the time, black and silent as the midnight zone of a cave. Life is out there in expected abundance. The jungle teems, but in a manner mostly beyond the reach of the human senses. Ninety-nine percent of the animals find their way by chemical trails laid over the surface, puffs of odor released into the air or water, and scents diffused out of little hidden glands and into the air downwind. Animals are masters of this chemical channel, where we are idiots. But we are geniuses of the audiovisual channel, equaled in this modality only by a few odd groups (whales, monkeys, birds). So we wait for the dawn, while they wait for the fall of darkness; and because sight and sound are the evolutionary prerequisites of intelligence, we alone have come to reflect on such matters as Amazon nights and sensory modalities.

I swept the ground with the beam from my headlamp for signs of life, and found—diamonds! At regular intervals of several meters, intense pinpoints of white light winked on and off with each turning of the lamp. They were reflections from the eyes of wolf spiders, members of the family Lycosidae, on the prowl for insect prey. When spotlighted the spiders froze, allowing me to approach on hands and knees and study them almost at their own level. I could distinguish a wide variety of species by size, color, and hairiness. It struck me how little is known about these creatures of the rain forest, and how deeply satisfying it would be to spend months, years, the rest of my life in this place until I knew all the species by name and every detail of their lives. From specimens beautifully frozen in amber we know that the Lycosidae have survived at least since the beginning of the Oligocene epoch, forty million years ago, and probably much longer. Today a riot of diverse forms occupy the whole world, of which this was only the minutest sample, yet even these species turning about now to watch me from the bare yellow clay could give meaning to the lifetimes of many naturalists.

The moon was down, and only starlight etched the tops of the trees. It was August in the dry season. The air had cooled enough to make the humidity pleasant, in the tropical manner, as much a state of mind as a physical sensation. The storm I guessed was about an hour away. I thought of walking back into the forest with my headlamp to hunt for new treasures, but was too tired from the day's work. Anchored again to my chair, forced into myself, I welcomed a meteor's streak and the occasional courtship flash of luminescent click beetles among the nearby but unseen shrubs. Even the passage of a jetliner 10,000 meters up, a regular event each night around ten o'clock, I awaited with pleasure. A week in the rain forest had transformed its distant rumble from an urban irritant into a comforting sign of the continuance of my own species.

But I was glad to be alone. The discipline of the dark envelope summoned fresh images from the forest of how real organisms look and act. I needed to concentrate for only a second and they came alive as eidetic images, behind closed eyelids, moving across fallen leaves and decaying humus. I sorted the memories this way and that in hope of stumbling on some pattern not obedient to abstract theory of textbooks. I would have been happy with *any* pattern. The best of science doesn't consist of mathematical models and experiments, as textbooks make it seem. Those come later. It springs fresh from a more primitive mode of thought, wherein the hunter's mind weaves ideas from old facts and fresh metaphors and the scrambled crazy images of things recently seen. To move forward is to concoct new patterns of thought, which in turn dictate the design of the models and experiments. Easy to say, difficult to achieve.

The subject fitfully engaged that night, the reason for this research trip to the Brazilian Amazon, had in fact become an obsession and, like all obsessions, very likely a dead end. It was the kind of favorite puzzle that keeps forcing its way back because its very intractability makes it perversely pleasant, like an overly familiar melody intruding into the relaxed mind because it loves you and will not leave you. I hoped that some new image might propel me past the jaded puzzle to the other side, to ideas strange and compelling.

Bear with me for a moment while I explain this bit of personal esoterica; I am approaching the subject of central interest. Some kinds of plants and animals are dominant, proliferating new species and spreading over large parts of the world. Others are driven back until they become rare and threatened by extinction. Is there a single formula for this biogeographic difference, for all kinds of organisms? The process, if articulated, would be a law or at least a principle of dynastic succession in evolution. I was intrigued by the circumstance that social insects, the group on which I have spent most of my life, are among the most abundant of all organisms. And among the social insects, the dominant subgroup is the ants. They range 20,000 or more species strong from the Arctic Circle to the tip of South America. In the Amazon rain forest they compose more than 10 percent of the biomass of all animals. This means that if you were to collect, dry out, and weigh every animal in a piece of forest, from monkeys and birds down to mites and roundworms, at least 10 percent would consist of these insects alone. Ants make up almost half of the insect biomass overall and 70 percent of the individual insects found in the treetops. They are only slightly less abundant in grasslands, deserts, and temperate forests throughout the rest of the world.

It seemed to me that night, as it has to others in varying degrees of persuasion many times before, that the prevalence of ants must have something to do with their advanced colonial organization. A colony is a superorganism, an assembly of workers so tightly knit around the mother queen as to act like a single, well-coordinated entity. A wasp or other solitary insect encountering a worker ant on its nest faces more than just another insect. It faces the worker and all her sisters, united by instinct to protect the queen, seize control of territory, and further the growth of the colony. Workers are little kamikazes, prepared—eager—to die in order to defend the nest or gain control of a food source. Their deaths matter no more to the colony than the loss of hair or a claw tip might to a solitary animal.

There is another way to look at an ant colony. Workers foraging around their nest are not merely insects searching for food. They are a living web cast out by the superorganism, ready to congeal over rich food finds or shrink back from the most formidable enemies. Superorganisms can control and dominate the ground and treetops in competition with ordinary, solitary organisms, and that is surely why ants live everywhere in such great numbers.

I heard around me the Greek chorus of training and caution: *How can you prove that is the reason for their dominance? Isn't the connection just another shaky conclusion that because two events occur together, one causes the other? Something else entirely different might have caused both. Think about it—greater individual fighting ability? Sharper senses? What?*

Such is the dilemma of evolutionary biology. We have problems to solve, we have clear answers—too many clear answers. The difficult part is picking out the right answer. The isolated mind moves in slow circles and breakouts are rare. Solitude is better for weeding out ideas than for creating them. Genius is the summed production of the many with the names of the few attached for easy recall, unfairly so to other scientists. My mind drifted into the hourless night, no port of call yet chosen.

The storm grew until sheet lightning spread across the western sky. The thunderhead reared up like a top-heavy monster in slow motion, tilted forward, blotting out the stars. The forest erupted in a simulation of violent life. Lightning bolts broke to the front and then closer, to the right and left, 10,000 volts dropping along an ionizing path at 800 kilometers an hour, kicking a countersurge skyward ten times faster, back and forth in a split second, the whole perceived as a single flash and crack of sound. The wind freshened, and rain came stalking through the forest.

In the midst of chaos something to the side caught my attention. The lightning bolts were acting like strobe flashes to illuminate the wall of the rain forest. At intervals I glimpsed the storied structure: top canopy 30 meters off the ground, middle trees spread raggedly below that, and a lowermost scattering of shrubs and small trees. The forest was framed for a few moments in this theatrical setting. Its image turned surreal, projected into the unbounded wildness of the human imagination, thrown back in time 10,000 years. Somewhere close I knew spear-nosed bats flew through the tree crowns in search of fruit, palm vipers coiled in ambush in the roots of orchids, jaguars walked the river's edge; around them eight hundred species of trees stood, more than are native to all of North America; and a thousand species of butterflies, 6 percent of the entire world fauna, waited for the dawn.

About the orchids of that place we knew very little. About flies and beetles almost nothing, fungi nothing, most kinds of organisms nothing. Five thousand kinds of bacteria might be found in a pinch of soil, and about them we knew absolutely nothing. This was wilderness in the sixteenth-century sense, as it must have formed in the minds of the Portuguese explorers, its interior still largely unexplored and filled with strange, myth-engendering plants and animals. From such a place the pious naturalist would send long respectful letters to royal patrons about the wonders of the new world as testament to the glory of God. And I thought: there is still time to see this land in such a manner.

The unsolved mysteries of the rain forest are formless and seductive. They are like unnamed islands hidden in the blank spaces of old maps, like dark shapes glimpsed descending the far wall of a reef into the abyss. They draw us forward and stir strange apprehensions. The unknown and prodigious are drugs to the scientific imagination, stirring insatiable hunger with a single taste. In our hearts we hope we will never discover everything. We pray there will always be a world like this one at whose edge I sat in darkness. The rain forest in its richness is one of the last repositories on earth of that timeless dream.

That is why I keep going back to the forests forty years after I began, when I flew down to Cuba, a graduate student caught up in the idea of the "big" tropics, free at last to look for something hidden, as Kipling had urged, something lost behind the Ranges. The chances are high, in fact certain, of finding a new species or phenomenon within days or, if you work hard, hours after arrival. The hunt is also on for rare species already discovered but still effectively unknown—represented by one or two specimens placed in a museum drawer fifty or a hundred years ago, left with nothing but a locality and a habitat note handwritten on a tiny label ("Santarém, Brazil, nest on side of tree in swamp forest"). Unfold the stiff yellowing piece of paper and a long-dead biologist speaks: I was there, I found this, now you know, now move on.

There is still more to the study of biological richness. It is a microcosm of scientific exploration as a whole, refracting hands-on experience onto a higher plane of abstraction. We search in and around a subject for a concept, a pattern, that imposes order. We look for a way of speaking about the rough unmapped terrain, even just a name or a phrase that calls attention to the object of our attention. We hope to be the first to make a connection. Our goal is to capture and label a process, perhaps a chemical reaction or behavior pattern driving an ecological change, a new way of classifying energy flow, or a relation between predator and prey that preserves them both, almost anything at all. We will settle for just one good question that starts people thinking and talking: Why are there so many species? Why have mammals evolved more quickly than reptiles? Why do birds sing at dawn?

These whispering denizens of the mind are sensed but rarely seen. They rustle the foliage, leave behind a pug mark filling with water and a scent, excite us for an instant and vanish. Most ideas are waking dreams that fade to an emotional residue. A first-rate scientist can hope to capture and express only several in a lifetime. No one has learned how to invent with any consistent success the equations and phrases of science, no one has captured the metaformula of scientific research. The conversion is an art aided by a stroke of luck in minds set to receive them. We hunt outward and we hunt inward, and the value of the quarry on one side of that mental barrier is commensurate with the value of the quarry on the other side. Of this dual quality the great chemist Berzelius wrote in 1818 and for all time:

> All our theory is but a means of consistently conceptualizing the inward processes of phenomena, and it is presumable and adequate when all scientifically known facts can be deduced from it. This mode of conceptualization can equally well be false and, unfortunately, presumably is so frequently. Even though, at a certain period in the development of science, it may match the purpose just as well as a true theory. Experience is augmented, facts appear which do not agree wih it, and one is forced to go in search of a new mode of conceptualization within which these facts can also be accommodated; and in this manner, no doubt, modes of conceptualization will be altered from age to age, as

experience is broadened, and the complete truth may perhaps never be attained.

The storm arrived, racing from the forest's edge, turning from scattered splashing drops into sheets of water driven by gusts of wind. It forced me back to the shelter of the corrugated iron roof of the open-air living quarters, where I sat and waited with the *mateiros*. The men stripped off their clothing and walked out into the open, soaping and rinsing themselves in the torrential rain, laughing and singing. In bizarre counterpoint, leptodactylid frogs struck up a loud and monotonous honking on the forest floor close by. They were all around us. I wondered where they had been during the day. I had never encountered a single one while sifting through the vegetation and rotting debris on sunny days, in habitats they are supposed to prefer.

Farther out, a kilometer or two away, a troop of red howler monkeys chimed in, their chorus one of the strangest sounds to be heard in all of nature, as enthralling in its way as the songs of humpback whales. A male opened with an accelerating series of deep grunts expanding into prolonged roars and was then joined by the higher-pitched calls of the females. This far away, filtered through dense foliage, the full chorus was machine-like: deep, droning, metallic.

Such raintime calls are usually territorial advertisements, the means by which the animals space themselves out and control enough land to forage and breed. For me they were a celebration of the forest's vitality: *Rejoice! The powers of nature are within our compass, the storm is part of our biology!*

For that is the way of the nonhuman world. The greatest powers of the physical environment slam into the resilient forces of life, and nothing much happens. For a very long time, 150 million years, the species within the rain forest evolved to absorb precisely this form and magnitude of violence. They encoded the predictable occurrence of nature's storms in the letters of their genes. Animals and plants have come to use heavy rains and floods routinely to time episodes in their life cycle. They threaten rivals, mate, hunt prey, lay eggs in new water pools, and dig shelters in the rain-softened earth.

On a larger scale, the storms drive change in the whole structure of the forest. The natural dynamism raises the diversity of life by means of local destruction and regeneration.

Somewhere a large horizontal tree limb is weak and vulnerable, covered by a dense garden of orchids, bromeliads, and other kinds of plants that grow on trees. The rain fills up the cavities enclosed by the axil sheaths of the epiphytes and soaks the humus and clotted dust around their roots. After years of growth the weight has become nearly unsupportable. A gust of wind whips through or lightning strikes the tree trunk, and the limb breaks and plummets down, clearing a path to the ground. Elsewhere the crown of a giant tree emergent above the rest catches the wind and the tree sways above the rain-soaked soil. The shallow roots cannot hold, and the entire tree keels over. Its trunk and canopy arc downward like a blunt ax, shearing through smaller trees and burying understory bushes and herbs. Thick lianas coiled through the limbs are pulled along. Those that stretch to other trees act as hawsers to drag down still more vegetation. The massive root system heaves up to create an instant mound of bare soil. At yet another site, close to the river's edge, the rising water cuts under an overhanging bank to the critical level of gravity, and a 20-meter front collapses. Behind it a small section of forest floor slides down, toppling trees and burying low vegetation.

Such events of minor violence open gaps in the forest. The sky clears again and sunlight floods the ground. The surface temperature rises and the humidity falls. The soil and ground litter dries out and warms up still more, creating a new environment for animals, fungi, and microorganisms of a different kind from those in the dark forest interior. In the following months pioneer plant species take seed. They are very different from the young shade-loving saplings and understory shrubs of the prevailing old-stand forest. Fast-growing, small in stature, and short-lived, they form a single canopy that matures far below the upper crowns of the older trees all around. Their tissue is soft and vulnerable to herbivores. The palmate-leaved trees of the genus *Cecropia*, one of the gap-filling specialists of Central and South America, harbor vicious ants in hollow internodes of the trunk. These insects, bearing the appropriate scientific name *Azteca*, live in symbiosis with their hosts, protecting them from all predators except sloths and a few other herbivores specialized to feed on *Cecropia*. The symbionts live among new assemblages of species not found in the mature forest.

All around the second-growth vegetation, the fallen trees and branches rot and crumble, offering hiding places and food to a vast array of basidiomycete fungi, slime molds, ponerine ants, scolytid beetles, bark lice, earwigs, embiopteran webspinners, zorapterans, entomobryomorph springtails, japygid diplurans, schizomid arachnids, pseudoscorpions, real scorpions, and other forms that live mostly or exclusively in this habitat. They add thousands of species to the diversity of the primary forest.

Climb into the tangle of fallen vegetation, tear away pieces of rotting bark, roll over logs, and you will see these creatures teeming everywhere. As the pioneer vegetation grows denser, the deepening shade and higher humidity again favor old-forest species, and their saplings sprout and grow. Within a hundred years the gap specialists will be phased out by competition for light, and the tall storied forest will close completely over.

In the succession, pioneer species are the sprinters, old-forest species the long-distance runners. The violent changes and a clearing of space bring all the species briefly to the same starting line. The sprinters dash ahead, but the prolonged race goes to the marathoners. Together the two classes of specialists create a complex mosaic of vegetation types across the forest which, by regular tree falls and landslides, is forever changing. If square kilometers of space are mapped over decades of time, the mosaic turns into a riotous kaleidoscope whose patterns come and go and come again. A new marathon is always beginning somewhere in the forest. The percentages of successional vegetation types are consequently more or less in a steady state, from earliest pioneer species through various mixes of pioneer and deep-forest trees to stands of the most mature physiognomy. Walk randomly on any given day for one or two kilometers through the forest, and you will cut through many of these successional stages and sense the diversity sustained by the passage of storms and the fall of forest giants.

It is diversity by which life builds and saturates the rain forest. And diversity has carried life beyond, to the harshest environments on earth. Rich assemblages of animals swarm in the shallow bays of Antarctica, the coldest marine habitats on earth. Perch-like notothenioid fishes swim there in temperatures just above the freezing point of salt water but cold enough to turn ordinary blood to ice, because they are able to generate glycopeptides in their tissues as antifreeze and thrive where other fish cannot go. Around them flock dense populations of active brittlestars, krill, and other invertebrate animals, each with protective devices of its own.

In a radically different setting, the deep unlighted zone of caves around the world, blind white springtails, mites, and beetles feed on fungi and bacteria growing on rotting vegetable matter washed down through ground water. They are eaten in turn by blind white beetles and spiders also specialized for life in perpetual darkness.

Some of the harshest deserts of the world are home to unique ensembles of insects, lizards, and flowering plants. In the Namib of southwestern Africa, beetles use leg tips expanded into oarlike sandshoes to swim down through the shifting dunes in search of dried vegetable matter. Others, the swiftest runners of the insect world, race over the baking hot surface on bizarre stilt legs.

Archaebacteria, one-celled microorganisms so different from ordinary bacteria as to be candidates for a separate kingdom of life, occupy the boiling water of mineral hot springs and volcanic vents in the deep sea. The species composing the newly discovered genus *Methanopyrus* grow in boiling vents at the bottom of the Mediterranean Sea in temperatures up to 110°C.

Life is too well adapted in such places, out to the edge of the physical envelope where biochemistry falters, and too diverse to be broken by storms and other ordinary vagaries of nature. But diversity, the property that makes resilience possible, is vulnerable to blows that are greater than natural perturbations. It can be eroded away fragment by fragment, and irreversibly so if the abnormal stress is unrelieved. This vulnerability stems from life's composition as swarms of species of limited geographical distribution. Every habitat, from Brazilian rain forest to Antarctic bay to thermal vent, harbors a unique combination of plants and animals. Each kind of plant and animal living there is linked in the food web to only a small part of the other species. Eliminate one species, and another increases in number to take its place. Eliminate a great many species, and the local ecosystem starts to decay visibly. Productivity drops as the channels of the nutrient cycles are clogged. More of the biomass is sequestered in the form of dead vegetation and slowly metabolizing, oxygen-starved mud, or is simply washed away. Less competent pollinators take over as the best-adapted bees, moths, birds, bats, and other specialists drop out. Fewer seeds fall, fewer seedlings sprout. Herbivores decline, and their predators die away in close concert.

In an eroding ecosystem life goes on, and it may look superficially the same. There are always species able to recolonize the impoverished area and exploit the stagnant resources, however clumsily accomplished. Given enough time, a new combination of species—a reconstituted fauna and flora—will reinvest the habitat in a way that transports energy and materials somewhat more efficiently. The atmosphere they generate and the composition of the soil they enrich will resemble those found in comparable habitats in other parts of the world, since the species are adapted to penetrate and reinvigorate just such degenerate systems. They do so because they gain more energy and materials and leave more offspring. But the restorative power of the fauna and flora of the world as a whole depends on the existence of enough species to play that special role. They too can slide into the red zone of endangered species.

Biological diversity—"biodiversity" in the new parlance—is the key to the maintenance of the world as we know it. Life in a local site struck down by a passing storm springs back quickly because enough diversity still exists. Opportunistic species evolved for just such an occasion rush in to fill the spaces. They entrain the succession that circles back to something resembling the original state of the environment.

This is the assembly of life that took a billion years to evolve. It has eaten the storms—folded them into its genes—and created the world that created us. It holds the world steady. When I rose at dawn the next morning, Fazenda Dimona had not changed in any obvious way from the day before. The same high trees stood like a fortress along the forest's edge; the same profusion of birds and insects foraged through the canopy and understory in precise individual timetables. All this seemed timeless, immutable, and its very strength posed the question: how much force does it take to break the crucible of evolution?

Krakatau

K RAKATAU, earlier misnamed Krakatoa, an island the size of Manhattan located midway in the Sunda Strait between Sumatra and Java, came to an end on Monday morning, August 27, 1883. It was dismembered by a series of powerful volcanic eruptions. The most violent occurred at 10:02 A.M., blowing upward like the shaped explosion of a large nuclear bomb, with an estimated force equivalent to 100–150 megatons of TNT. The airwave it created traveled at the speed of sound around the world, reaching the opposite end of the earth near Bogotá, Colombia, nineteen hours later, whereupon it bounced back to Krakatau and then back and forth for seven recorded passages over the earth's surface. The audible sounds, resembling the distant cannonade of a ship in distress, carried southward across Australia to Perth, northward to Singapore, and westward 4,600 kilometers to Rodriguez Island in the Indian Ocean, the longest distance traveled by any airborne sound in recorded history.

As the island collapsed into the subterranean chamber emptied by the eruption, the sea rushed in to fill the newly formed caldera. A column of magma, rock, and ash rose 5 kilometers into the air, then fell earthward, thrusting the sea outward in a tsunami 40 meters in height. The great tidal waves, resembling black hills when first sighted on the horizon, fell upon the shores of Java and Sumatra, washing away entire towns and killing 40,000 people. The segments traversing the channels and reaching the open sea continued on as spreading waves around the world. The waves were still a meter high when they came ashore in Ceylon, now Sri Lanka, where they drowned one person, their last casualty. Thirty-two hours after the explosion, they rolled in to Le Havre, France, reduced at last to centimeter-high swells.

The eruptions lifted more than 18 cubic kilometers of rock and other material into the air. Most of this tephra, as it is called by geologists, quickly rained back down onto the surface, but a residue of sulfuric-acid aerosol and dust boiled upward as high as 50 kilometers and diffused through the stratosphere around the world, where for several years it created brilliant red sunsets and "Bishop's rings," opalescent coronas surrounding the sun.

Back on Krakatau the scene was apocalyptic. Throughout the daylight hours the whole world seemed about to end for those close enough to witness the explosions. At the climactic moment of 10:02 the American barque *W. H. Besse* was proceeding toward the straits 84 kilometers east northeast of Krakatau. The first officer jotted in his logbook that "terrific reports" were heard, followed by

a heavy black cloud rising up from the direction of Krakatoa Island, the barometer fell an inch at one jump, suddenly rising and falling an inch at a time, called all hands, furled all sails securely, which was scarcely done before the squall struck the ship with terrific force; let go port anchor and all the chain in the locker, wind increasing to a hurricane; let go starboard anchor, it had gradually been growing dark since 9 A.M. and by the time the squall struck us, it was darker than any night I ever saw; this was midnight at noon, a heavy shower of ashes came with the squall, the air being so thick it was difficult to breathe, also noticed a strong smell of sulfur, all hands expecting to be suffocated; the terrible noises from the volcano, the sky filled with forked lightning, running in all directions and making the darkness more intense than ever; the howling of the wind through the rigging formed one of the wildest and most awful scenes imaginable, one that will never be forgotten by any one on board, all expecting that the last days of the earth had come; the water was running by us in the direction of the volcano at the rate of 12 miles per hour, at 4 P.M. wind moderating, the explosions had nearly ceased, the shower of ashes was not so heavy; so was enabled to see our way around the decks; the ship was covered with tons of fine ashes resembling pumice stone, it stuck to the sails, rigging and masts like glue.

In the following weeks, the Sunda Strait returned to outward normality, but with an altered geography. The center of Krakatau had been replaced by an undersea crater 7 kilometers long and 270 meters deep. Only a remnant at the southern end still rose from the

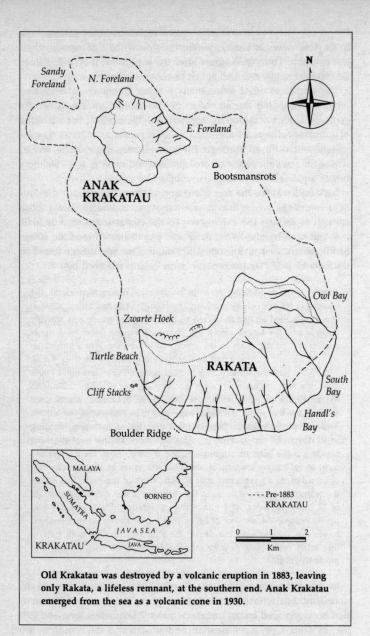

Old Krakatau was destroyed by a volcanic eruption in 1883, leaving only Rakata, a lifeless remnant, at the southern end. Anak Krakatau emerged from the sea as a volcanic cone in 1930.

sea. It was covered by a layer of obsidian-laced pumice 40 meters or more thick and heated to somewhere between 300° and 850°C, enough at the upper range to melt lead. All traces of life had, of course, been extinguished.

Rakata, the ash-covered mountain of old Krakatau, survived as a sterile island. But life quickly enveloped it again. In a sense, the spinning reel of biological history halted, then reversed, like a motion picture run backward, as living organisms began to return to Rakata. Biologists quickly grasped the unique opportunity that Rakata afforded: to watch the assembly of a tropical ecosystem from the very beginning. Would the organisms be different from those that had existed before? Would a rain forest eventually cover the island again?

The first search for life on Rakata was conducted by a French expedition in May 1884, nine months after the explosions. The main cliff was eroding rapidly, and rocks still rolled down the sides incessantly, stirring clouds of dust and emitting a continuous noise "like the rattling of distant musketry." Some of the stones whirled through the air, ricocheting down the sides of the ravines and splashing into the sea. What appeared to be mist in the distance turned close up into clouds of dust stirred by the falling debris. The crew and expedition members eventually found a safe landing site and fanned out to learn what they could. After searching for organisms in particular, the ship's naturalist wrote that "notwithstanding all my researches, I was not able to observe any symptom of animal life. I only discovered one microscopic spider—only one; this strange pioneer of the renovation was busy spinning its web."

A baby spider? How could a tiny wingless creature reach the empty island so quickly? Arachnologists know that a majority of species "balloon" at some point in their life cycle. The spider stands on the edge of a leaf or some other exposed spot and lets out a thread of silk from the spinnerets at the posterior tip of its abdomen. As the strand grows it catches an air current and stretches downwind, like the string of a kite. The spider spins more and more of the silk until the thread exerts a strong pull on its body. Then it releases its grip on the surface and soars upward. Not just pinhead-sized babies but large spiders can occasionally reach thousands of meters of altitude and travel hundreds of kilometers before settling to the ground to start a new life. Either that or land on the water and die. The voyagers have no control over their own descent.

Ballooning spiders are members of what ecologists, with the accidental felicity that sometimes pops out of Greek and Latin sources, have delightfully called the aeolian plankton. In ordinary parlance, plankton is the vast swarm of algae and small animals carried passively by water currents; aeolian refers to the wind. The creatures composing the aeolian plankton are devoted almost entirely to long-distance dispersal. You can see some of it forming over lawns and bushes on a quiet summer afternoon, as aphids use their feeble wings to rise just high enough to catch the wind and be carried away. A rain of planktonic bacteria, fungus spores, small seeds, insects, spiders, and other small creatures falls continuously on most parts of the earth's land surface. It is sparse and hard to detect moment by moment, but it mounts up to large numbers over a period of weeks and months. This is how most of the species colonized the seared and-smothered remnant of Krakatau.

The potential of the planktonic invasion has been documented by Ian Thornton and a team of Australian and Indonesian biologists who visited the Krakatau area in the 1980s. While studying Rakata they also visited Anak Krakatau ("Child of Krakatau"), a small island that emerged in 1930 from volcanic activity along the submerged northern rim of the old Krakatau caldera. On its ash-covered lava flows they placed traps made from white plastic containers filled with seawater. This part of the surface of Anak Krakatau dated from localized volcanic activity from 1960 to 1981 and was nearly sterile, resembling the condition on Rakata soon after the larger island's violent formation. During ten days the traps caught a surprising variety of windborne arthropods. The specimens collected, sorted, and identified included a total of 72 species of spiders, springtails, crickets, earwigs, barklice, hemipterous bugs, moths, flies, beetles, and wasps.

There are other ways to cross the water gaps separating Rakata from nearby islands and the Javan and Sumatran coasts. The large semiaquatic monitor lizard *Varanus salvator* probably swam over. It was present no later than 1899, feasting on the crabs that crawl along the shore. Another long-distance swimmer was the reticulated python, a giant snake reaching up to 8 meters in length. Probably all of the birds crossed over by powered flight. But only a small percentage of the species of Java and Sumatra were represented because it is a fact, curiously, that many forest species refuse to cross water gaps even when the nearest island is in full view. Bats, straying off course, made the Rakata landfall. Winged insects of larger size, especially butterflies and dragonflies, probably also traveled under their own power. Under similar conditions in the Florida Keys, I have watched such insects fly easily from one small island to another, as though they were moving about over meadows instead of salt water.

Rafting is a much less common but still important means of transport. Logs, branches, sometimes entire trees fall into rivers and bays and are carried out to sea, complete with microorganisms, insects, snakes, frogs, and occasional rodents and other small mammals living on them at the moment of departure. Blocks of pumice from old volcanic islands, riddled with enough closed air spaces to keep them afloat, also serve as rafts.

Once in a great while a violent storm turns larger animals such as lizards or frogs into aeolian debris, tearing them loose from their perches and propelling them to distant shores. Waterspouts pick up fish and transport them live to nearby lakes and streams.

Swelling the migration further, organisms carry other organisms with them. Most animals are miniature arks laden with parasites. They also transport accidental hitchhikers in soil clinging to the skin, including bacteria and protozoans of immense variety, fungal spores, nematode worms, tardigrades, mites, and feather lice. Seeds of some species of herbs and trees pass live through the guts of birds, to be deposited later in feces, which serves as instant fertilizer. A few arthropods practice what biologists call phoresy, deliberate hitchhiking on larger animals. Pseudoscorpions, tiny replicas of true scorpions but lacking stings, use their lobster-like claws to seize the hairs of dragonflies and other large winged insects, then ride these magic carpets for long distances.

The colonists poured relentlessly into Rakata from all directions. A 100-meter-high electrified fence encircling the island could not have stopped them. Airborne organisms would still have tumbled in from above to spawn a rich ecosystem. But the largely happenstance nature of colonization means that flora and fauna did not return to Rakata in a smooth textbook manner, with plants growing to sylvan thickness, then herbivores proliferating, and finally carnivores prowling. The surveys made on Rakata and later on Anak Krakatau disclosed a far more haphazard buildup, with some species inexplicably going extinct and others flourishing when seemingly they should have quickly disappeared. Spiders and flightless carnivorous crickets persisted almost miraculously on bare pumice fields; they fed on a thin diet of insects landing in the aeolian debris. Large lizards and some of the birds lived on beach crabs, which subsisted in turn on dead marine plants and animals washed ashore by waves. (The original name of Krakatau was Karkata, or Sanskrit for "crab"; Rakata also means crab in the old Javanese language.) Thus animal diversity was not wholly dependent on vegetation. And for its part vegetation grew up in patches, alternately spreading and retreating across the island to create an irregular mosaic.

If the fauna and flora came back chaotically, they also came back fast. In the fall of 1884, a little more than a year after the eruption, biologists encountered a few shoots of grass, probably *Imperata* and *Saccharum*. In 1886 there were fifteen species of grasses and shrubs, in 1897 forty-nine, and in 1928 nearly three hundred. Vegetation dominated by *Ipomoea* spread along the shores. At the same time grassland dotted with *Casuarina* pines gave way here and there to richer pioneer stands of trees and shrubs. In 1919 W. M. Docters van Leeuwen, from the Botanical Gardens at Buitenzorg, found forest patches surrounded by nearly continuous grassland. Ten years later he found the reverse: forest now clothed the entire island and was choking out the last of the grassland patches. Today Rakata is covered completely by tropical Asian rain forest typical in outward appearance. Yet the process of colonization is far from complete. Not a single tree species characterizing the deep, primary forests on Java and Sumatra has made it back. Another hundred years or more may be needed for investment by a forest fully comparable to that of old, undisturbed Indonesian islands of the same size.

Some insects, spiders, and vertebrates aside, the earliest colonists of most kinds of animals died on Rakata soon after arrival. But as the vegetation expanded and the forest matured, increasing numbers of species took hold. At the time of the Thornton expeditions of 1984–85, the inhabitants included thirty species of land birds, nine bats, the Indonesian field rat, the ubiquitous black rat, and nine reptiles, including two geckos and *Varanus salvator*, the monitor lizard. The reticulated python, recorded as recently as 1933, was not present in 1984–85. A large host of invertebrate species, more than six hundred in all, lived on the island. They included a terrestrial flatworm, nematode worms, snails, scorpions, spiders, pseudoscorpions, centipedes, cockroaches, termites, barklice, cicadas, ants, beetles, moths, and butterflies. Also present were microscopic rotifers and tardigrades and a rich medley of bacteria.

A first look at the reconstituted flora and fauna of Rakata, in other words Krakatau a century after the apocalypse, gives the impression of life on a typical small Indonesian island. But the community of species remains in a highly fluid state. The number of resident bird species may now be approaching an equilibrium, the rise having slowed markedly since 1919 to settle close to thirty. Thirty is also about the number on other islands of Indonesia of similar size. At the same time, the *composition* of the bird species is less stable. New species have been arriving, and earlier ones have been declining to extinction. Owls and flycatchers arrived after 1919, for example, while several old residents such as the bulbul *(Pycnonotus aurigaster)* and gray-backed shrike *(Lanius schach)* disappeared. Reptiles appear to be at or close to a similar dynamic equilibrium. So are cockroaches, nymphalid butterflies, and dragonflies. Flightless mammals, represented solely by the two kinds of rats, are clearly not. Nor are plants, ants, or snails. Most of the other invertebrates are still too poorly explored on Rakata over sufficiently long periods of time to judge their status, but in general the overall number of species appears to be still rising.

Rakata, along with Panjang and Sertung, and other islands of the Krakatau archipelago blasted and pumice-coated by the 1883 explosion, have within the span of a century rewoven a semblance of the communities that existed before, and the diversity of life has largely returned. The question remains as to whether endemic species, those found only on the archipelago prior to 1883, were destroyed by the explosion. We can never be sure because the islands were too poorly explored by naturalists before Krakatau came so dramatically to the world's attention in 1883. It seems unlikely that endemic species ever existed. The islands are so small that the natural turnover of species may have been too fast to allow evolution to attain the creation of new species, even without volcanic episodes.

In fact the archipelago has suffered turbulence that destroyed or at least badly damaged its fauna and flora every few centuries. According to Javanese legend, the volcano Kapi erupted violently in the Sunda Strait in 416 A.D.: "At last the mountain Kapi with a tremendous roar burst into pieces and sunk into the deepest of the earth. The water of the sea rose and inundated the land." A series of smaller eruptions, burning at least part of the forest, occurred during 1680 and 1681.

Today you can sail close by the islands without guessing their violent history, unless Anak Krakatau happens to be smoldering that day. The thick green forest offers testimony to the ingenuity and resilience of life. Ordinary volcanic eruptions are not enough, then, to break the crucible of life.

CHAPTER THREE

The Great Extinctions

W HAT WAS the greatest blow ever suffered by life through all time? Not the 1883 explosions at Krakatau, which were not even the worst in recorded history. An 1815 eruption at Tambora, 1,400 kilometers to the east of Krakatau on the Indonesian island of Sumbawa, lifted five times as much rock and ash as Krakatau. It inflicted more environmental destruction and killed tens of thousands of people. About 75,000 years ago a still greater eruption occurred in the center of northern Sumatra. It blew out a phenomenal 1,000 cubic kilometers of solid material, creating an oval depression 65 kilometers long that filled with fresh water and persists to this day as Lake Toba. Paleolithic people lived on the island then. We can only imagine what they felt in the presence of an eruption one hundred times the magnitude of Krakatau, and what stories of gods and apocalypse proliferated in the culture afterwards.

Great eruptions are likely to have occurred repeatedly across long stretches of geological time. A simple form of statistical reasoning leads to this conclusion. The frequency curve of the intensity of volcanic eruptions around the world, like so many chance phenomena, peaks near the low end and tapers off for a long distance toward the high end. This means that most eruptions are relatively minor perturbations, consisting of a plume of vapor from a fumarole here, a minor lava flow there. Lava fountains and big flows, the next step up, are less common but still occur on a yearly basis somewhere in the world. An event the size of the Krakatau explosion happens once or twice a century. An eruption as big as the one at Toba is far rarer but, over millions of years, probably inevitable.

The same statistical reasoning applies to the fall of meteorites. A large number ranging in size from dust particles to pebbles reach the earth's surface each year, streaking in at 15 to 75 kilometers a second. A much smaller number range in size from baseballs to soccer balls. They account for the majority of the thirty or so meteorites worldwide that can be seen traveling all the way down and are then located by searchers on foot. A very few are much more massive. The largest ever observed in the United States was a 5,000-kilogram meteorite that fell in Norton County, Kansas, on February 18, 1948. Over millions of years only a few truly gigantic meteorites reach the earth's surface. One with a diameter of 1,250 meters gouged out Canyon Diablo in Arizona. Another monster, 3,200 meters in diameter, created the Chubb Depression at Ungava, Quebec.

By extrapolation upward along the scale of violence, it is conceivable and even likely that a volcanic eruption or a meteorite strike occurs once every 10 million or 100 million years so great as to literally shake the earth, drastically change its atmosphere, and as a result extinguish a substantial portion of the species then living. Something of that kind might have happened at the end of the Mesozoic era 65 million years ago, when dinosaurs and a few other prevailing groups of animals were set back or extinguished altogether. So concluded Luis Alvarez and three other physical scientists at Berkeley in 1979. They found abnormally high concentrations of iridium, an element of the platinum group, in a thin geological deposit separating the older Mesozoic era from the younger Cenozoic era. More precisely, the layer separates rocks of the Cretaceous period, youngest of the Mesozoic periods, and the Tertiary period, oldest of the Cenozoic. Proceeding upward across this thin line, called the K-T boundary (the two single-letter symbols given the two periods respectively), the fossils shift from a prevalence of dinosaurs and a few small mammals to no dinosaurs and a prevalence of mammals. Iridium has a strong affinity for iron; as a result, during the formation of the planet most of it had been drawn deep into the iron-rich core of the earth. Its presence in the K-T boundary, so close to the surface, was a mystery.

The Berkeley team noticed that iridium is also abundant in some meteorites. This anomaly and some mathematical modeling led them to the following scenario: 65 million years ago a meteorite 10 kilometers in diameter crashed into the earth at 72,000 kilometers an hour. The impact conveyed a force greater than the detonation of all the nuclear weapons in the world. It rang the earth like a bell, ignited wildfires, washed the shores with giant tsunamis, and kicked up an immense dust cloud that enshrouded the planet and then either cooled the atmosphere by blocking out the sun or else warmed it by trapping heat as in a greenhouse. As the dust settled it formed a silt layer half a centimeter thick, laced with iridium. Afterward acid rain washed the surface residue for months or years. All these effects combined, according to the Alvarez scenario, to kill the dinosaurs and a medley of other plants and animals.

If a strike anywhere near this magnitude did occur, it should have left telltale signs in addition to iridium enrichment of the earth's surface. In the intense discussion and research that followed Alvarez's proposal, one key piece of new evidence came to light. Geochemists know that when quartz is subjected to extreme pressures, such as those at an impact site, it is "shocked": the crystal lattice is disrupted so that irregular planes appear in thin sections of the mineral examined microscopically between crossed polarizing filters. Such planes were indeed found in quartz grains in some parts of the K-T boundary. At this point the case for the meteorite hypothesis looked very good.

First rule of the history of science: when a big, new, persuasive idea is proposed, an army of critics soon gathers and tries to tear it down. Such a reaction is unavoidable because, aggressive yet abiding by the rules of civil discourse, this is simply how scientists work. It is further true that, faced with adversity, proponents will harden their resolve and struggle to make the case more convincing. Being human, most scientists conform to the psychological Principle of Certainty, which says that when there is evidence both for and against a belief, the result is not a lessening but a heightening of conviction on both sides. During the 1980s, hundreds of experts wrote over two thousand articles for and against the meteorite hypothesis. Tensions rose at scientific conferences, arguments and counterarguments flowed through the pages of *Science*, a small industry grew up in the laboratories and seminar rooms of research universities.

Rule number two: the new idea will, like mother earth, take some serious hits. If good it will survive, probably in modified form. If bad it will die, usually at the time of death or retirement of the last original proponent. As Paul Samuelson once said of the science of economics: funeral by funeral, theory advances. In this case, the antimeteorite critics had a powerful competing hypothesis. They said that every few tens of millions of years huge volcanic eruptions, either one-time gargantuan Krakataus or concerted volleys of ordinary Krakataus, could produce the effects observed at the K-T boundary. Some present-day volcanoes do bring up elevated levels of iridium in their ash. They might also generate enough pressure to shock quartz, although field tests in progress (as I write) have not resolved the matter either way.

The volcanists and other critics raised another even more troubling piece of evidence to undermine the meteorite hypothesis: many extinctions did occur at the end of the Cretaceous period, no question about that, but not all at once. The times of extinction of various groups were smeared out over millions of years on either side of the K-T boundary. Dinosaurs, for example, declined noticeably during the last ten million years before the end of the Cretaceous period. In Montana and southern Alberta, about thirty species were present 10 million years before the end. The number decreased gradually to thirteen just before the end, with the horned dinosaur *Triceratops* remaining most abundant in the final group. A similar pattern was followed by the ammonoids, mollusks with a chambered shell like that of the modern pearly nautilus. It was also followed by inoceramid pelecypods, bivalve mollusks that included giant species with shells a meter wide, and by rudists, other bivalves that built reefs from the mass of their shells. Many groups of foraminiferans, amoeba-like marine creatures that secrete elaborate and exquisitely designed calcareous skeletons, phased out in steps across a million years. Some disappeared before the end of the Cretaceous, others later at differing times, all to be replaced by new kinds of foraminiferans emerging during several hundred thousand years. Insects passed through the K-T boundary relatively unscathed. All of the orders, the highest-ranking taxonomic groups, survived, including the Coleoptera (beetles), Diptera (flies), Hymenoptera (bees, wasps, and ants), and Lepidoptera (moths and butterflies). Most if not all of the families, the next highest-ranking groups, also came through, including the Formicidae (ants), Curculionidae (weevils), and Stratiomyidae (soldier flies). The fossil record is still too poor on the Cretaceous side to estimate extinctions at the species level, of particular kinds such as the modern-day house fly *(Musca domestica)* or the cabbage white butterfly *(Pieris rapae)*.

In order to accommodate the staggered schedule of extinctions across the K-T boundary brought into focus by the controversy, some paleontologists conceived of a series of violent eruptions over millions of years toward the end of the Cretaceous, creating periodic global dust veils, wildfires, acid rain, and climatic cooling. These malign events conspired to reduce population levels of all kinds of organisms and to shrink their geographical ranges to limited regions of the world. A few kinds of animals, such as the dinosaurs, ammonoids, and foraminiferans, were hit hard. Insects and plants persisted more or less intact, perhaps because of their ability to function at low physiological levels for months or even years at a time.

Some of the scientists favoring the meteorite hypothesis, but impressed by the new evidence on extinction, adjusted their model by abandoning the hypothesis of a single cataclysmic event. They postulated a series of smaller meteorite strikes across the million-year transition. Many such events, they said, could have smeared out the timing of the extinctions widely on either side of the K-T boundary.

Not all paleontologists were so ready to abandon both the mega-Krakatau and the single-strike hypotheses. They redoubled their efforts to locate fossils close to the K-T boundary in order to fix more precisely the time of the mass extinctions. Now the balance appears to be shifting somewhat to the single-event hypothesis. With more fossils in hand, it does seem more plausible that dinosaurs and ammonoids were cut down suddenly at the time of the supposed meteorite strike or the mega-Krakatau. The data on foraminiferans remain ambiguous and disputed. Plants offer clearer evidence of a single catastrophe. Their fossils are more abundant and easily interpreted, especially the pollen grains incorporated into lake-bottom silt year by year. Western North America saw a sudden and severe reduction in the pollen grains of flowering plants at the K-T boundary, followed by an equally abrupt jump in fern spores—the "fern spike" of the fossil record—followed soon afterward by a return of the pollen of flowering plants, this time representing a different assemblage of species. The temporary decline of flowering plants and the rise of ferns is consistent with a boundary-event winter, a darkening and cooling of the climate caused by clouds of dust and smoke and lasting for a year or two. Some plant species became extinct, especially broad-leaved evergreens in the general category represented today by magnolias and rhododendrons. Others came back after a time, the descendants of scattered survivors, but as part of a different mix of the post-Mesozoic era. In the southern hemisphere the effect on the vegetation was less severe.

Most paleontologists are now leaning cautiously toward the hypothesis of a sudden, catastrophic closure of the Mesozoic era. Meanwhile, the search goes on for the kind of evidence most coveted during all scientific odysseys: one easily understood discovery that definitely implicates a single major cause while dismissing the alternatives. The most obvious candidate is the smoking gun from a great meteorite strike, a giant crater somewhere on earth that could be dated precisely to the K-T boundary. Since two-thirds of the earth is covered by water, the remnant of a large blowout might lie undiscovered on the ocean floor. In 1990 two candidate craters were proposed on the basis of the distribution of shocked quartz and distinctive geological formations in accessible strata: one in the Caribbean southwest of Haiti, the other just south of western Cuba 1,350 kilometers from the first site. The evidence is not yet strong enough to be accepted. The geological conformations are under study, and the search continues in other ocean basins.

A compromise may be in order. Both the meteorite and the volcano explanations could be correct. The two events might have occurred at the same time. A meteorite 10 kilometers in diameter striking the surface at thousands of kilometers an hour would not only shake the earth's surface and darken the atmosphere but also trigger volcanic eruptions over all the planet. Alternatively, an unprovoked volcanic activity might be the key, with a meteorite strike delivering the coup de grace to the dinosaurs and the most sensitive marine animals at the time we have come to call the K-T boundary.

This brings us to the important fact that the Cretaceous extinction was only one of five such catastrophes that occurred over the last half-billion years, and it was not the most severe. Furthermore, the earlier spasms appear not to have been associated with meteorite strikes or unusually heavy volcanism. The five mass extinctions occurred in this order, according to geological period and time before the present: Ordovician, 440 million years; Devonian, 365 million years; Permian, 245 million years; Triassic, 210 million years; and Cretaceous, 65 million years. There have been a great many second- and third-order dips and rises, but these five are at the far end of the curve of violence, and they stand out. They are to other episodes as a catastrophe is to a misfortune, a hurricane to a summer squall.

The organisms that most clearly display extinction rates were animals that lived in the sea, from mollusks and arthropods to fishes, for the simple artifactual reason that their remains settled quickly to the bottom to be silted over and turned into fossils before they fully decomposed. It is further true that the units of taxonomic measure are families of related species because if any species in the group were alive at the time of the deposit, there is a good chance that at least one will now come to light in fossil form. To rely on individual species, many of which were likely to be rare or spottily distributed at any given time, is to introduce a large statistical error.

Consider the large amount of data on marine animals collected and analyzed by John Sepkoski and David Raup of the University of Chicago and others. The loss of families on which reliable data are available was just about the same, roughly 12 percent, in each of the spasms except the Permian, which saw a staggering loss of 54 percent. Statistical methods exist by which it is possible to count the number of extinguished families and make a sound educated guess as to the loss of the species that composed the families. The great Permian crash is estimated to have resulted in the loss of between 77 and 96 percent of all marine animal species. Raup has remarked that "if these estimates are reasonably accurate, global biology (for higher organisms at least) had an extremely close brush with total destruction." Trilobites and placoderm fishes, two highly distinctive and dominant groups in earlier periods, did in fact come to an end. On land, the mammal-like reptiles, distant ancestors of humanity, were devastated, with only a few survivors squeezing through. Insects and plants were less affected; they somehow acquired the invisible shield that was to surround them through all the later episodes.

No iridium has been found in deposits dating to the time of the first four spasms. Thus there was evidently no meteorite strike of sufficient magnitude to cause first-order extinction spasms. There were massive volcanic eruptions in north-central Siberia about the time of the Permian extinctions, perhaps sufficient to alter the global climate, but the connection to the decline of life is far from proved. So what did happen? In the view of Steven Stanley and some other paleobiologists, the primary agent of destruction was long-term climatic change. The evidence is circumstantial but persuasive. It includes a general retreat of tropical organisms toward the equator, reaching a peak at the time of the crises. Reef-building organisms, including algae and calcareous sponges, were especially vulnerable. They vanished over large portions of the earth. The nonliving, skeletal parts of the reefs were then either eroded by wave action or silted over. (One fossil reef, formed in Western Australia 350 million years ago, somehow resisted erosion and is still a prominent feature of the landscape.) The ranges of surviving tropical organisms were compressed toward the equator during the crises. Glaciation was more extensive.

The earth thus appears to have cooled dramatically during the first four crises, eliminating many species and forcing others into smaller ranges, rendering them more vulnerable to extinction from other causes.

I have begged the question of ultimate causation. If global cooling was the killing event, what caused the cooling? The most likely answer deduced by geologists is the movement of land masses and the fringing seas during continental drift. At the time of the first major extinction spasms—Ordovician, Devonian, and Permian—the land was moving to form the supercontinent called Pangaea. As its southern block, Gondwanaland, edged over the South Pole in Late Ordovician and Devonian times, extensive glaciation resulted, and the biological crises occurred more or less concurrently. During Permian times, Pangaea drifted farther northward, and glaciers spread over both the northern and southern ends. As ice formed, the sea level fell, drastically reducing the extent of the warm inland seas in which most of marine life then lived.

Continental drift does not appear to have been a cause of global cooling at the end of the Mesozoic period, so our attention is justifiably fixed on meteorites and volcanoes. Today the land mass of the world is arrayed in a configuration that favors high levels of diversity: widely separated continents with long shorelines and stretches of shallow tropical water dotted with lots of islands. No evidence exists of meteorite showers or volcanic blasts of world-altering proportions for the past 65 million years, at least none strong enough to have collapsed the house of cards we know as biodiversity.

To summarize: life was impoverished in five major events, and to lesser degree here and there around the world in countless other episodes. After each downturn it recovered to at least the original level of diversity. How long did it take for evolution to restore the losses after the first-order spasms? The number of families of animals living in the sea is as reliable a measure as we have been able to obtain from the existing fossil evidence. In general, five million years were enough only for a strong start. A complete recovery from each of the five major extinctions required tens of millions of years. In particular the Ordovician dip needed 25 million years, the Devonian 30 million years, the Permian and Triassic (combined because they were so close together in time) 100 million years, and the Cretaceous 20 million years. These figures should give pause to anyone who believes that what *Homo sapiens* destroys, Nature will redeem. Maybe so, but not within any length of time that has meaning for contemporary humanity.

In the chapters to follow I will describe the formation of life's diversity as it is understood—with traversals—by most biologists. I will give evidence that humanity has initiated the sixth great extinction spasm, rushing to eternity a large fraction of our fellow species in a single generation. And finally I will argue that every scrap of biological diversity is priceless, to be learned and cherished, and never to be surrendered without a struggle.

Biodiversity
Rising

The Fundamental Unit

THE MOST WONDERFUL mystery of life may well be the means by which it created so much diversity from so little physical matter. The biosphere, all organisms combined, makes up only about one part in ten billion of the earth's mass. It is sparsely distributed through a kilometer-thick layer of soil, water, and air stretched over a half billion square kilometers of surface. If the world were the size of an ordinary desktop globe and its surface were viewed edgewise an arm's length away, no trace of the biosphere could be seen with the naked eye. Yet life has divided into millions of species, the fundamental units, each playing a unique role in relation to the whole.

For another way to visualize the tenuousness of life, imagine yourself on a journey upward from the center of the earth, taken at the pace of a leisurely walk. For the first twelve weeks you travel through furnace-hot rock and magma devoid of life. Three minutes to the surface, five hundred meters to go, you encounter the first organisms, bacteria feeding on nutrients that have filtered into the deep water-bearing strata. You breach the surface and for ten seconds glimpse a dazzling burst of life, tens of thousands of species of microorganisms, plants, and animals within horizontal line of sight. Half a minute later almost all are gone. Two hours later only the faintest traces remain, consisting largely of people in airliners who are filled in turn with colon bacteria.

The hallmark of life is this: a struggle among an immense variety of organisms weighing next to nothing for a vanishingly small amount of energy. Life operates on only 10 percent of the sun's energy reaching earth's surface, that portion fixed by the photosynthesis of green plants. The free energy is then sharply discounted as it passes through the food webs from one organism to the next: very roughly 10 percent passes to the caterpillars and other herbivores that eat the plants and bacteria, 10 percent of that (or 1 percent of the original) to the spiders and other low-level carnivores that eat the herbivores, 10 percent of the residue to the warblers and other middle-level carnivores that eat the low-level carnivores, and so on upward to the top carnivores, which are consumed by no one except parasites and scavengers. Top carnivores, including eagles, tigers, and great white sharks, are predestined by their perch at the apex of the food web to be big in size and sparse in numbers. They live on such a small portion of life's available energy as always to skirt the edge of extinction, and they are the first to suffer when the ecosystem around them starts to erode.

A great deal can be learned quickly about biological diversity by noticing that species in the food web are arranged into two hierarchies. The first is the energy pyramid, a straightforward consequence of the law of diminishing energy flow as noted: a relatively large amount from the sun's energy incident on earth goes into the plants at the bottom, tapering to a minute quantity to the big carnivores on top. The second pyramid is composed of biomass, the weight of organisms. By far the largest part of the physical bulk of the living world is contained in plants. The second largest amount belongs to the scavengers and other decomposers, from bacteria to fungi and termites, which together extract the last bit of fixed energy from dead tissue and waste at every level in the food web, and in exchange return degraded nutrient chemicals to the plants. Each level above the plants diminishes thereafter in biomass until you come to the top carnivores, which are so scarce that the very sight of one in the wild is memorable. Let me stress that point. No one looks twice at a sparrow or squirrel, or even once at a dandelion, but a peregrine falcon or mountain lion is a lifetime experience. And not just because of their size (think of a cow) or ferocity (think of a house cat), but because they are rare.

The biomass pyramid of the sea is at first glance puzzling: it is turned upside down. The photosynthetic organisms still capture almost all the energy, which is discounted in steps by the 10 percent rule, but they have less total bulk than the animals that eat them. How is this inversion possible? The answer is that the photosynthetic organisms are not plants in the traditional landbound sense. They are phytoplankton, microscopic single-celled algae carried passively by water currents. Cell for cell, planktonic algae fix more solar energy and manufacture more protoplasm than plants on the land, and they grow, divide, and die at an immensely faster pace. Small animals, particularly copepods and other small crustaceans carried in the sea currents, hence called zooplankton, consume the algae. They harvest huge quantities without exhausting the standing photosynthetic crop in the water. Zooplankton in turn are eaten by larger invertebrate animals and fish, which are then eaten by still larger fish and marine mammals such as seals and porpoises, which are hunted by killer whales and great white sharks, the top carnivores. The inversion of the biomass pyramid is why the waters of the open ocean are so clear, why you can look into them and spot an occasional fish but not the green plants—algae—on which all the animals ultimately depend.

We have arrived at the question of central interest. The larger organisms of earth, composing the visible superstructures of the energy and biomass pyramids, owe their existence to biological diversity. Of what then is biodiversity composed? Since antiquity biologists have felt a compelling need to posit an atomic unit by which diversity can be broken apart, then described, measured, and reassembled. Let me put the matter as strongly as this important issue merits. Western science is built on the obsessive and hitherto successful search for atomic units, with which abstract laws and principles can be derived. Scientific knowledge is written in the vocabulary of atoms, subatomic particles, molecules, organisms, ecosystems, and many other units, including species. The metaconcept holding all of the units together is hierarchy, which presupposes levels of organization. Atoms bond into molecules, which are assembled into nuclei, mitochondria, and other organelles, which aggregate into cells, which associate as tissues. The levels then progress on upward as organs, organisms, societies, species, and ecosystems. The reverse procedure is decomposition, the breaking of ecosystems into species, species into societies and organisms, and so on downward. Both theory and experimental analysis in science are predicated on the assumption—the trust, the faith—that complex systems can be cleaved into simpler systems. And so the search proceeds relentlessly for natural units until, like the true grail, they are found and all rejoice. Scientific fame awaits those who discover the lines of fracture and the processes by which lesser natural units are joined to create larger natural units.

So the species concept is crucial to the study of biodiversity. It is the grail of systematic biology. Not to have a natural unit such as the species would be to abandon a large part of biology into free fall, all the way from the ecosystem down to the organism. It would be to concede the idea of amorphous variation and arbitrary limits for such intuitively obvious entities as American elms (species: *Ulmus americana*), cabbage white butterflies *(Pieris rapae)*, and human beings *(Homo sapiens)*. Without natural species, ecosystems could be analyzed only in the broadest terms, using crude and shifting descriptions of the organisms that compose them. Biologists would find it difficult to compare results from one study to the next. How might we assess, for example, the thousands of research papers on the fruit fly, which form much of the foundation of modern genetics, if no one could tell one kind of fruit fly from another?

I will try to cut to the heart of the matter with the "biological-species concept": *a species is a population whose members are able to interbreed freely under natural conditions*. This definition is an idea easily stated but filled with exceptions and difficulties, all interesting, all reflective of the range of complexity in evolutionary biology itself. My opinion is that the grail, though nicked and tarnished, is in our possession. The chalice sits on the shelf. I must add at once that not all biologists accept the biological-species concept as sound or as the pivotal unit on which the description of biological diversity can be based. They look to the gene or the ecosystem to play these roles, or they are happy to live with conceptual anarchy. I think they are wrong, but in any case will return shortly to the difficulties of the biological species to give voice to their misgivings.

For the moment let me go on to expand the definition, which is accepted at least provisionally by a majority of evolutionary biologists. Notice the qualification it contains, "under natural conditions." This says that hybrids bred from two kinds of animals in captivity, or two kinds of plants cultivated in a garden, are not enough to classify them as a member of a single species. To take the most celebrated example, zookeepers have for years crossed tigers with lions. The offspring are called tiglons when the father is a tiger and ligers when the father is a lion. But the existence of these creatures proves nothing, except perhaps that lions and tigers are genetically closer to each other than they are to other kinds of big cats. The still unanswered question is, do lions and tigers hybridize freely where they meet under natural conditions?

Today the two species do not meet in the wild, having been driven back by the expansion of human populations into different corners of the Old World. Lions occur in Africa south of the Sahara and in one small population in the Gir Forest of northwestern India. Tigers live in small, mostly endangered populations from Sumatra north through India to southeastern Siberia. In India, no tigers are found near the Gir Forest. It would seem at first that the test of the biological-species concept, free interbreeding in nature, cannot be applied. But this is not so: during historical times the two big cats overlapped across a large part of the Middle East and India. To learn what happened in these earlier days is to find the answer.

At the height of the Roman Empire, when North Africa was covered by fertile savannas—and it was possible to travel from Carthage to Alexandria in the shade of trees—expeditions of soldiers armed with net and spear captured lions for display in zoos and in colosseum spectacles. A few centuries earlier, lions were still abundant in southeastern Europe and the Middle East. They preyed on humans in the forests of Attica while being hunted themselves for sport by Assyrian kings. From these outliers they ranged eastward to India, where they still thrived during British rule in the nineteenth century. Tigers ranged in turn from northern Iran eastward across India, thence north to Korea and Siberia and south to Bali. To the best of our knowledge, no tiglons or ligers were recorded from the zone of overlap. This absence is especially notable in the case of India, where under the British Raj trophies were hunted and records of game animals kept for more than a century.

We have a good idea why the two species of big cats, despite their historical proximity, failed to hybridize in nature. First, they liked different habitats. Lions stayed mostly in open savanna and grassland and tigers in forests, although the segregation was far from perfect. Second, their behavior was and is radically different in ways that count for the choice of mates. Lions are the only social cats. They live in prides, whose enduring centers are closely bonded females and their young. Upon maturing, males leave their birth pride and join other groups, often as pairs of brothers. The adult males and females hunt together, with the females taking the lead role. Tigers, like all other cat species except lions, are solitary. The males produce a different urinary scent from that of lions to mark their territories and approach one another and the females only briefly during the breeding season. In short, there appears to have been little opportunity for adults of the two species to meet and bond long enough to produce offspring.

Each biological species is a closed gene pool, an assemblage of organisms that does not exchange genes with other species. Thus insulated, it evolves diagnostic hereditary traits and comes to occupy a unique geographic range. Within the species, particular individuals and their descendants cannot diverge very far from others because they must reproduce sexually, mingling their genes with those of other families. Over many generations all families belonging to the same biological species are by definition tied together. Linked as one by the chains of ancestry and future descent, they all evolve in the same general direction.

The biological-species concept works best if used in a single locality, such as a state or county or small island, over a short period of time. Consider any group of organisms in such a place and time. Select one haphazardly: say the hawks of Harris County, Texas. Walk through the remaining natural habitats around the city of Houston looking for accipiters, buzzard hawks, harriers, ospreys, and falcons, and you will eventually find sixteen species. Some, such as the red-shouldered hawk *(Buteo lineatus)* and kestrel *(Falco sparverius)*, are relatively common. Others, Harlan's hawk *(Buteo harlani)* and the prairie falcon *(Falco mexicanus)*, are rare. In the end, after enough visits to fields, scrub-pine woods, and timbered swamps, you will have compiled the same list as other veteran birdwatchers, and your characterizations will coincide with those in Roger Tory Peterson's *Field Guide to the Birds of Texas and Adjacent States*. Each hawk species possesses a diagnostic combination of anatomical traits, call, favored prey, flight pattern, and geographical range. Some of these qualities, such as mating behavior, can be seen to contribute to the reproductive isolation of the sixteen species. Hybrids in nature are next to none.

It might immediately have come to your mind that the general agreement on hawk species is only a cultural artifact, a convention about anatomy and scientific names that arose in the same fashion as the autonomous evolution of common law, from intuition and historical accident—dependent on who first used plumage color to classify types, who first applied a Latin name to some recognizable form or other, and so on until a classification emerged with which a sufficient number of people were able to feel comfortable and, finally, to which Roger Tory Peterson gave his imprimatur. You would be wrong. There is a test that can distinguish cultural artifacts from natural units: the comparison of classifications made by human societies that have never been in contact. In 1928 the great ornithologist Ernst Mayr traveled as a young man to the remote Arfak Mountains of New Guinea to make the first thorough collection of birds, including hawks. Before departing, he visited key bird collections already deposited in European museums. By studying specimens gathered from western New Guinea, he estimated that a little more than one hundred bird species would be likely to occur in the Arfak Mountains. His species concept was that of a European scientist looking at dead birds, who then sorts the specimens in piles according to their anatomy, as a bankteller stacks nickels, dimes, and quarters. Once settled in camp, after a long and hazardous trek, Mayr hired native hunters to help him collect all the birds of the region. As the hunters brought in each specimen, he recorded the name they used in their own classification. In the end he found that the Arfak people recognized 136 bird species, no more, no less, and that their species matched almost perfectly those distinguished by the European museum biologists. The only exception was a single pair of closely similar species that Mayr, the trained scientist, was able to separate but that Arfak mountain people, although practiced hunters, lumped together.

Many years later, when I was twenty-five years old, about Mayr's age at the time of his Arfak adventure, I made a long trek through the Saruwaget Mountains of northeastern New Guinea to collect ants. I repeated the cross-cultural test and found that the Saruwaget people could not tell one ant from another. An ant was an ant was an ant. This should have come as no surprise. It was not that Saruwaget ants and natives failed the test, only that Papuans have no practical need to classify ants. The Arfak people are hunters who use their knowledge of bird diversity to make a living, just as European ornithologists do. In Mayr's time at least, wild birds were their principal source of meat.

To the same end, the Amerindian tribes of the Amazon and Orinoco basins have an intimate knowledge of the plants of the rain forest. A few shamans and tribal elders are able to put names on a thousand or more species of plants. Not only do the botanists of Europe and North America generally agree with these species distinctions, but they have learned a great deal from their Amerindian colleagues about the habitat preferences, flowering seasons, and practical uses of the different plants. It is a remarkable fact that the only crop used widely by developed countries not already known to native peoples is the macadamia nut, which originated in Australia. Unfortunately, much of the indigenous knowledge is being lost as European culture continues to intrude and the last preliterate native

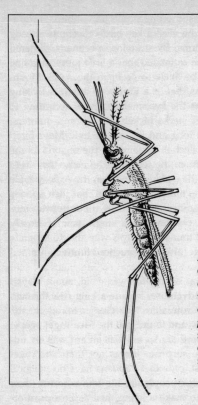

A malaria-carrying mosquito of Europe. The female depicted here resting on a wall belongs to a sibling species of the *Anopheles maculipennis* group—one of several forms so similar that even experts have difficulty identifying among them.

cultures in tropical countries weaken and disappear. We are losing irretrievably what is in a real sense scientific knowledge.

In all cultures, taxonomic classification means survival. The beginning of wisdom, as the Chinese say, is calling things by their right names. Following the discovery in 1895 that malaria is carried by *Anopheles* mosquitoes, governments around the world set out to eradicate those insect vectors by draining wetlands and spraying infested areas with insecticides. In Europe the relation between the malarial agent, protozoan blood parasites of the genus *Plasmodium*, and the vector mosquito, *Anopheles maculipennis*, seemed at first inconsistent, and control efforts lacked pinpoint accuracy. In some localities the mosquito was abundant but malaria rare or absent, while in others the reverse was true. In 1934 the problem was solved. Entomologists discovered that *A. maculipennis* is not a single species but a group of at least seven. In outward appearance the adult mosquitoes seem almost identical, but in fact they are marked by a host of distinctive biological traits, some of which prevent them from hybridizing. The first such "characters" detected were the size and shape of the rafts of eggs laid by female mosquitoes on the water surface. Those of two species were found to produce no clusters at all, instead laying each egg separately. Entomologists were alerted, and other parts of the puzzle soon fell into place. More characters were quickly added: color of the eggs, gross structure of the chromosomes, hibernation versus continuous breeding in winter, and geographical distribution. Most important of all, some of the species distinguished by these traits were found to feed on human blood and thus to carry the malarial parasites. Once identified, the dangerous members of the *A. maculipennis* complex could be targeted and eradicated. Malaria virtually disappeared from Europe.

Systematists are often able to solve biological problems in this manner by breaking species into characters of sibling species. In the reverse direction, they also frequently lump other forms previously considered good species into larger, variable species by demonstrating the existence of only one population of freely breeding individuals. When done correctly, splitting and lumping open the door to a secure interpretation of the organisms on which the analysis was performed.

Still, the biological-species concept has chronic deep problems. From the beginning of its first clear formulation at the turn of the century, it has been corroded by exceptions and ambiguities. The fundamental reason is that each species defined as a reproductively isolated population or set of populations is in some stage or other of evolution that makes it different from all other species. It is moreover a unique individual, not merely one unit in a class of identical units such as a hydrogen atom or a molecule of benzene. This qualification makes it different from a concept in physics or chemistry, which is a summary term for a set of measurable quantities. An electron, for example, is a postulated unit with 4.8×10^{-10} units of negative charge and 9.1×10^{-28} grams of mass. Of course no one has actually seen electrons, but physicists believe in them because the properties attributed to them permit a precise explanation of cathode rays, electromagnets, the photoelectric effect, electricity, and chemical bonds. A large part of physics and chemistry depends on a precise visualization of the way electrons are stripped from atoms and molecules to create positive ions and free electrons. In the language of physics, they are virtual; they have undoubted corporeal existence. At Cambridge University in the 1930s Lord Rutherford and his research group sang of these invisible bodies at the annual Cavendish dinner, to the tune of the popular song "My Darling Clementine":

> There the atoms in their glory,
> Ionize and recombine.
> Oh my darlings, oh my darlings,
> Oh my darlings, ions mine.

But all the members of a given class are identical, and the class is forever absolute and unalterable.

If an electron is truly an electron and an ion is an ion, and all in a class are interchangeable, a species is a thing-unto-itself that merely shares certain properties with most other species, most of the time. For species are always evolving, which means that each one perpetually changes in relation to other species. In some cases sibling species are so similar that only biochemical tests or mating experiments can tell them apart, bringing despair to practicing biologists who have to sort organisms quickly. In the eastern United States slipper animalcules, the familiar little protozoans of the genus *Paramecium* widely used in high-school biology classes, used to be separated into three common "species," *P. aurelia*, *P. bursaria*, and *P. caudatum*, on the basis of anatomical differences that can be readily seen under the light microscope. Close study, however, has disclosed that no fewer than 20 species exist, separable at the least by their mating behavior and hence constituting independently evolving populations. There is a strong temptation just to ignore the biological complexity and stay with the three old, easy species, but the malaria example counsels otherwise. Biologists know in their hearts that there can be no compromise on matters of such importance, that they must soldier on until all the true closed-gene pools have been defined. Every one of the atomic units must be given a name.

Sibling species present no more than technical problems. They threaten no crisis for biological theory. More serious conceptual problems are created by "semispecies," populations that partially interbreed—not enough to constitute one big freely interbreeding gene pool, but enough to produce a good many hybrids under natural conditions. The problem is acute in many plants, especially in those that are pollinated by the wind so that pollen is broadcast scattershot and often settles on flowers of the wrong species. Along the Pacific coast of North America, about a third of the oak and pine species are actually semispecies. Yet somehow the semispecies stay apart as breeding systems. They are recognizable as discrete entities in the field even while exchanging genes by occasional hybridization. They can be distinguished by the anatomy of their foliage and flowers and the habitats in which they preferentially grow. Alan Whittemore and Barbara Schaal, after completing a study of DNA differences among the native white oaks of the eastern United States, concluded:

> The genus *Quercus* (the oaks) is outstanding for the very poor development of sterility barriers between its species. Oak species are interfertile in many combinations, and natural hybrids may be formed between pairs of species that are very different from one another both morphologically and physiologically. Although some pairs of interfertile species show strong ecological separation, many interfertile species pairs show extensive ecological overlap.

Yet somehow the white oaks remain distinct. Hybridization between the species is much less frequent than breeding within them, and as a result the gene pools stay partly closed.

It is further true that an abundance of hybrids, and with it the maintenance of semispecies in a state of ambiguous tension, may not be a global phenomenon in plants. Tropical species appear to exchange genes less extensively than those in the temperate zones. In other words, they "behave" more like animals in the maintenance of a stricter pattern of species diversity. Since the great majority of plant species occur in the tropics, this evolutionary conservatism may prove to be a more general botanical trait than hybridization at the intensity displayed by oaks. A conspicuous exception is the large tree genus *Erythrina*, whose species commonly hybridize. But the genetic study of hybridization and species formation in tropical plants has hardly begun, and caution is advised.

It is the nature of species formation, to be explained in the next chapter, that for a while after the splitting of a single species into two species—call them A and B—some members of species A can be more closely related to some members of species B than they are to other As, and vice versa. These relatives in A and B have a close common ancestry, but they have acquired one or a few crucial differences that prevent them from exchanging genes. They are like sisters who live in different countries and are unable to cross the national boundary. Some biologists have argued that such individuals should be put in the same species, a single "phylogenetic species," regardless of their inability to interbreed. Other biologists, and I am one of them, firmly disagree. The idea of the phylogenetic species is an interesting and useful but not fatal challenge to the biological species. To seek genealogical information within and between populations does not require the overthrow of reproductive isolation as the salient process of diversification at the population level. Sisterhoods are important, but nations more so.

We must now confront an even more serious conceptual difficulty for the biological-species concept. The idea of the closed-gene pool has no meaning at all for the minority of organisms that are either obligatorily hermaphroditic—those that have both ovaries and testes and fertilize themselves—or parthenogenetic, producing offspring from unfertilized eggs. By one or the other of these devices, various microorganisms, fungi, plants, mites, tardigrades, crustaceans, insects, and even lizards simply forgo the inconvenience and perils and gambler's excitement of sexual reproduction.

How to solve the dilemma? Asexual and self-fertilizing forms tend to maintain a remarkable integrity. The vast majority, even though freed from the evolutionary constraints of sexual compatibility, do not vary in all directions, do not fan out to create wide continuous variation and taxonomic confusion. The gene combinations of the organisms are prone to exist in clusters, enabling systematists to place most specimens with ease. It is widely believed that the clustering is due to the lower survival and reproduction rate of vagrant intermediate forms. Only those organisms with anatomy and behavior close to the norm are able to do well. Also, many of the nonsexual species have recently evolved from sexually reproducing ancestors and therefore have not had enough time to diverge or expand. In the end, however, the lines drawn by the biologist around such species must be arbitrary.

The closed-pool concept also loses its meaning in the case of chronospecies, which are stages in the evolution of the same species through time. Consider our own species, *Homo sapiens*, which evolved in a straight line from *Homo erectus*, found through Africa and Eurasia roughly a million years ago. Obviously we cannot know whether *H. sapiens* and *H. erectus* would interbreed freely if brought together in nature. The question is hollow when presented away from context; it is a scientist's koan, the equivalent of the sound of one hand clapping, the length of a string. Yet paleontologists, driven by practical necessity, go on distinguishing and naming chrono-species. They are right to do so. It would be irresponsible to call *H. sapiens* and *H. erectus* the same species, and even more so to add their immediate antecedent *H. habilis* and—still further back—the primitive australopithecine man-apes.

Searching for an anchor, willing to compromise in order to find some process shared by a large fraction of organisms, biologists keep re-turning to the biological-species concept. In spite of its difficulties, regardless of the fact that it can never be employed as an abstract entity like the electron to make exact quantitative calculations, the concept is likely to continue to hold center stage for the simple reason that it works well enough in enough studies on most kinds of organ-isms, most of the time.

The vast majority of species are in fact sexual; they do exist as closed-gene pools. The biological-species concept works very well in the study of local faunas and floras, such as the hawks of Texas and the mosquitoes of Europe and the primates of the Old World includ-ing *Homo sapiens*, and maximally so in well-demarcated communities on islands and isolated habitat patches, circumstances, in short, found in a large part of the real world.

For years I have suffered through seminars and hallway debates on the biological species. I have read through a library of opinions and witnessed the concept wax and wane in the minds of evolution-ary biologists. The core problem seems to be the democratic process of science, in the sense that the concept has a weak constituency: people really don't need it most of the time. Systematists perform their sorting largely on the basis of differences they find between museum specimens. Asked if the differences are maintained by re-productive isolation, they respond: probably. But they don't care enough to go to the trenches. They are satisfied that the anatomical gaps exist and leave it to population biologists to find out why this is so. Population biologists for their part are enchanted by the dy-namism of the speciation process and the many problems thrown up to embarrass the biological-species concept in the early stages of species separation. What is wrong, many ask, with disorder or even chaos, for a while? Why not play around with many species concepts, each designed to fit ad hoc circumstances? Content with the thrill of the race they are running, and with scattered applause along the way, few see profit in sprinting to the finish line.

Unlike the systematists, population biologists do not have to classify a million species. And they forget that reproductive isolation between breeding populations is the point of no return in the creation of biological diversity. During the earliest stages of divergence, there may be less difference between the two species than exists as variation within them. A surge of hybrids may yet occur to erase the barrier and confuse the picture even more. But in most cases the two species are embarked on an endless journey that will carry them further and further apart. The differences between them will in time far exceed anything possible among the members of their own breeding populations. In the real world, the great range of biological diversity has been generated by the divergence of species that were created in turn by the defining step expressed in the biological-species concept.

Someday biologists may come up with a single concept that unifies sexual species, nonsexual species, and chronospecies into a single, theoretically powerful natural unit. But I doubt it. The dynamism of the evolutionary process and the individuality of species make it unlikely that a completely universal species definition will ever be fashioned. Instead, two to several concepts will continue to be recognized, like the waves and particles of physics, as optimal in different circumstances. Of these, the biological species is likely to remain central to the explanation of global diversity. But whatever the outcome, the imperfections of the concept, and thereby of our classificatory system, reflect the idiosyncratic essence of biological diversity. They give even more reason to cherish each species in turn as a world unto itself, worthy of lifetimes of study.

New Species

WHAT IS THE ORIGIN of biological diversity? This profoundly important problem can be most quickly solved by recognizing that evolution creates two patterns across time and space. Think of a butterfly species with blue wings as it evolves into another species with purple wings. Evolution has occurred but leaves only one kind of butterfly. Now think of another butterfly species, also with blue wings. In the course of its evolution it splits into three species, bearing purple, red, and yellow wings respectively. The two patterns of evolution are vertical change in the original population and speciation, which is vertical change plus the splitting of the original population into multiple races or species. The first blue butterfly experienced pure vertical change without speciation. The second blue butterfly experienced pure vertical change plus speciation. Speciation requires vertical evolution, but vertical evolution does not require speciation. The origin of most biological diversity, in a phrase, is a side product of evolution.

Vertical change is mostly what Darwin had in mind when he published his 1859 masterwork. The full title tells the story: *On the Origin of Species by Means of Natural Selection, or the Preservation of Favoured Races in the Struggle for Life.* In essence, Darwin said that certain hereditary types within a species (the "favoured races") survive at the expense of others and in so doing transform the makeup of the entire species across generations. A species can be altered so extensively by natural selection as to be changed into a different species, said Darwin. Yet no matter how much time elapses, no matter how much change occurs, only one species remains. In order to create diversity beyond mere variation among the competing organisms, the species must split into two or more species during the course of vertical evolution.

Darwin understood in a general way the difference between vertical evolution and the splitting of species, but he lacked a biological-species concept based on reproductive isolation. As a result he did not discover the process by which multiplication occurs. His thinking on diversity remained fuzzy. In that sense, the short title *On the Origin of Species* is misleading.

The distinction between the two patterns of evolution is illustrated in concrete detail by human evolution. The earliest known hominid in the fossil record is the man-ape *Australopithecus afarensis*. A hominid, to be technically explicit, is a member of the family Hominidae, which includes modern *Homo sapiens* and earlier human and human-like species. When *A. afarensis* lived in the savannas and woodlands of Africa five to three million years ago, it was, so far as the evidence shows, the only species of its kind. The scant fossils reveal that adults walked in a bipedal manner broadly similar to that of *Homo sapiens*. This posture was and is unique among all the mammals. It freed the man-apes to carry burdens in their arms and hands. They or their descendants were able to transport infants, tools, and food for long distances. Perhaps they established campsites (although evidence so far back is lacking), and from that habit divided labor among those who stayed at home to tend the camp and those who foraged abroad. In brains *A. afarensis* was not conspicuously endowed. Its cranial capacity was no greater than that of modern chimpanzees, about 400 cubic centimeters. But the stage was set for the evolutionary advance toward humankind.

As time passed, to no later than two million years ago, the early man-ape populations both evolved and split into at least three distinct species. Two of the species, the advanced man-apes *Australopithecus boisei* and *Australopithecus robustus*, were 1.5 meters (five feet) tall, with a gorilla-like bony crest down the midline of the skull that anchored enormous jaw muscles. They were probably vegetarians, using molar teeth up to 2 centimeters long to crush seeds and shred tough vegetation, much as gorillas do today. The third species derived from the primitive man-apes, *Homo habilis*, was more nearly a true human in the contemporary sense, enough so for anthropologists to remove it from the genus *Australopithecus* and place it in *Homo*. It stood just under 1.5 meters and weighed about 45 kilograms

(100 pounds). Its form was essentially that of modern *H. sapiens*, with one outstanding exception: the brain volume was between 600 and 800 cubic centimeters, still only half as much as in modern *H. sapiens* but considerably larger than in chimpanzees.

During the next million years the man-apes disappeared, and with them most hominid diversity. *H. habilis* survived and continued to evolve in total size and cranial capacity, slowly at first and then at an accelerating rate. It metamorphosed into the intermediate species *Homo erectus*, reaching that level, or "evolutionary grade" as biologists like to call it, about 1.5 million years ago. Sometime during its early history *H. erectus* expanded its range from Africa to Europe and Asia.

Fossils found at Zhoukoudian, near Beijing, China, bear witness to the ensuing evolution over a quarter million years. The brain size rose steadily from 915 to 1,140 cubic centimeters, while the stone tools designed by these "Peking men" grew increasingly sophisticated. Across the far-flung *H. erectus* populations of the Old World, the brain size and dentition progressed to essentially modern levels, by half a million years before the present. The chronospecies *H. sapiens*, or the modern species derived from the archaic species by vertical evolution, had arrived. Its taxonomic diagnosis is extraordinary: brain 3.2 times larger than in an ape of human size, housed in a wobbly spherical skull; jaw and teeth feeble; body borne erect on elongated hindlegs; skin mostly hairless except for patches that warm the head and display the genitalia; internal organs supported by a basin-shaped pelvis; thumb abnormally long for a primate, turning the hand into a specialized device for handling tools; mind fashioned from symbolic language and semantic memory with the aid of elaborate speech-control centers located in the parietal cortex.

Now to recapitulate human evolution as a display of the two patterns of evolution: the hominids experienced a modest amount of speciation during the early man-ape period, followed by extinction of all the lines except the rapidly evolving single species of *Homo*. The ancestral *Australopithecus afarensis* probably had a generalized diet. The hominids that succeeded it combined vertical evolution and full species formation enforced by reproductive isolation. The species deployed into different niches in the manner typical of successful, expanding animal groups. The man-apes *Australopithecus boisei* and *A. robustus* became increasingly vegetarian. *Homo habilis*, which had already diverged enough to be placed in a genus of its own, added meat to its diet by means of both hunting and scavenging. At the same time, it evidently consumed enough vegetable material to retain what today would be called a balanced diet. Then the man-apes disappeared, possibly driven to extinction by *H. habilis* or its descendant, *H. erectus*. What followed was mostly the evolution of a single species, traversing the series *H. habilis* to *H. erectus* to *H. sapiens*.

Most animal and plant species, including *H. sapiens*, retain their identity simply by not breeding with other species. How does this segregation come about in the first place? The process as we understand it is surprisingly simple. Any evolutionary change whatsoever that reduces the chances of producing a fertile hybrid can yield a new species. The reason is that the launching of fertile hybrids is a complicated and delicate procedure. It is somewhat like putting a space vehicle into orbit. A vast number of parts must work correctly, and the timing of their operation must be nearly perfect. Otherwise the mission fails. Consider a male of species A and a female of species B trying to create a fertile hybrid offspring. Because they are genetically different from one another, things can go wrong. The two individuals might want to mate in different places. They might try to breed at different seasons or times of the day. Their courtship signals could be mutually incomprehensible. And even if the representatives of the two species actually mate, their offspring might fail to reach maturity or, attaining maturity, turn out to be sterile. The wonder is not that hybridization fails but that it ever works. *The origin of species is therefore simply the evolution of some difference—any difference at all—that prevents the production of fertile hybrids between populations under natural conditions.*

Biologists speak weightily of all the things that fail as "intrinsic isolating mechanisms." By intrinsic they mean hereditary, in other words differences prescribed by the genes of the opposing populations. They do not mean something extrinsic, such as a river or mountain range, that keeps populations A and B apart. Step by step through the reproductive process, no intrinsic isolating mechanism must intrude if two populations are to remain in the same species. Step by step, the appearance of even one such mechanism will cleave them into two distinct species. Distinct, that is, if you accept the biological-species concept, which you must do if we are to avoid chaos in general discussions of evolution.

By all means let us avoid chaos. Take any set of sexually reproducing species that live together in the same geographical region. They are reproductively isolated from one another by their own distinctive isolating mechanisms. This is just a formal way of saying that some hereditary difference between the species, examined pair by pair, prevents them from producing large numbers of fertile hybrids.

Consider flycatchers of the genus *Empidonax*, little birds that perch on tree limbs or powerlines and dart out from time to time to snatch flying insects. Five species occur together in the northern United States. They remain genetically distinct in part because they prefer different habitats. For example:

Least flycatcher *(E. minimus)*, open woods and farmland

Alder flycatcher *(E. alnorum)*, alder swamps and wet thickets

Yellow-bellied flycatcher *(E. flaviventris)*, coniferous woods and cold bogs

In addition, each species uses its own identifying call during the breeding season, so distinctive in combination with habitat choice as to leave little room for mistakes and the creation of hybrids.

The possibility for error has no limit, and so intrinsic isolating mechanisms are endless in variety. Examples worked out by field researchers are not merely the substance of academic biology. They also explain a great deal of wondrous and otherwise indecipherable natural history. Some examples:

• The giant silkworm moths of North America (family Saturniidae) fly and mate at various times during late afternoon and through the night. The females call in the males over distances of up to several kilometers by releasing a powerful chemical scent. They pop out an eversible sac folded into the rear tips of their bodies, exposing the attractant to the air and allowing it to evaporate and disperse into a downwind plume. The males are extremely responsive to the sex attractant. When they detect only a few molecules, they fly upwind, bringing them close to the females. Each species of giant silkworm moth is sexually active only during a limited span of time in each daily cycle, as follows:

Females of promethea moths *(Callosamia promethea)* call from about 1600 to 1800 hours

Females of polyphemus moths *(Antheraea polyphemus)* call from about 2200 to 0400 hours

Females of cecropia moths *(Hyalophora cecropia)* call from about 0300 to 0400 hours

And so on through the 69 species of North American Saturniidae, each of which so far as we know has its own *heure d'amour*. Agitated in flight, the males select females of their own species not only by the time of the calls but by the chemical composition of the sex attractants. Few if any mistakes are made thanks to the combination of timing and scent, and thus few if any hybrids come into being.

• Male jumping spiders (family Salticidae) recognize females of their own species by sight. The courting male faces the female. His visage, which is strikingly colored according to species, is instantly recognizable to the sharp-eyed female spider and to human observers alike. In the woods and fields of New England lives a species with a red brow, white face, and black fangs. Another can be found bearing gray brow, red face, and white fangs; still another with black brow and face and snowy white hair that envelop the fangs like ermine muffs; and yet another with tufts of hair elevated behind the head like black-speckled fairy wings. The males posture and dance before the females. Those of one species supplement the yellow and black vertical striping of their fangs by raising black-tipped yellow forelegs above their heads in a gesture resembling surrender. Those of others variously bob up and down, weave side to side, bend their abdomens over their heads or twist them to the side, or raise their forelegs and wave them side to side like semaphore flags. When biologists give females a choice between males, the spiders use the colors and movements to select mates of their own species.

• The heliconias (family Heliconiaceae) are mostly tropical American plants pollinated by hummingbirds, which they attract with huge flower-like structures called bracts and copious offerings of nectar. Hummingbirds respond to this extravagant generosity by carrying the pollen of the heliconias with speed and efficiency. They serve the plants well, but there is a problem: the birds like to visit more than one heliconia species, creating a risk of hybridization. The heliconias have solved the problem by the evolution of flower parts of differing length. Each species plants its pollen on a particular zone of the hummingbird's body. The hummingbird in turn deposits the pollen only on the stigmas of flowers of the same length, in other words only on flowers belonging to the same species.

It is possible to proceed down the catalogue of evolved procedures known to biologists by which species avoid hybridization and seldom see one repeated in exact detail. Many of the most elaborate and beautiful displays of nature function as intrinsic isolating mechanisms, from bright colors to beguiling odors and melodious songs.

But wait: I have been speaking of the origin of species in paradoxical language. In the traditional language of biology, the "mechanisms" have "functions." Yet they represent whatever can go wrong, not what can go right. In other words, beauty arises from error. How can both of these apparently contradictory perceptions be true? The answer, based on studies of many populations in the wild, is this: *the differences between species ordinarily originate as traits that adapt them to the environment, not as devices for reproductive isolation.* The adaptations may also serve as intrinsic isolating mechanisms, but the result is accidental. Speciation is a by-product of vertical evolution.

To see why this strange relationship holds, consider the special but widespread mode of diversification called *geographical speciation.* Start with an imaginary population of birds—say, flycatchers—that was split by the last glacial advance in North America. Over several thousand years, the population living in what would today be the southwestern United States adapted to life in open woodland, while the other population, in the southeastern United States, adapted to life in swamp forests. These differences were independently acquired and functional. They allowed the birds to survive and reproduce better in the habitats most readily available to them south of the glacial front. With the retreat of the ice, the two populations expanded their ranges until they met and intermingled across the northern states. One now breeds in open woodland, the other in swamps. The differences in their preferred habitats, based on hereditary differences acquired during the period of enforced geographical separation, makes it less likely that the two newly evolved populations will closely associate during the breeding season and hybridize. The adaptive difference in habitat thus accidentally came to serve as an isolating mechanism.

Other traits might well have diverged between two such bird populations during their period of geographical separation, including songs used by the males to attract females and places within the forest favored for the construction of nests. Any of these hereditary differences might lessen the chances that adults of the woodland and swampland species mate. If mating does occur, the hybrids will be intermediate in the newly evolved traits. They will not be well suited to either open woodland or swampland, and hence have less chance to survive. Any kind of barrier, from habitat differences to maladapted hybrids, will work as an intrinsic barrier if it is strong enough. The two populations have turned into distinct species because they are reproductively isolated where they meet under natural condi-

A species multiplies into two or more daughter species when its populations are isolated from one another long enough by geographical barriers. In this example, based on a composite of real cases, the parental species is at first widely distributed across terrain composed mostly of grassland *(top)*. The local climate grows wetter, and a river splits the species in two, so that now one population lives in grassland and the other in woodland *(bottom)*.

In time the two populations evolve apart, until they attain the level of new species *(top)*. When the river barrier disappears, the two forms are able to live together without interbreeding *(bottom)*.

tions. The single ancestral, pre-glaciation species has been split into two species, an entirely incidental result of the vertical evolution of its populations while they were separated by a geographical barrier.

The fission of the flycatcher species is a simplified composite of real speciation events that occurred in North America during the last glacial advance. Something like it has been repeated in many parts of the world, among many kinds of plants and animals, over hundreds to thousands of years. The process is initiated by geographical barriers that rise and fall, promoting the origin of a random array of hereditary isolating mechanisms sufficient to keep newly formed species from interbreeding when they do come in contact. Evolutionary biologists have discovered an extreme diversity of such barriers, as seen in these examples:

• In the Amazon Basin, droughts break up continuous forest into scattered patches. Some of the plant and animal populations newly isolated within them start to diverge. Other populations are fragmented by shifts in river courses that repeatedly open and close rainforest corridors linking the patches.

• Along the coast of New Guinea, stretches of the continental shelf are partially submerged as the sea level rises. Those parts with the highest elevation remain above water as islands, and populations restricted to them start to diverge.

• The Hawaiian islands are colonized by birds, crickets, wasps, damselflies, beetles, snails, flowering plants, and other kinds of organisms arriving as occasional windblown waifs. As the first colonists multiply and spread, they evolve in response to the distinctive environment of the islands and hence diverge from the ancestral populations left behind on the mainlands of North America and Asia. They also spread from island to island within the archipelago and from valley to valley and along the ridge tops on each island's mountainous interior, generating new isolated and diverging populations as they go. A single species that colonized Hawaii 100,000 years ago could easily have given rise to hundreds of species, each endemic at the present time to a particular island, valley, or mountain ridge.

I have stressed the imperfections of the species as a natural unit. The flaws that plague it are the inevitable consequence of the particularities of history. Every population of animals exists in a highly dynamic state, growing in size if it can, thrusting into new terrain when permitted, evolving in new directions as opportunity arises. Raw chance weighs heavily in guiding its evolutionary trajectory.

Consider a biologically diverse environment such as a forested valley on Kauai, a shallow-water shelf along the shores of Lake Victoria, or a cypress swamp in northern Florida. Some of the resident species are individually specialized for narrow niches and limited to small geographical ranges. Their dispersal ability is poor, their close relatives few or nonexistent. Their vertical evolution creeps along, and speciation is stalled. They have no geographic races and little prospect for multiplication. At the opposite extreme, other species possess flexible food habits and are excellent dispersers. They form new populations readily and evolve quickly into new niches, from diet to habitat to season of activity. Their potential for diversification is high, and they pile up species in the same localities by repeated cycles of dispersal and reinvasion.

In focusing on this last group of actively evolving populations, we are most likely to encounter all the stages of geographical speciation as interpreted by prevailing theory. At the earliest stage, the population is spread continuously across its range, and all the organisms in it freely interbreed. Few if any differences occur from one end of the range to the other. At the next stage, the population is still continuously distributed but divided into subspecies. Where the subspecies meet, they freely crossbreed. Imagine such a race of butterflies in Texas with large spots on their wings, and another in Mississippi lacking spots. Where the two subspecies meet in Louisiana and interbreed, the butterflies have spots of intermediate size. Those near Texas have larger spots, tending toward full Texas size; those near Mississippi have very small spots, tending toward the unadorned Mississippi condition.

Time passes, and at a more advanced stage the subspecies are still able to interbreed freely if they meet, but by now they have diverged in many genetic qualities. Butterflies from the same populations might differ not only in wing pattern but also in size, preferred food plant, growth rate of the caterpillars, and so on through any combination of hundreds of traits subject to genetic variation. The divergence of the subspecies will be hastened if some physical barrier, such as a broad river or dry grassland corridor, separates the two populations and restricts the flow of genes between them.

Finally, the two populations have diverged so far that they do not interbreed when they meet. They have been transformed into full-fledged biological species. Our two butterfly species now coexist in Louisiana, kept apart by differences in breeding season, courtship behavior, or some other intrinsic isolating mechanism singly or in combination—an inherited failure, in other words, to mesh their reproductive activity. Few if any small-spotted hybrids occur in the zone of overlap.

This classical model of geographical speciation makes a tidy picture. It has a core of truth, but real evolution is much messier. In fact, evolution is so messy that a faithful description of real cases converts the science into natural history, in which unique details are as important as the principles by which they are explained.

Consider the subspecies. The category seems an inevitable intermediate step in an Aristotelian progression running from no subspecies to subspecies to species. What exactly is a subspecies? The textbooks define it as a geographical race, a population with distinctive traits occupying part of the range of the species.

What then is a population? We are in immediate trouble. It is easy to say that a clearly defined population, one recognizable by everyone at a glance, occupies an exclusive part of the range of a species. And geneticists like to add, for purposes of mathematical clarity but not as an absolute requirement, that the population is a "deme": its members interbreed at random, and any member is equally likely to mate with any other member in the population, regardless of its location.

Few such objectively definable populations exist in nature. Most that do look like textbook examples are endangered species, with so few organisms left that there is no doubt as to the boundaries of the population they compose. The last surviving ivory-billed woodpeckers, found in one mountain forest of eastern Cuba, belong to this parlous category. So do the Devil's Hole pupfish, barely hanging on in a tiny desert spring at Ash Meadows, Nevada. You can stand at the entrance to Devil's Hole, look down 15 meters to where the water laps over a sunlit ledge, and see the entire species swimming around like goldfish in a bowl.

Most species are not confined this stringently, which is fortunate for conservation and unfortunate for textbook theory. Take the redbacked salamander (*Plethodon cinereus*), one of the most widespread and abundant salamanders of North America. The species ranges from Nova Scotia and Ontario south to Georgia and Louisiana. Redbacks occur almost continuously through the northern three quarters of this range. It is tempting to classify the whole northern ensemble as one huge population. Salamander taxonomists do just that and call it a subspecies, *Plethodon cinereus cinereus* (a formal rule

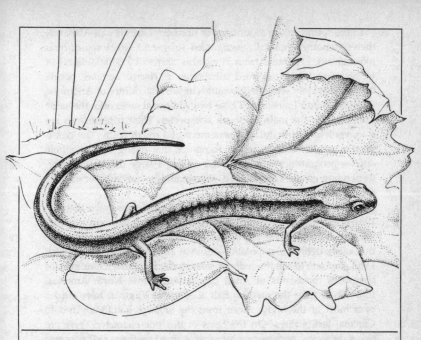

The red-backed salamander *(Plethodon cinereus)* of eastern North America, a wide-spread species with a pattern of racial variation that is typically ambiguous, such that subspecies can be defined only by subjective criteria.

of nomenclature: to designate a subspecies, add a third name). But redbacks are far from continuously distributed. They are largely confined to moist lowland forest, which is not a continuous habitat but an irregularly broken filigree laid on the land. Even within habitable forest, the population is divided into local aggregates that slowly expand, contract, and reform into new configurations across the generations. The rate of interbreeding among local demes in forested valleys and woodlots is unresearched and unknown. In short, with more data biologists might be able to distinguish thousands of populations across the vast range of *P. cinereus cinereus*. A diligent taxonomist might legitimately break the one formally recognized subspecies into large numbers of subspecies with smaller ranges.

To the south, in the mountains of northern Georgia and Alabama, there is another generally recognized subspecies, *Plethodon cinereus polycentratus*, separated from *P. cinereus cinereus* by 80 kilometers of redback-free terrain. A third subspecies, *P. cinereus serratus*, occurs in several widely separated localities in the hill country of Arkansas, Oklahoma, and Louisiana. These two additional races offer the same difficulties as the main northern subspecies. Their triple names are a convenient shorthand, the statement of a rough truth. The classification works so long as we recognize that dicing up the whole species geographically is imprecise and to a large degree arbitrary. Depending on the criteria used, there could be one subspecies of *P. cinereus*, or there could be hundreds.

An even more fundamental difficulty of subspecies is the discordance of the traits by which the subspecies are defined. Suppose that we agreed to ignore the population problem for the moment. Imagine that easily definable populations exist in an idealized species (which I will continue to call the redback salamander for clarity). One species comprises thousands of small populations across North America. Individuals from the *southern* half, Georgia to Virginia, have stripes over most of the body; those from the *northern* half, Maryland to Canada, lack stripes. On the basis of this one character, there are two subspecies, two geographic races: striped southern redbacks and plain northern redbacks. We notice, however, that *western* individuals of the species are larger. These two characters, stripedness and size, are obviously discordant—they break along different geographic lines. They can be used to define four subspecies: big striped in the *southwest*, little striped in the *southeast*, little unstriped in the *northeast*, and big unstriped in the *northwest*. Next we find that the eyes of juveniles are amber southwest of a line running from the Great Lakes to Georgia and yellow to the northeast of the line. Two more subspecies are added to produce a total of six overall. Looking still more closely we find . . .

Here is the point of this exercise in geometry: most traits varying geographically in a given species are discordant. They change at different places and in different directions. It follows that subspecies are recognized according to whatever traits taxonomists choose to study. It also follows that the greater the number of traits, the larger the number of the subspecies that must be recognized.

The uncertainty of the limits of populations combined with the discordance of traits means that the subspecies is an arbitrary unit of classification. That uncertainty is reflected in the confusion over human races. In past years anthropologists struggled hopelessly with attempts to define human races. Estimates of the number of races made by researchers during the 1950s ranged from six to more than sixty. The variation in numbers is due precisely to the fact that *Homo sapiens* is a typical evolving species.

Anthropologists, like biologists, have now largely abandoned the formal subspecies concept. They prefer a convenient shorthand to designate a certain part of a population with reference to one or two traits. They say, for example, "northern Asians tend to have more prominent canthal folds," knowing full well that canthal folds differ in geography from blood types, which deviate in turn from average height, lactose intolerance, Tay-Sachs syndrome, eye color, hair structure, infant passivity, and so on through hundreds of other more or less discordant traits prescribed or at least influenced by 200,000 or so human genes scattered through 46 chromosomes. The emphasis in research in anthropology and biology has passed from the description of subspecies to the analysis of the geography of separate traits and their respective contributions, singly or in combination, to survival and reproduction.

The demotion of the subspecies should carry with it a word of caution, in the service of moderation. Real populations do exist, however difficult to define. Genetic traits still vary. It may be artificial to divide up and label redback salamanders from the southern United States as subspecies, but they nevertheless differ in many genetic traits and compose a reservoir of unique genes. It is further true that some populations of widespread animals and plants are sufficiently isolated and genetically distinct to compose objective subspecies even in the abstract textbook sense. It is useful to label such populations formally as subspecies. Stephen O'Brien and Ernst Mayr, for example, have proposed guidelines to that end for use by conservation biologists and policy makers. They suggest that subspecies be defined as individuals occupying a particular part of the range of the species, with genes and natural history distinct from those of other subspecies. Members of different subspecies can freely interbreed. They can arise either as populations adapting to local conditions or as hybrids between subspecies.

The delimitation of subspecies, an occasional bureaucratic necessity when the U.S. Endangered Species Act or its equivalent is invoked, will usually be difficult and even controversial. Evolution, to repeat, is messy. The Florida panther offers a case in point. Once the panther, also known as the mountain lion or cougar, occurred throughout the southern United States. Now it is down to about fifty individuals in southern Florida, the subspecies *Felis concolor coryi*. Biochemical tests have revealed that this tiny population was derived from two stocks: the final survivors of the original North American panthers that once roamed Florida, and seven animals of mixed North American and South American origin released from captivity into the Everglades between 1957 and 1967. The present population is thus of hybrid origin, but it contains a unique ensemble of genes of partial North American origin and deserves protection as a native mammal.

The ambiguity of the subspecies as a taxonomic unit creates an interesting dilemma in evolutionary reasoning. We have before us an idealized sequence that starts with a geographically isolated population still identical to other populations of the same species. The population then evolves into a subspecies, still capable of interbreeding with the other populations, if they could somehow breach the geographical barrier and meet along the edges of their ranges. The subspecies finally evolves into a full species, meaning that if it meets the other populations it would no longer freely interbreed with them. The dilemma is this: if subspecies are usually amorphous and cannot be defined by a single objective criterion, how can such an arbitrary unit give rise to the species, which is sharply defined and objective?

The answer to the puzzle tells us a great deal about the origin of diversity. In order to spring forth as a species, a group of breeding individuals need only acquire one difference in one trait in their biology. This difference, the innate isolating mechanism, prevents them from freely interbreeding with other groups. It does not matter if the limits of the populations as a whole are poorly marked. Nor does it matter if all the other traits vary in a crazy-quilt manner across the populations that are splitting off as species. What counts is that somehow a group of individuals occupying some part of the total range evolves a different sex attractant, nuptial dance, mating season, or any other hereditary trait that prevents them from freely interbreeding with other populations. When that happens, a new species is born. The truly objective unit, the closed-gene pool of future generations, is the group of individual organisms that acquires the isolating trait. This new species can be defined by a single isolating trait. Other characters that vary geographically—hairiness, color, cold hardiness, whatever—can show any pattern of geographical variation whatever, either concordant with the isolating trait or totally different, without changing the outcome. Once segregated in this manner, the species will inevitably evolve away from other species, becoming ever more different in a steadily enlarging suite of traits as time passes.

The decisive isolating change can be further based on only a slight alteration in the genes or chromosomes. Some species of leafroller moths in the family Tortricidae, for instance, are segregated by relatively minor deviations in the chemistry of their female sex attractants. The variation they manifest from one species to the next is the kind commonly based on mutations in only one gene. The segregation can in fact be even more elementary. Some leafroller species are separated by close to the absolute conceivable minimum: a difference not in the organic structure of the substances composing the sex attractants but in the percentages of the substances, in this case a variety of acetates, that go into the blend. The females of each moth species have their own delicate bouquet that the males sample, and from this signal alone they choose either to press forward or to depart. In theory at least, it is possible for large numbers of tortricid moth species to arise quickly through minor changes in their genes that alter either the chemistry or the proportions of the sex attractants.

Geographical speciation is supplemented in nature by a rich medley of other modes of speciation. By far the best documented is polyploidy, an increase in the number of chromosomes. More precisely, a polyploid individual has twice as many chromosomes in each cell as an ordinary individual—or three times or four times or any other exact multiple. Polyploidy is virtually instantaneous in its effect, potentially isolating a group of individuals from its ancestors in one generation. This immediate isolation is caused by the inability of the hybrids of polyploid individuals and nonpolyploid individuals to develop in a normal manner or, if full development is achieved, to reproduce. Polyploidy is responsible for the origin of almost half of the living species of flowering plants and of a smaller number of animal species.

The stage for polyploid speciation is set by the passage of all sexually reproducing species through a two-part life cycle. In the haploid phase of the cycle, there is a single set of chromosomes in each cell. That is followed by the diploid phase, in which there are two sets of chromosomes in each cell. The diploid phase is ended by a reduction of the chromosomes to a single set, returning the organisms to the haploid phase, and so on. The haploid phase includes the sperm and eggs, each of which is a little organism with a single set of chromosomes. In higher plants and animals, the haploid phase consists exclusively of this ephemeral period. When a sperm and an egg unite, the chromosome number is doubled and the diploid phase begins. Two little organisms become one little organism, which is then capable of growing into a very big organism consisting of billions of diploid cells. The haploid number of chromosomes in human beings (the number in each sex cell) is 23; the diploid number, found in the remaining tissues after fertilization, is 46. If the base number is somehow tripled, thus creating a triploid organism (a triploid human being, for example, would have 69 chromosomes), the organism is marked for trouble. Difficulties are likely to be encountered during development of the embryo and beyond, into adult life. Down's syndrome is one of many defects caused by triple numbers, in this case three copies of chromosome number 21 (so called because biologists give each of the 23 chromosomes its own number for quick labeling).

When a triploid starts to manufacture sex cells, it will encounter difficulty. The two chromosomes of a normal diploid plant or animal can go through meiosis—the reduction to one chromosome per cell—with relative ease. In the first cell division of meiosis, when reduction takes place, the two related chromosomes pair up (in human beings, for example, all the chromosomes together create 23 pairs), then separate to different cells each with the haploid chromosome number. The three chromosomes of a triploid get tangled up during the pairing and separation (in this, the ultimate sense, three *is* a crowd), and either abort the process or produce large numbers of abnormal sex cells.

Triploids play the pivotal role in species isolation by polyploidy, simply because they are the unworkable products created when polyploid organisms try to breed with their diploid relatives. In full, the process of speciation by polyploidy proceeds as follows:

• Start with a polyploid plant that has newly originated within a diploid population. It is usually a tetraploid, in which the ordinary diploid number was accidentally doubled in early embryonic development. Each ordinary cell therefore has four chromosomes of each kind in each ordinary cell, rather than the usual two. As a result the tetraploid plant—the new species—places two chromosomes of a kind in each sex cell instead of the usual one.

• Suppose that plants of the ancestral species have 10 chromosomes in each ordinary cell and 5 in each sex cell. The polyploid plants have

20 and 10, respectively, in the two kinds of cell. Diploid plants of the ancestral species can interbreed with one another, and polyploid plants can interbreed with other polyploid plants.

• Some of the diploid and polyploid plants cross-breed to produce hybrids. When an ordinary sex cell (5 chromosomes) is fused with a polyploid sex cell (10 chromosomes), the hybrid is a triploid (15 chromosomes). As a result the hybrid plant may experience difficulties during growth. Even if it is able to attain sexual maturity, however, it cannot produce normal sex cells and is sterile. The diploid ancestor and polyploid derivative are reproductively isolated, and hence the polyploid derivative is a new species, created in a single generation.

There is another, even more innovative way that polyploidy can create species: by multiplying the number of chromosomes in hybrids of two species already in existence. The ordinary hybrids of many species of plants are sterile, even when they have the same number of chromosomes and growth is untroubled up to the time of flowering. The reason for the sterility is the incompatibility of the parental chromosomes during the formation of the sex cells. Let us call the two species that hybridize A and B. When a chromosome from species A attempts to line up with its counterpart from species B, a normal procedure for the exchange of blocks of genes during the production of sex cells, the A and B chromosomes differ too much from each other to complete the maneuver.

The way out of the impasse is to double the number of chromosomes in the hybrid. Then during the production of its sex cells, each A chromosome can be matched with an identical A chromosome, and each B chromosome can be matched with an identical B chromosome. Now the hybrid polyploid is fertile. It can breed with other hybrid polyploids of identical type, but not with either of the diploid parents from which it arose. Such hybridization followed by doubling does occur spontaneously in nature from time to time.

New species can be created in the laboratory and garden by joining old species in this Frankensteinian manner. The most celebrated example is the polyploid hybrid of the radish *(Raphanus sativus)* and the cabbage *(Brassica oleracea)*, which are genetically similar but reproductively isolated members of the mustard family Brassicaceae. Radishes and cabbages both have 9 chromosomes in their sex cells and 18 chromosomes in their diploid tissues. Hybrid radish-cabbage plants can be readily produced by cross-fertilization. They too have 18 chromosomes in their diploid tissues, 9 from each parent. But the two sets of 9 cannot pair with each other and complete the formation of sex cells during the reduction division, a defect that renders the hybrids sterile. When the chromosome numbers of the hybrids are doubled, bringing the diploid number to 36, the plants are fertile. Now each radish chromosome and each cabbage chromosome has an exact counterpart with which it can pair, and the production of sex cells can proceed normally. The rabbage or cabbish—choose the name you like—is self-sustaining as a species. It cannot be bred back to either of the parent species.

In spite of its clear importance in the origin of plant diversity, polyploidy may not be the most prevalent rapid process for organisms as a whole. Another, possibly even more widespread mode is nonpolyploid sympatric speciation. The term *sympatric speciation* refers to the origin of a new species in the same place as the parent species (literally, "of the same country"). It is contrasted with geographic speciation or, more formally, allopatric ("of different countries") speciation, in which the new species originates in a different place while isolated by a physical barrier.

Speciation by polyploidy is sympatric because the new polyploid form arises as a few plants directly from diploid plants, in one generation. Nonpolyploid sympatric speciation is merely sympatric origin by some other means. The most persuasively modeled and documented process in this category is through the intermediate host races of insects that feed on plants. Here are the key steps suggested by Guy Bush and others who have developed and tested the theory in recent years:

• Members of the parental insect species live and mate on one kind of plant. This degree of specificity is widespread among insects. It might characterize millions of species of plant-eating and parasitic insects and other small creatures that spend most or all of their lives on single plants.

• Some individuals of a given species then move to a second kind of plant, where they begin to feed and mate. The new host plant grows in the immediate vicinity of the old, so closely that individual plants of the two might even be intermingled. The insect shift is accompanied by enough genetic change to alter preference to the new plant species and improve the chance of survival on it. Later the evolved insects seek out the new host species whenever they disperse from plant to plant.

• When the evolution of the new insect strain proceeds far enough to settle it firmly on the adopted host, but not enough to isolate it reproductively in full from the old strain, it becomes by definition a host race. When the host race diverges still further, picking up enough differences to forestall even the potential of interbreeding, it becomes by definition a full species.

Host races can originate and then evolve to species rank within a few generations. Some fruit flies of the genus *Rhagoletis* appear to make the transition that quickly. In North America, species that infest hawthorns have occasionally spread to domestic fruit trees. The colonizing flies are prone to turn into host races because they breed only while the host fruit is available, and different tree species come into fruit at different times of the growing season. In 1864 *Rhagoletis pomonella*, which lives on native hawthorns, invaded apples in the Hudson River Valley and subsequently spread over most of the apple-growing regions of North America. A second host race of *R. pomonella* colonized cherries in Door County, Wisconsin, in the mid-1960s. The three races are partially separated by the season in which their host fruit matures, in the following order through the spring and summer: cherry, apple, hawthorn.

Speciation in the sympatric mode, transiting no more than an eyeblink of geological time, might easily have created vast numbers of insects and other invertebrates known to be specialized on particular plant species. It could also have underwritten the proliferation of parasite species that spend most or all of their lives on one kind of animal host. The theory looks good, but we can only guess at the purity of its truth. The early stages are difficult to detect, and few studies have been initiated in the invertebrates most likely to display them. *Rhagoletis* fruit flies are exceptional in the attention they receive, because of the economic damage they cause.

In conclusion, species can be created quickly, and diversity can therefore expand explosively. Our knowledge of evolution, though imperfect, tells us at the very least why life has that potential. Given the right circumstances, a new species can arise in one to several generations.

This vision of the origin of diversity raises a troubling question with ethical overtones: if evolution can occur rapidly, with the number of species quickly restored, why should we worry about species extinction? The answer is that new species are usually cheap species. They may be very different in outward traits, but they are still genetically similar to the ancestral forms and to the sister species that surround them. If they fill a new niche, they probably do so with relative inefficiency. They have not yet been fine-tuned by the vast number of mutations and episodes of natural selection needed to insert them solidly into the community of organisms into which they were born. Pairs of newly created sister species are often so close in their diet, nest preference, susceptibility to particular diseases, and other biological traits that they cannot coexist. Each tends to push out the other by competition. They then occupy different ranges, so that local communities are not enriched by the presence of both.

Great biological diversity takes long stretches of geological time and the accumulation of large reservoirs of unique genes. The richest ecosystems build slowly, over millions of years. It is further true that by chance alone only a few new species are poised to move into novel adaptive zones, to create something spectacular and stretch the limits of diversity. A panda or a sequoia represents a magnitude of evolution that comes along only rarely. It takes a stroke of luck and a long period of probing, experimentation, and failure. Such a creation is part of deep history, and the planet does not have the means nor we the time to see it repeated.

The Forces of Evolution

WHAT DRIVES EVOLUTION? This is the question that Darwin answered in essence and twentieth-century biologists have refined to produce the synthesis, called neo-Darwinism, with which we now live in uneasy consensus. To answer it in modern idiom is to descend below species and subspecies to the genes and chromosomes, and thence to the ultimate sources of biological diversity.

The fundamental evolutionary event is a change in the frequency of genes and chromosome configurations in a population. If a population of butterflies shifts through time from 40 percent blue individuals to 60 percent blue individuals, and if the color blue is hereditary, evolution of a simple kind has occurred. Larger transformations are accomplished by a great many such statistical changes in combination. Shifts can occur purely in the genes, with no effect on wing color or any other outward trait. But whatever their nature or magnitude, the changes in progress are always expressed in percentages of individuals within or among populations. Evolution is absolutely a phenomenon of populations. Individuals and their immediate descendants do not evolve. Populations evolve, in the sense that the proportions of carriers of different genes change through time. This conception of evolution at the population level follows ineluctably from the idea of natural selection, which is the core of Darwinism. There are other causes of evolution, but natural selection is overwhelmingly dominant.

Evolution by natural selection as we understand it today is a continuous cycle that can be stopped only by the death of the entire population. The starting point is the origin of variation by mutations, which are random changes in the chemical makeup of the genes, in the positions of the genes on the chromosomes, and in the numbers of chromosomes themselves. Genes are the portions of the DNA that ultimately prescribe outward traits, as simple as the color of wings and as complex as the power of flight. Each gene is composed of up to several thousand nucleotide pairs, or genetic "letters." Three nucleotide pairs in a row specify an amino acid. The amino acids in turn are assembled into proteins; proteins are the building blocks of the cells, and cells are the building blocks of organisms.

The number of genes in a typical larger organism, such as a human being, is on the order of 100,000. At least five genes on different chromosome positions affect variation in quantitative traits such as the date of flowering in plants, fruit size, the eye diameter of fish, and skin color in human beings. As many as one hundred genes work together to prescribe traits as complex as ear structure or skin texture.

A large number of molecular steps translate the nucleotide code into the assembly of the distinctive qualities of a species. The precise order of march leads from the triplet letters of DNA to messenger RNA and from there in sequence to transfer RNA and amino acids; the amino acids bond together into proteins; some of the proteins assemble into cell structures, and others into enzymes that catalyze the cell construction itself; additional enzymes accelerate metabolism; and finally the whole self-organizing ensemble projects to the world those properties of anatomy, physiology, and behavior by which the organism lives or dies, reproduces or not.

The commonest and most elementary kind of mutation is an alteration in the chemistry of a gene, specifically a chance substitution of one nucleotide pair for another. Sickle-cell anemia in human beings

The life cycle and the gene pool. The diploid organisms, each with two genes of a kind per cell, produce sex cells that individually carry only one gene of a kind. The sex cells, composed of sperm and eggs, are the haploid generation in the cycle. Sperm combine with eggs to create the next crop of diploid organisms. Thus the genes in a population—collectively the gene pool— repeatedly separate and recombine to create new variations to be acted on by natural selection. (The animal modeled here is the salt-marsh harvest mouse, a threatened species of California's wetlands.)

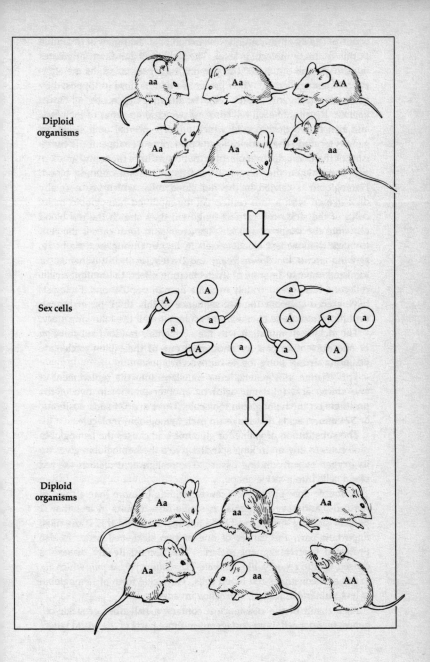

Diploid organisms

Sex cells

Diploid organisms

is one of the most thoroughly researched of all examples of evolution at this ultimate molecular level. The sickle-cell condition originates in perhaps one out of 100,000 persons each generation, as the alteration of a single gene. When present in double dose (unopposed by a normal counterpart in each cell), the mutant gene causes full-blown anemia. Recall that each cell contains two chromosomes of each type and hence two positions in which either a normal or a sickle-cell gene is located. The sickle-cell gene prescribes a change in the chemistry of the hemoglobin molecules that crystallizes them into an elongate form when the oxygen level falls in the surrounding blood. Hemoglobin is carried in the red blood cells, which are normally disk-shaped with a thin center. As the modified hemoglobin molecules of the sickle-cell carriers lengthen, they stretch the red blood cells into the shape of a sickle. The change in form causes the cells to block the smallest blood vessels in tissues throughout the body, slowing circulation downstream and thereby inducing ischemia, or localized anemia. In spite of its debilitating effect, the mutant sickle-cell gene has spread widely in some human populations. Biologists have pieced together the full sequence of this small bit of human evolution from gene chemistry to ecology to tell the following story.

The sickle-cell mutation was the accidental, random substitution of one nucleotide pair for another at one of the billion nucleotide positions strung along the 46 human chromosomes.

This change in a genetic letter translates into the replacement of one amino acid (glutamic acid) by another (valine) in two of the positions on the hemoglobin molecule. There are 574 such positions, or 574 amino acids that make up each hemoglobin molecule.

The substitution of valine for glutamic acid causes the hemoglobin molecules to line up in long spindles when the hemoglobin gives up its oxygen to surrounding tissues. The realignment distorts the red blood cells into a sickle shape.

A double dose of the gene causes sickling in more than a third of the cells, and severe anemia. A single dose results in less than 1 percent sickling and at most mild anemia. But, and this is the most important part, the carrier of one or two sickle-cell genes is also protected from malignant malaria. This second deadly disease is caused by the amoeba-like parasite *Plasmodium falciparum*, which invades and consumes red blood cells. The sickle form of hemoglobin is less vulnerable to the *Plasmodium* invaders.

As a result of the resistance it confers, a half dose of sickle-cell genes, one per cell, is advantageous in those parts of the world where malignant malaria is common. Until recent historical times, this danger zone included tropical Africa, the eastern Mediterranean, the Arabian peninsula, and India. Over most of this area, natural selection favored the sickle-cell gene. Its frequency commonly hovered above 5 percent, and it rose to as high as 20 percent in a few regions of Mozambique, Tanzania, and Uganda. The natural selection is balanced. When the gene becomes common, more people acquire a double dose and die off from hereditary anemia. When it becomes rare, more people die of malaria, a parasite-induced anemia. Over centuries the percentages of the gene observed in Africa and elsewhere have moved up or down according to the frequency with which they encounter malignant malaria.

Of the multitude of gene mutations and chromosomal rearrangements that arise throughout a population in each generation, many are so minor as to be neutral in effect, neither favoring nor disfavoring survival and reproduction. Either that, or they affect such quantitative traits as height and longevity, adding or subtracting from the traits in ways that are difficult to detect. The vast majority of genetic changes whose effects are large enough to be easily detected are also harmful. By definition they are opposed by natural selection and therefore scarce. In human beings these genetic defects are called genetic diseases. They include Down's syndrome, Tay-Sachs disease, cystic fibrosis, hemophilia, and sickle-cell anemia, as well as thousands of other abnormalities. When on the other hand a new mutant or novel combination of rare preexisting alleles (different forms of the same gene) happens to be superior to the ordinary, "normal" allele, it tends to spread through the population over many generations. In time it becomes the new genetic norm. If human beings were to move into a new environment that somehow gave sickle-cell hemoglobin a total (as opposed to just partial) Darwinian advantage over ordinary hemoglobin, then in time the sickle-cell trait would predominate and be regarded as the norm.

The sickle-cell trait puts a twist on moral reasoning that is worth a moment's reflection. Natural selection, it reminds us, is ethically neutral. Malarial anemia is balanced by hereditary anemia through the mindless agency of differential survival. Those who die of malaria are victims of a harsh environment. Those who die of a double dose of sickle genes are Darwinian wreckage, cast off as the accidental side product of a chance mutation. The tragedy of the hereditary loss is repeated relentlessly in high numbers because in this case natural selection happens to have been balanced rather than directional. No gods decreed it, no moral precept emerges from it. The sickle-cell gene happens to be common in a few parts of the world because the hemoglobin molecule defeats a parasite through the agency of one of its conveniently available mutant forms, and it does so in an inept manner.

The process of evolution by natural selection can be summarized as follows. Random nucleotide substitutions in the gene yield corresponding changes in anatomy, physiology, or behavior. The process sprinkles multiple forms of the gene created this way through the population. Genetic change is also initiated when genes shift position on the chromosomes, or when the number of chromosomes (and hence the number of genes) is raised or lowered. In biological phrasing, the genotype has been altered by one of these forms of mutation or another, and as a result there is now a different phenotype. New phenotypes, or the altered traits in anatomy, physiology, or behavior, usually have some effect on survival and reproduction. If the effect is favorable, if they confer higher rates of survival and reproduction, the mutant genes prescribing them proceed to spread through the population. If the effect is unfavorable, the prescribing genes decline and may disappear altogether.

It can be easily seen why Darwinism is both the greatest idea in nineteenth-century science and the simplest. Its power arises from the fact that natural selection is protean in form. In some cases selection is lethal, mediated by predation, disease, and starvation. In others it is benign, arising from differences in family size, without increasing mortality in the least. Its products range in magnitude from fixing the number of hairs on a fly's wing to the creation of the human brain. Like the old god Proteus it is endless in the form it takes and is therefore filled with the information of realized Nature. Natural selection has these near magical properties because, in one sense, it is a creation of our language. It is nothing more than the active-voice metaphor of all the differences in survival and reproduction among genotypes arising from the effects of the genotypes on organisms. But what it represents is real, and very powerful.

The environment is the theater, as the ecologist G. Evelyn Hutchinson once remarked, and evolution is the play. And more: genetic prescription of the developmental process is the language, and mutation invents the words—but like an idiot spouting gibberish. Finally, natural selection is the editor and principal driving, creative force. Guided by no vision, bound to no distant purpose, evolution composes itself word by word to address the requirements of only one or two generations at a time.

Evolution is blinkered still more by the fact that the frequency of genes and chromosomes can be shifted by pure chance. The process, an alternative to natural selection called genetic drift, occurs most rapidly in very small populations. It proceeds faster when the genes are neutral, having little or no effect on survival and reproduction. Genetic drift is a game of chance. Suppose that a population of organisms contained 50 percent A genes and 50 percent B genes at a particular chromosome site, and that in each generation it reproduced itself by passing on A and B genes at random. Imagine that the population comprises only five individuals and hence 10 genes on the chromosome site. Draw out 10 genes to make the next generation. They can all come from one pair of adults or as many as five pairs of adults. The new population could end up with exactly 5 A and 5 B genes, duplicating the parental population, but there is a high probability that in such a tiny sample the result instead will be 6 A and 4 B, or 3 A and 7 B, or something else again. Thus in very small populations the percentages of alleles can change significantly in one generation by the workings of chance alone. That in a nutshell is genetic drift, about which mathematicians have published volumes of sophisticated and usually incomprehensible calculations.

But let us go on. Population size is critical in genetic drift. If the population were 500,000 individuals with 500,000 A genes and 500,000 B genes respectively, the picture could be entirely different. At this large number, and given that even a small percentage of the adults reproduced—say 1 percent reproduced—the sample of genes drawn would remain very close to 50 percent A and 50 percent B in each generation. In such large populations genetic drift is therefore a relatively minor factor in evolution, meaning that it is weak if opposed by natural selection. The stronger the selection, the more quickly the perturbation caused by drift will be corrected. If drift leads to a high percentage of B genes but A genes are superior in nature to B genes, the selection will tend to return the B genes to a lower frequency.

An important version of genetic drift is the founder effect, believed by some evolutionary biologists to accelerate the formation of new species. Suppose that we started with the same large population containing a mix of A and B alleles. Again, for simplicity make it 50 percent of each. A small group of individuals strays to an offshore island or some other remote locality previously unoccupied by the species. Say a mated pair of birds flies to the new location. The genes they carry are four in number at each chromosome position, including the one that carries gene forms A and B. By sheer accident the founder population may turn out to contain 2 A and 2 B, preserving the ancestral ratios. But there is also an excellent chance that they will contain 3 A and 1 B or 1 A and 3 B, or all A or all B. In other words, because founder populations are so likely to be small in size, they are also likely to differ genetically from the parent population by chance alone. That initial difference, combined with geographical isolation and the exigencies of a new and different environment, can propel populations into new ways of life, new adaptive zones. It can also lead them more quickly to the formation of reproductive barriers and full species status.

Three features of evolution conspire to give it great creative potential. The first is the vast array of mutations, including nucleotide-pair substitutions, shifts in positions of genes on chromosomes, changes in the numbers of chromosomes, and transfers of pieces of chromosomes. All populations are subject to a continuous rain of such new genetic types that test the old.

A second source of evolutionary creativity is the speed at which natural selection can act. Selection does not need geological time, spanning thousands or millions of years, to transform a species. The point is best argued with explicit examples from the theory of population genetics. Take a dominant gene, one whose expression overrides recessive genes at the same chromosome position. For example, the dominant gene for normal blood clotting overrides the one for hemophilia, and the dominant gene for the ability to roll the tongue into a tube overrides the one for lack of that ability. When a dominant gene occurs in the same cells as the recessive gene, a combination called the *heterozygous condition*, it is the one expressed in the phenotype. Only when the recessive gene is alone, in other words the double-dose or homozygous condition, does its phenotype find expression. A dominant gene whose phenotype enjoys a 40 percent advantage in survival or reproduction over the recessive phenotype can largely replace it within the population in twenty generations, passing from a frequency of 5 percent to a frequency of 80 percent in that interval. Twenty generations amount to as little as four or five hundred years in human beings, forty years or less in dogs, and one year in fruit flies. A recessive gene with the same degree of advantage requires sixty generations to traverse the same frequency range, still a very short period by geological standards. If dominance is incomplete—both genes are expressed when together—and if the advantage of the winning gene is total, the changeover can, in theory and laboratory populations at least, be accomplished in a single generation.

The final creative feature of natural selection is the ability to assemble complicated new structures and physiological processes, including new patterns of behavior, with no blueprint and no force behind them other than natural selection itself acting on chance mutations. This is a key point missed by creationists and other critics of evolutionary theory, who like to argue that the probability of assembling an eye, a hand, or life itself by genetic mutations is infinitesimally small—in effect, impossible. But the following thought experiment shows that the opposite is true. Suppose that a new trait emerges if two new mutations, call them C and D, occur simultaneously on different chromosome sites. The chance of C occurring is one in a million per individual organism, a typical mutation rate in the real world, and the chance of D occurring is also one in a million. Then the chances of both C and D occurring in the same individual is a million times a million, or a trillion, a near impossibility—as the critics have insisted. But natural selection subverts the process. If C confers even a slight advantage all by itself, it will become the prevailing gene through the population at its chromosome position. Now the chance of CD appearing is one in a million. In moderate to large populations of plants and animals, which often contain more than a million individuals, the changeover to CD is a virtual certainty.

The emerging picture of evolution at the level of the gene has altered our conception of both the nature of life and the human place in nature. Before Darwin it was customary to use the vast complexity of living organisms as proof of the existence of God. The most famous exposition of this "argument from design" came from the Reverend William Paley, who in his 1802 *Natural Theology* introduced the watchmaker analogy: the existence of a watch implies the existence of a watchmaker. In other words, great effects imply great causes. Common sense would seem to dictate the truth of this deduction, but common sense is merely unaided intuition, and unaided intuition is reasoning performed in the absence of instruments and the tested knowledge of science. Common sense tells us that massive satellites cannot hang suspended 36,000 kilometers above one point on the earth's surface, but they do, in synchronous equatorial orbits.

Phenotypic evolution, based on gene action but expressed in the outward traits of organisms, can be correspondingly swift. If a single gene is easily substituted in fewer than a hundred generations by moderate selection pressure, a single gene thus inserted might also exercise profound effects on the biology of a species. One gene can change the shape of a skull. It can lengthen lifespans, restructure the color pattern on a wing, or create a race of giants.

This point is made most compellingly by allometry, when parts of the body grow at different rates. A familiar example is the slower growth of the human skull relative to the body in children, causing adults to end up with heads not much larger than those of infants but atop large muscular bodies. If the allometry of a species is strong, small adults can be strikingly different from large adults in many biological traits, even if they are all identical genetically in the trait under consideration. Among animals the process can be taken to bizarre extremes. In some stag-beetle species, such as the European *Lucanus cervus*, little males have relatively short, simple mandibles, while large males have more massive mandibles half as long as the rest of the body, an armament that gains them superiority in combat. What is inherited in the males is not one of a series of body types, and not necessarily even a particular body size, but rather the allometric growth pattern common to all the males. Males that obtain less food or terminate growth early end up small and feminized; those that reach large size become hulking, top-heavy supermales. The allometry itself is relatively simple, dependent only on differences in rates of growth among certain patches of tissue. It is easy to imagine a rapid switch of a magnitude often associated with the origin of species that is nevertheless based on the simplest hereditary change. Minor mutations in one or several genes might easily alter the allometric pattern, so that all of the males come to more closely resemble females. Alternatively, the change could push the pattern the other way, so that all stag-beetle males sprout huge mandibles.

The social systems of ants illustrate the power of allometry even more dramatically. The caste system of each ant colony, from queens to big-headed soldiers to small-headed workers, is based on a single allometric pattern common to all female members of the colony. Depending on the food and chemical stimuli she receives as a larva, a female ant becomes a queen or a soldier or a minor worker. All fall

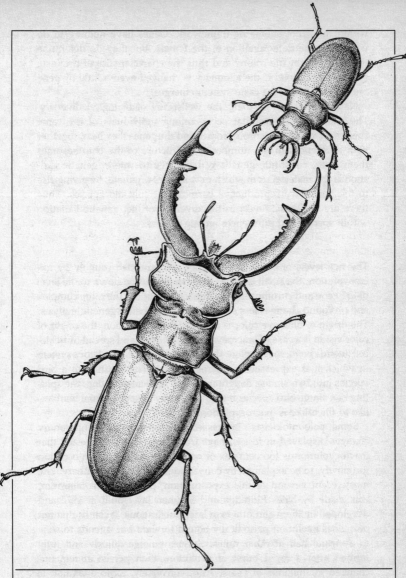

Two males of the European stag beetle prepare for combat. Their different body forms are due to allometry, the more rapid growth of certain parts of the body relative to others. In this species the head and mandibles grow fastest, so that the larger male is much more heavily armored than its smaller opponent.

within the same allometric framework. Genes have nothing to do with the caste determination of the female, but they do determine the allometry of the colony and thus the characteristics of the caste system as a whole. If the allometry is changed even a little by gene mutations, a different caste system emerges.

Natural selection is, then, the wellspring of biological diversity. The allelic differences that occur among individuals of the same species, across all the chromosomes and the genes they bear, together with differences in the number and structure of the chromosomes themselves, constitute genetic variation. Furthermore, genetic variation is the material from which new species originate, by giving rise to the hereditary reproductive barriers that split old species. Thus there are two basic levels in the diversity of life: genetic variation within species and differences among species.

The two levels of biological diversity are paralleled roughly by microevolution, the small changes that can be tracked down to the level of the gene and chromosome, and macroevolution, the more complex and profound changes less susceptible to immediate genetic analysis. The origin of blue eye color is a case of microevolution; the origin of color vision is a case of macroevolution. The rise and spread of sickle-cell anemia is microevolution; the formation of the circulatory system in which it is expressed is macroevolution. The splitting of a bird species into two similar daughter species is microevolution; the splitting of a single bird species into a wide array of species from warbler-like to finch-like is macroevolution.

Some paleontologists, impressed by the striking evolutionary changes displayed in fossils, have from time to time suggested that macroevolution is too complex or occurs too rapidly, or on occasion too slowly, to be explained by conventional evolutionary theory. The most recent version of this expostulation, punctuated equilibrium, was made by Niles Eldredge and Stephen Jay Gould in 1972 and developed by them and others in later publications. It claims that not only does evolution periodically bound forward but it tends to slow to a virtual halt at other times. Species emerge quickly and fully formed after a rapid burst of evolution, then persist almost unchanged for millions of years. And, conversely, rapid evolution is driven mostly or entirely during species formation. The alternation between leaps and pauses creates a jerky pattern, a punctuated equilibrium, so extreme as to point to novel processes of evolution beyond the natural selection of genes and chromosomes. Macroevolution, the reasoning in its most radical form concludes, is in some fashion unique, not the same as microevolution.

The punctuated-equilibrium thesis received a great deal of attention because it was at first promoted as a challenge to the neo-Darwinian theory of evolution; in effect, a new theory of evolution. That claim has been abandoned by most of its proponents. The fossil evidence for the widespread occurrence of jerky patterns proved weak, and most examples put forth at the outset were discredited.

More to the point, the possibility of swift evolution was already a cornerstone of traditional evolutionary theory and therefore in no sense a challenge to it. The models of population genetics, the foundation of quantitative theory, predict that evolution by natural selection can be so rapid as to seem nearly instantaneous in geological time. The models also allow for stasis, or long periods with little or no evolution of a kind detectable in fossils. These predictions of population genetics have been upheld by decades of careful study in the laboratory and field, across a wide range of animals, plants, and microorganisms. They have illustrated gradual transitions between closely related species from a single tick in microevolution to large advances in macroevolution, from the earliest beginnings of geographical variation to the origin of species and panoramic expansions of species into multiple adaptive zones.

In general, the continuity between microevolution and macroevolution has been upheld. Neo-Darwinian theory was not challenged in substance, only semantically—a renaming, so to speak, as opposed to a reinventing of the wheel. Punctuated equilibrium is now used mostly as a descriptive term for a pattern of alternating rapid and slow evolution, especially when the rapid phase is accompanied by species formation. Its fate illustrates the principle that in science failed ideas live on as ghosts in the glossaries of the survivors. The value of the punctuated-equilibrium challenge lay not in its claims but in the research it stimulated on evolutionary rates and in the favorable public attention it brought to evolution studies as a whole.

To say that microevolution shades into macroevolution by degree instead of kind does not mean, however, that all we know about evolution is written in the narrow script of modern genetics. It only affirms that nothing learned so far is inconsistent with that canon, in the same sense that nothing learned so far about the molecular processes of the cell is inconsistent with contemporary physics and chemistry. There is still a great deal more to evolution than its genetic mechanisms.

A case in point is species selection, a process that has begun to be profitably explored by paleontologists, who study fossils, as well as by neontologists, who work on modern organisms. The subject has been muddled somewhat by the fact that the two groups favor different vocabularies, but the process itself is easily explained. A newly evolved species, like a newborn organism, comes into the world bearing its own distinctive set of traits. Depending on which traits it possesses, the species can last a long time or a short time before it becomes extinct. It is also prone either to split into multiple species or to remain intact as a single species throughout its lifespan. These influential hereditary traits are emergent properties of species but are not invested by some mysterious macroevolutionary process. They are the properties evolved by the organisms that compose the species. They originate by microevolution, in other words by changes in the frequencies of genes and chromosome arrays, and are translated upward into the species-level patterns we call macroevolution.

This translation has two key properties. It is blind, and it echoes back down to speed or to slow the evolution of organisms. Organisms, in the course of their struggle for survival and reproduction, are unconcerned in a Darwinian sense with the persistence of the species as a whole. They are also unaffected by the degree to which the species multiplies. Thus their genes are inserted into the next generation or perish by their own idiosyncratic actions, no matter whether the species is expanding and multiplying or dwindling toward extinction. The traits they possess nevertheless cause the species to last a long time or a short time and to remain a single species or to multiply. That influence is what has been identified as the upward translation from microevolution to macroevolution. Conversely, and this is the essence of species selection, the longevity of a species, together with its tendency to form new species, affects how rapidly the crucial traits are spread through the fauna or flora as a whole. This is the downward echo that makes the theory of species selection more than just a boring statement of the obvious.

Consider a set of species that might be subject to selection at the higher level. By a *set* I mean multiple species of common ancestry, such as the cichlid fishes of Lake Victoria or the lycaenid butterflies of tropical America. Natural selection among the species can reinforce organism-level selection that occurs within each species in turn. Hence the evolution of the trait proceeds more rapidly within the set of species as a whole. The character of the fauna and flora shifts accordingly. Alternatively, species selection can oppose organism-level selection, slowing the rate of spread.

How important is species selection? If the group is defined broadly enough, such as all vascular plants or all land vertebrates, it is of overwhelming importance. In the late Mesozoic era, cycads and conifers gave way to the poleward-spreading flowering plants. After the catastrophic end of the Mesozoic, mammals took over from dinosaurs and crocodilians. But we knew all that; the fact sheds little light on the biological process of species selection. To link species selection to natural selection at the level of individuals and populations, we need smaller groups of species and sharper insights into the fine details of ecology and adaptation. It is easy to imagine the existence of such groups but difficult to find them in nature. We have been limited so far to a scattering of cases, of which the following appear to be the most promising.

• Among insects, a shift from predatory behavior or scavenging to plant feeding increases the rate of species formation. The reason is that more species can specialize on particular kinds of plants or even different parts of the same plant. They can radiate into these niches even more rapidly by forming host races, which are believed to be precursors to full-fledged species. In such insects, to put the matter as simply as possible, individual selection and species selection conspire to increase the rate of evolution.

• During the latter half of the Mesozoic era 100 to 65 million years ago, oysters, clams, and other mollusks varied from one set of species to another in dispersal ability and hence in the breadth of their geographical ranges. Those with large ranges also survived for longer periods of geological time. The study of present-day mollusk species shows that dispersal ability is probably the result of natural selection at the organism level. If true, it follows that individual and species selection worked together to increase the average geographical range and longevity of mollusks.

• Ants, beetles, lizards, and birds display what has been called the *taxon cycle*. Some of the species adapt—more precisely, their member organisms adapt—to habitats from which it is easy to disperse. Such places include the seashore, edges of rivers, and windswept grasslands. Quite coincidentally, these habitats are also the best staging areas for long-distance dispersal. The species concentrated there attain the widest geographical spread and potential for speciation. When some of the far-flung populations penetrate more sheltered habitats nearby, they "settle down"—lose dispersal power and thereby become more prone to species formation. Eventually, they decline to extinction. The question arises: do they decline to extinction more rapidly, as was the case in the Mesozoic mollusks? If so, organisms that adapt for life in restricted habitats can be said to improve their own Darwinian fitness at the expense of the longevity of their species. In short, the two levels of selection, individual and species, are countervailing.

• A process similar to the taxon cycle has been occurring for millions of years in the rich fauna of antelopes, buffalo, and other bovid mammals of Africa. Species that are generalized, able to occupy more than one habitat—to shift from forest to grassland, say, and back again—survive for longer periods of time. Those specialized to live in particular habitats are more likely to be trapped there and to decline to extinction as the climate changes and the forest alternately advances and retreats. With their populations prone to fragmentation, the bovid specialists are also more likely to generate new species, and thus they gain and lose species more rapidly than bovid generalists. Overall, natural selection of individual animals leads to the natural selection of species, by either increasing their longevity or decreasing it according to circumstances.

• Desert plants such as *Dedeckera eurekensis* of the Mojave Desert may experience another category of individual-level natural selection that conflicts with species selection. During droughts, few seeds have a chance to germinate. Natural selection might easily lead to a strategy in which individual plants shut down seed production and concentrate their resources on survival. (The alternative strategy, not used by *Dedeckera*, is to produce many seeds, which then wait for rainfall.) If hard times drag on, a premium is put on longevity rather than on reproductive ability. The species whose members are pushed by natural selection to adopt the longevity strategy will end up limited to a small number of long-lived but nearly sterile individuals. The surviving individual plants are winners in the game of organism-level selection, but their success brings the species to the edge of extinction.

The picture emerging of natural selection at the level of the individual, whether or not it is enhanced by species selection, is one of exuberance, power, and a potential for quickness. If enough raw hereditary material exists in the first place, and if the selection pressure (differences in survival and reproduction) are strong, one gene or chromosome type can be substituted for another in fewer than a hundred generations. The possibility is there for rapid microevolution and even the early stages of macroevolution.

This capacity is well understood in theory and has been realized in laboratory experience. It is also displayed in wild populations when species are subjected to new selection pressures, such as the threat from a new parasite or access to a new food source. There has been time enough and more for natural selection alone to create radical new types of organisms. Consider: the Age of Reptiles lasted for 100 million reptilian generations, and the Age of Mammals succeeding it passed more than 10 million mammalian generations before the human species emerged. Earth took hundreds of millions of years to produce the first unicellular organisms, which were assembled out of an astronomical number of potent molecules.

What we understand best about evolution is mostly genetic, and what we understand least is mostly ecological. I will go further and suggest that the major remaining questions of evolutionary biology are ecological rather than genetic in content. They have to do with selection pressures from the environment as revealed by the histories of particular lineages, not with genetic mechanisms of the most general nature. I could be very wrong. Molecular biology is so vigorous and rapidly growing that new mechanisms driving evolution one way or the other may be discovered. We have a vast amount to learn about the way functional genes, the exons of DNA, originated and how they were reshuffled and elaborated to set the basis for the full flowering of biological diversity. It is further possible that extragenetic constraints on embryonic development, such as fundamental physical limits on cell size and tissue organization, play a guiding role. Competition and interference among cells and tissues might entail still undiscovered principles of a novel kind. Many surprises await us in the study of development. Sometime soon discoveries in the two key domains of the genetic code and embryonic development could shake neo-Darwinism to its foundations. But I doubt it. I think the greatest advances in evolutionary biology will be made in ecology, explaining more fully in time why the diversity of life is of such and such a nature and not some other.

Adaptive Radiation

EVOLUTION on a large scale unfolds, like much of human history, as a succession of dynasties. Organisms possessing common ancestry rise to dominance, expand their geographic ranges, and split into multiple species. Some of the species acquire novel life cycles and ways of life. The groups they replace retreat to relict status, being diminished in scattershot fashion by competition, disease, shifts in climate, or any other environmental change that serves to clear the way for the newcomers. In time the ascendant group itself stalls and begins to fall back. Its species vanish one by one until all are gone. Once in a while, in a minority of groups, a lucky species hits upon a new biological trait that allows it to expand and radiate again, reanimating the cycle of dominance on behalf of its phylogenetic kin.

When viewed in one slice of geological history, all contemporary dynastic successions taken together present a complex and strikingly beautiful pattern across the surface of the earth. Now the comparison is to a palimpsest, an ancient parchment on which the current dominant groups are boldly spread and past rulers survive as faded traces in spaces between the lines, in shrunken niches. Mammals, the dominant large vertebrates on the land today, are accompanied by turtles and crocodilians, among the last survivors of yesteryear's ruling reptiles. Forests of flowering plants shelter scattered ferns and cycads, remnants of the prevailing vegetation of the Age of Reptiles. And on a smaller scale the air is filled with flies, wasps, moths, and butterflies, relative newcomers of insect evolution. They are preyed upon by dragonflies, Paleozoic relics that still possess stiff outstretched wings and other archaisms that date back to the dawn of flight. Dragonflies are the Fokkers and Sopwith Camels of the insect world that somehow stayed aloft all those years.

Adaptive radiation is the term applied to the spread of species of common ancestry into different niches. *Evolutionary convergence* is the occupation of the same niche by products of different adaptive radiations, especially in different parts of the world. The Tasmanian wolf of Australia, a marsupial, outwardly resembles the "true" wolf of Eurasia and North America, a placental mammal. The first is a product of adaptive radiation in Australia, the second the product of a parallel adaptive radiation in the northern hemisphere. The two species converged to occupy similar niches within independent adaptive radiations on the different continents.

Adaptive radiation and evolutionary convergence are displayed in textbook fashion in distant archipelagoes around the world, including the Galápagos, Hawaii, and the Mascarenes. They are also sharply defined in ancient lakes such as Baikal and the Great Lakes of East Africa's Rift Valley. These places are so isolated that only a few kinds of plants and animals have been able to reach them. The fortunate colonists originated in large, crowded faunas and floras, pressed by competition, predators, and disease, restricted in habitat and diet. They arrived in a new and mostly empty world where, initially at least, opportunity was spread before them in abundance.

The archipelagoes and lakes are not only isolated but also small and young enough, in comparison with the continents and oceans, to keep the patterns of adaptive radiation and convergence simple and hence decipherable. That is why biologists see Hawaii as one of the prime laboratories of evolution. It is an archipelago rather than a single island, setting the stage for the splitting of populations into full-blown species. It is geographically the most remote of all archipelagoes, so that relatively few colonists have reached its shores. It is large enough to provide niches for the radiation of large numbers of new species, yet small enough to constrict and clearly display the patterns of speciation and adaptive radiation. Finally, although youthful relative to continents, it is still old enough, about 5 million years in the case of Kauai, for adaptive radiation to have attained an impressive degree of maturity.

The 10,000 known endemic species of insects on Hawaii are believed to have evolved from only about 400 immigrant species. Some have made unique shifts in habitat and lifestyle. Around the world, for example, almost all larvae of damselflies (small, delicate relatives of dragonflies) are aquatic, feeding on insects around them in ponds and other bodies of fresh water. But on Hawaii the nymphs of one species, *Megalagrion oahuense*, have left the water entirely and now hunt insect prey on the floor of wet mountain forests. Displaying an even more radical shift, caterpillars of the moth genus *Eupithecia* have abandoned their habit of feeding on plants to become ambush predators. These bizarre wormlike larvae lie in wait on vegetation for passing insects and seize their victims with sudden strikes of their forelegs. A cricket in the genus *Caconemobius* has gone from life on the land to a partial marine existence, living among boulders in the wave-splash zone and feeding on flotsam that washes ashore. Another *Caconemobius* species lives on bare lava flows, where it browses on windblown vegetation debris. Still other crickets of the same genus, all blind, live in caves. The killer caterpillars and entrepreneurial crickets have all been discovered within the past twenty years. Hawaii, familiar as it may seem to the casual visitor, is still a paradise full of surprises for the explorer naturalist.

Radiation and convergence on distant archipelagoes are marked by disharmony, defined in evolutionary biology as the wildly disproportionate representation of some major groups and the absence of others. When a few species split swiftly and massively in response to exceptional opportunity, they and their descendants seize a large part of the environment and hold it thereafter with a lion's share of the total diversity. The fauna and flora as a whole are thus unbalanced in comparison with those of continents, whose great biological diversity originated from many stocks over longer spans of time.

Hawaii harbors the most disharmonic bird fauna in the world. Until recent historical times, more than one hundred of the known species were endemic, by which is meant they are native and found nowhere else in the world. Their numbers include sixty species extinguished successively by the Polynesian and then the European colonists, and forty still surviving. More than half are or were honeycreepers, composing a unique tribe of their own, the Drepanidini, which in formal classifications is a branch of the finch subfamily Carduelinae, which in turn is a branch of the larger finch and sparrow family Fringillidae. All honeycreepers are descended from a single pair or small flock of colonists, most likely blown to the islands by a storm many thousands of years ago. This ancestral species was a relatively primitive cardueline bird, probably small, slender, and with a bill resembling that of a goldfinch. Its diet probably consisted of seeds and insects. Non-Hawaiian carduelines include goldfinches, canaries, and crossbills, which occur around the northern hemisphere but are concentrated in temperate Europe and Asia. It seems likely, then, that the first Hawaiian colonists flew or were blown in by a storm from either North America or eastern Asia. The honeycreepers, their populations expanding, then achieved an explosive adaptive radiation. They penetrated many new niches and diversified their anatomy and behavior in concert. As ecological conquerors of the first rank, they offer a textbook display of radiation and convergence on a scale small enough to be dissected and explained with reasonable certitude.

I should say that their *memory* offers such a display. Before the coming of the Polynesians 2,000 years ago, and before the arrival of European merchantmen and settlers eighteen centuries later, the forests of Hawaii swarmed with a riot of the sparrow-sized honeycreepers. The many species were distinguished variously by red and yellow and olive green plumages, slashed with wing bands in black, gray, and varying nuances of white. Even today the scarlet arapane *(Himatione sanguinea)* occurs in populations of one thousand to a square kilometer in some localities. Walk among them in a grove of ohia lehua trees, watch for their scintillas of bright color, listen for their thin whistling songs, and you will be granted a last snapshot of old Hawaii, as it was before the first Tahitian canoes touched shore.

Most of the honeycreepers are gone now. They retreated and vanished under pressure from overhunting, deforestation, rats, carnivorous ants, and malaria and dropsy carried in by exotic birds introduced to "enrich" the Hawaiian landscape. They disappeared as vanished species usually do, not in a dramatic cataclysm but unnoticed, at the end of a decline when those who knew them could acknowledge that none had been seen around for a while, that perhaps there were still a few to be found in such-and-such a valley, where in fact a predator had already snatched the last living individual, a lonely male, let us suppose, from its nocturnal perch. In old Polynesian times several generations might pass, the last tattered feather panache would be replaced in a ceremonial headdress laid aside for good, and the species would be as consigned and as forgotten, in the words of the Catholic liturgy, as the unremembered dead.

Still the sweep of the radiation even among the surviving honeycreepers is the greatest of any closely related group of birds in the world. The Maui parrotbill *(Pseudonestor xanthophrys)* has an anatomy somewhat resembling that of a true parrot but feeds on insects rather than fruit and seeds. It wields its stout bill to chew and rip open twigs and branches in order to reach beetle grubs and other insects burrowing through the wood. The ou *(Psittirostra psittacea)*, the finch analogue, possesses a thick bill with which it feeds on seeds primarily and insects secondarily, in the generalized finch manner. The akepa *(Loxops coccinea)* represents a partial approach to crossbills of the northern hemisphere. The tips of its bill reach past each other in a lateral direction, allowing it to twist open leaf buds and legume pods in search of insect prey. Other species of *Loxops* and *Himatione* resemble warblers, with small delicate bodies and short thin bills. Like typical warblers, which are dominant birds on most continents, they hunt insects that fly in the open or rest exposed on vegetation. The iiwi *(Vestiaria coccinea)* and several species of *Hemignathus* are closely convergent to the sunbirds of tropical Africa and Asia. They use long, slender, downward curving bills like siphons to extract nectar from flowers.

The species of *Hemignathus* have achieved a miniature adaptive radiation within the larger radiation, a secondary deployment into major niches. In addition to the primary nectar feeders with full, curved bills, they include the nukupuu, *Hemignathus lucidus*, whose lower bill is shortened to only a little more than half the length of the upper bill. This curious form has evolved partway to the status of a woodpecker. In addition to nectar feeding, accomplished with the upper bill, it uses the lower bill to tap tree trunks and branches, pry up pieces of bark, probe into crevices, and chase out insect prey caught unaware by the maneuver.

A second, even more remarkable species, the akiapolaau *(Hemignathus wilsoni)*, has traveled almost the full distance into the woodpecker niche. It uses its lower bill, which is short and completely straight, to hammer and chisel open bark and wood. This woodpeckerish behavior is a straightforward extension of the gentler tapping behavior of the nukupuu. Walter Bock has described it: "The decurved upper jaw is raised out of the way of the mandible when the bird is pecking. After a hole has been cut and the insect exposed, the longer decurved upper jaw is then used to probe for the insects. The combination of straight chisel-like mandible and a long decurved upper jaw for probing is an unusual, and perhaps unique, example of the two jaws in a single avian species being adapted for two quite

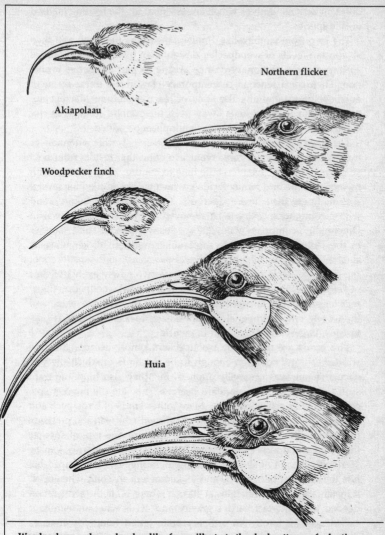

Labels within image:
Northern flicker
Akiapolaau
Woodpecker finch
Huia

Woodpeckers and woodpecker-like forms illustrate the dual patterns of adaptive radiation and evolutionary convergence. During radiations of birds in different parts of the world, separate lines evolved to fill the woodpecker niche: the akiapolaau, a honeycreeper of Hawaii; the common flicker of North America, one of many "true" woodpeckers; the woodpecker finch of the Galápagos; and the huia (female above, male below) of New Zealand.

different actions, both of which are essential for the feeding method of this species."

The step from sunbird-like *Hemignathus* to the nukupuu and then akiapolaau levels of woodpecker design is an instructive example of an important evolutionary change arising as part of species formation. Their coexistence in contemporary Hawaii is a freeze-frame of microevolution attaining the scale of macroevolution. Macroevolution, seen in the two steps away from the sunbird level, is microevolution writ large, with species multiplication added.

The ersatz woodpeckers of Hawaii deserve further attention as exemplars of imperfect convergent evolution that results from bold adaptive radiation given too little time to mature. They contrast with members of the bird family Picidae, which we are entitled for several reasons to call the true woodpeckers. The Picidae, a well-knit group whose common ancestry is far removed from that of the Hawaiian Drepanidini, comprises about 200 species worldwide. The 19 species of the United States include the familiar northern flicker *(Colaptes auratus)*, downy woodpecker *(Picoides pubescens)*, and sapsuckers of the genus *Sphyrapicus*. They also include two newly extinct victims of deforestation in North America, the ivory-billed woodpecker *(Campephilus principalis)*, largest of all the picids of the Nearctic realm, and the closely related imperial woodpecker of Mexico *(Campephilus imperialis)*, largest woodpecker in the world.

The picids are called true woodpeckers simply because they are widespread and common enough to be the birds to which the vernacular name was originally applied. But they also have the right stuff to be the nominal standard bearers. They are the premier specialists in their ecological class. Many other kinds of birds peck and pry wood to expose insects, but none do it with the élan and precision of a picid. The details of the hunting strike of one typical species, the acorn woodpecker *(Melanerpes formicivorus)* of California, can be seen with the aid of slow-motion photography. The awl-shaped bill hits the wood at between 20 and 25 kilometers an hour, whereupon it instantaneously decelerates at 1000G, where 1G is the acceleration needed to counteract earth's gravity and 4G is what an astronaut experiences on liftoff. An ordinary brain jarred hundreds of times daily by a blow to the head of this magnitude would be reduced to pulp. The woodpecker survives because it possesses two unusual features. The brain case is made of unusually dense spongy bone joined with sets of opposing muscles that appear to act as shock absorbers. And the woodpecker brings its head up and down like a metronome in a single plane, avoiding the rotational forces that would skew the brain from side to side and tear it loose from its moorings.

Jackhammer feeding is only one adaptation of the picid woodpeckers. Many of the species have a stiff wedge-shaped tail that braces them against tree trunks and bristlelike feathers over the nostrils that shield the air passages from wood dust. They employ cylindrical, sticky tongues that can be extruded beyond the tip of the bill by as much as twenty centimeters and snaked through the insect burrows to seize their prey, then retracted and coiled in a cavity encircling the inner surface of the skull.

On the other hand, picid woodpeckers are not very good at traveling over open water, and therein lies our tale. They never colonized Hawaii during the millions of years its bird fauna was evolving. The honeycreepers were free to fill the woodpecker niche, and they did so with the ingenious innovation of the akiapolaau. The species has a jerry-built look to it when put alongside one of the sophisticated, wood-pulverizing picids. The akiapolaau could never have survived in competition, or originated in the first place, had native Hawaiian picids been hammering in the forests when the first honeycreepers flew ashore. To exist as a woodpecker with full dependence on dead and dying trees as a hunting ground requires space: an ivorybill breeding pair used about 8 square kilometers of old-growth swamp forest. When its habitat was drastically reduced by lumbering in the southern United States, the species was doomed to extinction. The ivorybill population was never very large. It declined precipitously, and the last authenticated sightings were made in the 1970s. Today a tiny remnant population lives on in the mountain forests of eastern Cuba. Woodpeckers therefore compete intensely for the relatively scarce resource on which they depend, and they would almost certainly displace any akiapolaaus they encountered.

Picid woodpeckers are also absent from the Galápagos. This volcanic archipelago, located in deep water 800 kilometers west of Ecuador, is the site of extensive adaptive radiations in many kinds of plants and animals. The productions are not as rich as those on Hawaii, but they are conspicuous enough to have inspired Darwin with the idea of evolution. Among those he found most interesting were Darwin's finches, or the subfamily Geospizinae in technical classification. A single colonizing ancestor expanded into thirteen contemporary species overall, which fill some of the same feeding niches as the Hawaiian honeycreepers. They shout the truth of evolution, and a naturalist of Darwin's caliber could not have missed it. He wrote in his 1842 *Journal of Researches* the words that foretold his theory: "The most curious fact is the perfect gradation in the size of the beaks of the different species of Geospiza—Seeing this gradation and diversity of structure in one small, intimately related group of birds, one might really fancy that, from an original paucity of birds in this archipelago, one species had been taken and modified for different ends."

Some of the Darwin's finch species are warbler-like, using their slender bills to capture insects and drink nectar. Others act like "true" finches, wielding relatively thick bills to tear apart fruit and crack open seeds. The larger the bird and thicker the beak, the wider the range of food objects consumed. In hard times, those with the thickest bills are able to specialize on the largest and toughest fruits and seeds.

Adaptive radiation is never complete, neither on archipelagoes nor on continents. Perhaps because there are fewer flowers in the dry Galápagos forests, Darwin's finches have not entered the sunbird niche, so expertly filled by several species of Hawaiian honeycreepers. None possesses a long, curved bill or long tongue of the kind needed to collect nectar from the deep recesses of flowers. On the other hand, the Galápagos radiation has produced an adaptive type unique among all the birds of the world: vampires. On the small and remote islands of Darwin and Wolf, ground finches alight on the backs of boobies, large seabirds of the genus *Sula*, and peck at the feather roots on the wings and tail, drinking the blood that flows out. Not content with this villainy, the vampire finches also crack open the eggs of seabirds by pushing them against rocks, then drink the contents.

Two geospizines, the woodpecker finch *(Cactospiza pallida)* and the mangrove finch *(Cactospiza heliobates)*, have entered the woodpecker niche, again in a manner wholly new for birds. Their bills are shaped like those of conventional insect feeders. The birds peck at the surface of trunks and branches and pull up loose pieces of bark, but they do not strike the wood with vertical hammer blows. In this respect they are ordinary Darwin's finches, close to a few other species of similar mien. They neither pluck insects out with a long curved bill, like Hawaii's akiapolaau, nor fish for them with an extrusible tongue in the picid woodpecker manner. Their innovation is purely behavioral. Woodpecker finches pick up a cactus spine, twig, or leaf petiole, adjust it so that it sticks out in front of their head like a stiff extruded tongue, then insert this makeshift probe into crevices here and there to stir up insects and winkle them out for capture. This trick is one of the few uses of tools known among animals. Seeing them in action, it is hard not to think of the woodpecker finches as intelligent. They have in fact been observed to correct mistakes during their hunting procedure. One individual was seen trying to break in two a stick that had proved too long to use as a tool. Another picked up a forked stick, tried unsuccessfully to probe with the forked tip, then turned the stick around to work with the unforked tip, this time successfully.

How did the birds conceive such an innovation? Peter Grant, who has watched Darwin's finches in the wild longer than any other person, believes that the birds achieved tool use by accident rather than by thought and then relied on operant conditioning. "I can imagine," he wrote, "a frustrated Woodpecker Finch failing to drop a piece of bark it had just removed from around the entrance to a crevice in a branch, accidentally pushing it into the crevice and touching the prey, and being rewarded by the prey moving toward the entrance and within reach of the bird's beak." Evolution by genetic assimilation might follow. Birds with a greater capacity for such trial-and-error learning would imitate those who invented the technique, and thus survive at a higher rate. In time the population would then contain not only brighter birds but those with an instinct hardened to pick up and manipulate sticks in the first place. Evolutionary biologists believe that genetic assimilation of this kind can on occasion greatly accelerate evolution, with behavioral flexibility leading the way.

If necessity is the mother of invention, then opportunity is its mother's milk. Tool use in Darwin's finches, no less than the fantastical double-functioned bill of Hawaii's akiapolaau, arose in a remote place in the absence of competition from the dominant picid woodpeckers. There is an even more bizarre twist to illustrate this principle, provided by the huia (*Heteralocha acutirostris*) on New Zealand. In the absence of competition from native picids, this odd crowlike species evolved a division of labor between the male and female that enabled them to work together as a kind of compound woodpecker. The species is extinct now, having been last seen on North Island in 1907, but enough observations were made in its last days to give a picture of the foraging techniques, again unique among birds. The male was armed with a straight, stout bill, similar in shape to a picid's. He chiseled open both dead wood and green saplings, then snatched up the first beetle grubs and other insects exposed. His mate, in contrast, had a long, slender, curved bill like that of many Hawaiian honeycreepers. She worked closely with the male, probing into deeper crevices and plucking out insects beyond his reach.

The archives of natural history are filled with other cases of species formation exploding as a response to ecological opportunity. On the Galápagos, Rarotonga, Juan Fernandez, and other remote oceanic islands, members of the plant family Compositae have radiated repeatedly to fill a large fraction of the niches available to vegetation. Composites as a whole rank among the most diverse and widespread flowering plants in the world. They include such familiar plants as the asters, sunflowers, thistles, marigolds, and lettuce. Their flowers are actually flower heads, tight clusters of many small flowers surrounded by leaflike structures (bracts). In addition to gracing gardens and banks of wildflowers, they are encountered everywhere as pretty weeds, such as dandelions and goldenrods, indomitable in the summer, gone with the winter frost.

On the most distant forested islands, many composite species also dominate the native shrubs and trees, having evolved from small herbaceous plants into what may be loosely called tree-asters and tree-lettuces. St. Helena is one of the most isolated islands in the world, located in the South Atlantic midway between Africa and South America. Before being completely settled first by Dutch and then British colonists, a process completed by the late 1800s, St. Helena's volcanic slopes were covered by forests of woody composites. Among them grew additional composite species and other plants of herbaceous form, the whole flora comprising thirty-six endemic species of flowering plants. Living in the forests were 157 or more St. Helenan beetle species, evolved from as few as twenty colonizing stocks, feeding on vegetation, dead wood, fungi, and each other. Seventy percent of these insects were weevils, a proportion totally out of line with coleopterous faunas in the rest of the world. Yet the strange ensemble worked. St. Helena was a nearly closed ecosystem, a biosphere functioning in great isolation, one step removed from a satellite colony in space.

The floras of each such composite-invested islands around the world contain among them every principal step in the transition from herbs to shrubs to trees. Each island is a contemporary laboratory of macroevolution, its flora an independent evolutionary experiment in progress, waiting for evolutionary biologists to pick up the clues and tell the story. The experiments are made even more persuasive by the fact that they have been replicated in still other groups of herbaceous plants, including especially members of the family Lobeliaceae. "The metamorphosis of these lettuces into shrubs or trees," Sherman Carlquist wrote in an account of island biology, "invites comparison with what has happened on other islands, to other plants. The Hawaiian lobeliads provide an almost exact parallel . . . Each growth form and leaf type can be roughly matched, showing that islands with a particular climate and a particular degree of isolation tend to promote these forms, these sizes."

What selection force drives the herbs to larger dimensions and assembles the island forests? Evidence from many sources suggests that it is ecological opportunity afforded by the absence of conventional trees. The vast majority of both temperate and tropical tree species have limited dispersal powers. Beech mast, dipterocarp seeds, and citrus fruits cannot travel far from the mother trees or survive immersion in salt water. But composites, among the dominant weedy plants of the world, are superb dispersers. When islands such as St. Helena and Oahu emerged as volcanic cones from the sea, these plants along with grasses were evidently among the first to arrive. They were also among the pioneers of Krakatau after the 1883 explosion. Around the world the long-distant emigrants entered an environment mostly or entirely devoid of shrubs and trees. They had the chance to evolve into shrubs and trees and preempt the land before traditional woody plants arrived, assuming this later event were even possible. Darwin correctly deduced the process in *On the Origin of Species*, using the new idiom of natural selection:

> Trees would be little likely to reach distant oceanic islands; and an herbaceous plant, though it would have no chance of successfully competing in stature with a fully developed tree, when established on an island and having to compete with herbaceous plants alone, might readily gain an advantage by growing taller and taller and overtopping the other plants. If so, natural selection would often tend to add to the stature of herbaceous plants when growing on an island, to whatever order they belonged, and thus convert them first into bushes and ultimately into trees.

The arborescence of island weeds raises the larger question of why certain groups of organisms undergo radiation, and others do not.

The example of the Compositae shows that superior dispersal ability empowers at least some organisms some of the time. A species able to invade a new island, lake, or other empty environment, to fill it, and to divide into multiple specialized species is likely to control the land by preempting invasion and diversification by other species. On the Galápagos Islands a small assemblage of flycatchers, mocking-birds, and warblers coexists with the thirteen species of Darwin's finches, but none has achieved a comparable adaptive radiation. Is it possible that the finches, or more precisely the ancestral finch, simply got to the Galápagos first and closed off opportunities for later arrivals? Dominance could have been the reward of nothing more than superior dispersal power. Since we do not know the date of arrival of these birds, we cannot say for sure.

Alternatively, perhaps the ancestral Darwin's finch had qualities that allowed it to evolve and radiate more decisively than its rivals, no matter when it arrived. It might have possessed a generalized anatomy and behavior that adapted it quickly to a partially empty environment. If that is true, would it be possible to deduce the nature of the original species in this respect? Not with certainty, but we can make a good guess because, amazingly, something like the ancestral species still exists. One more Darwin's finch, number fourteen, lives on Cocos Island, a 47-square-kilometer speck of land 580 kilometers to the northeast of the Galápagos. The island, owned by Costa Rica, is hilly, uninhabited by human beings, and densely clothed in tropical rain forest. The Cocos Island finch, *Pinaroloxias inornata*, coexists with only three other breeding species of landbirds—a cuckoo, a fly-catcher, and a yellow warbler. This scarcity of competitors has al-lowed it to engage in what biologists call *ecological release*, the expan-sion of a single species into multiple habitats.

Ecological release is a common phenomenon on remote islands with small faunas and floras—I have observed it many times in ants, for example—but it occurred to spectacular degree in the Cocos Is-lands finch. The birds, all members of a single freely breeding spe-cies, occupy niches ordinarily divided among species, genera, and even entire families of birds. They range from shore to hilltop, forage within the forest from ground to canopy, and feed variously on insects, spiders, and other arthropods, mollusks, small lizards, seeds, fruit, and nectar. In this respect the Cocos finch far exceeds any one species of Darwin's finch in the Galápagos. Most striking of all, each individual bird specializes on a particular kind of food and maintains the habit for at least several weeks, perhaps for its entire life. This microcosmic adaptive radiation appears to be based on observational learning. During a ten months' visit to Cocos, Tracey Werner and Thomas Sherry, the biologists who discovered the release phenomenon, watched juvenile finches approach and imitate the distinctive feeding behavior of yellow warblers and sandpipers. No doubt the young birds also copy older birds of their own species. Like medieval apprentices selecting masters within specialized guilds, the young birds appear to prosper with personalized instruction.

Cocos finches have bills roughly intermediate in size and shape to those of warblers and finches. With the tendency of the birds to diversify in feeding habits individually, the stage is set for rapid species formation and adaptive radiation in the Galápagos mode, if circumstances permitted it. But circumstances do not permit it. The place is too small and too remote from other islands to allow the formation of new species. The Cocos finch radiation therefore remains stalled in an embryonic state, one species to the island.

Certain kinds of plants and animals, by virtue of distinctive biological traits they already possess, seem poised to expand and preempt many niches in sparsely inhabited environments. If the new home is complex enough to allow species formation and ecological specialization, the radiation proceeds to term. A second example of radiation proneness, as graphic as that of Darwin's finches, is found in the freshwater Cichlidae. This prolific fish family occurs from Texas to South America in the New World, from Egypt to the Cape Province in the Old World. A suite of primitive species also lives on Madagascar, and three more species are endemic to southern India and Sri Lanka.

The Great Lakes of East Africa, the necklace of fresh water strung along the Rift Valley from Uganda to Mozambique, swarm with cichlids. These fishes dominate the aquatic fauna, having radiated to fill almost all major niches available to freshwater fishes as a whole. Cichlids are the lacustrine equivalents of the Hawaiian honeycreepers. The three hundred or more species of Lake Victoria alone, for example, include the following major adaptive types:

Astatotilapia elegans, perch-like in shape, a generalized bottom feeder

Paralabidochromis chilotes, large mouth with thickened lips; preys on insects

Macropleurodus bicolor, small mouth; uses pebble-shaped pharyngeal teeth to crush snails and other mollusks

Lipochromis obesus, possesses a heavier body and somewhat enlarged mouth; preys on the young of other fish

Prognathochromis macrognathus, resembles a pike, with slender body, disproportionately large head and jaws, sharp teeth; preys on other fish

Pyxichromis parorthostoma, constricted head, upturned mouth with thickened lips; probably a feeding specialist but habits still unknown

Haplochromis obliquidens, teeth expanded and flat at the tips; grazes on algae

The Lake Victoria array is the greatest found anywhere in the world in a single group of fishes limited to a single body of water. Equally remarkable are the graduated steps that link the species in each of the adaptive classes, from the earliest stages of anatomical modification to the most extreme specialization in body form. Among the mollusk feeders, to illustrate the point, are found some species with only a few slightly enlarged pharyngeal teeth used by the fishes to crush the shells of their prey. Others, somewhat more advanced, possess a greater number of such teeth, many pebble-shaped, which are ground together on the mollusk shells with somewhat enlarged throat muscles. Still others, the extreme mollusk specialists, use pharyngeal bones packed with pebble teeth and powered by heavier throat muscles. Comparable morphoclines—series of species arrayed from the most generalized to the most specialized—occur among the cichlid algal feeders and predators on other fish.

All of the Lake Victoria cichlids appear to have descended from a single ancestral species that colonized Lake Victoria from nearby older lakes. The evidence, presented by Axel Meyer and his coworkers in 1990, is based on the degree of similarity in the genetic codes of the fishes. More specifically, fourteen of the species, representing nine genera, display very little variation in the nucleotide sequences of mitochondrial DNA, less diversity than occurs within the entire human species.

Most of the Lake Victoria cichlids belong to a larger group called *haplochromines*, an informal designation used in past years to suggest a common recent ancestry, a hypothesis now supported by the molec-

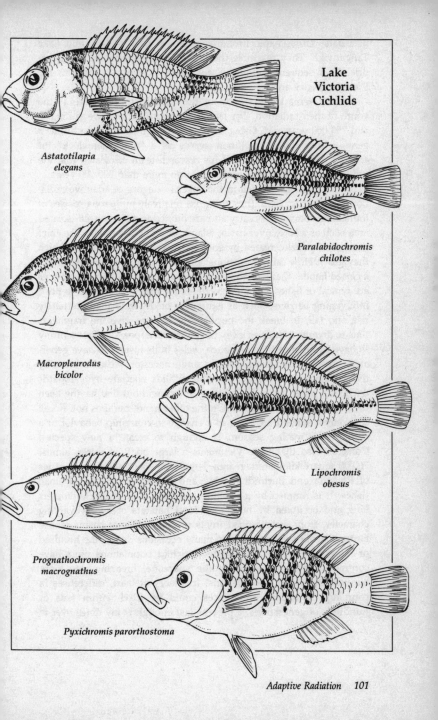

Lake Victoria Cichlids

Astatotilapia elegans

Paralabidochromis chilotes

Macropleurodus bicolor

Lipochromis obesus

Prognathochromis macrognathus

Pyxichromis parorthostoma

ular data. Other haplochromines occur in Lake Malawi and Lake Tanganyika. They resemble the Lake Victoria species in mitochondrial DNA sequences, but are not nearly so close to them as the Victoria species are to one another.

Another remarkable feature of the Lake Victoria cichlids is the youth of their radiation. The lake is estimated to be between 250,000 and 750,000 years old. Using DNA sequences from the cytochrome *b* gene, where steady evolution serves as a "molecular clock" for animals generally, Meyer and his research team calculated that the full cichlid evolution was achieved in no more than 200,000 years.

The Victoria cichlids fall in the special category of adaptive radiation called *species flocks*: they comprise relatively numerous species of immediate common ancestry and are limited to a single well-isolated area such as a lake, river basin, island, or mountain range. The chief theoretical puzzle created by species flocks is the process by which they grow. How can populations split repeatedly into species within a closed habitat that has no geographical barriers? If haplochromines are typical of fishes and other vertebrates, they would seem to need intervening barriers, such as isthmuses of dry land that alternately rise and fall, to break the populations up and give the fragments time to diverge to species level. Lake Victoria appears on first scrutiny to have experienced too few such cycles in its history to have generated three hundred species from a single ancestor. We are pressed by the evidence to conclude that the cichlids speciate by sympatric means, in other words by splitting in two without first having been divided by a physical barrier. On the other hand, perhaps not. Recall that only a single trait, such as a change in courtship behavior or a shift in the mating season, is enough to create a new species. Consider also that Lake Victoria is a large body of water, almost 70,000 square kilometers in area, larger than the combined countries of Rwanda and Burundi nearby, and home for millions of small fishes. It is rimmed by a twisting shoreline over 24,000 kilometers long and occupied by numerous local habitats of widely varying character, from wave-lapped inlets to deep offshore basins whose bottoms never see sunlight. On many occasions during the hundred or more millennia of their history, cichlid populations must have contracted their ranges along the shoreline, breaking into local, temporarily isolated populations. In theory at least, differences in courtship or habitat preferences could be fixed within tens or hundreds of generations, a process fast enough many times over to have generated three hundred cichlid species during the lifespan of Lake Victoria.

The evolutionary explosion might have occurred all the more easily if cichlid fishes are prone to rapid evolution in the demonstrated manner of Darwin's finches. The clue to look for is a cichlid species somewhere that is the equivalent of the Cocos Island finch, that is, opposed by few if any competitors, highly variable, and multipurpose in its life habits. And, as if created for the delectation of life scientists and textbook writers, such an example exists. It does not occur in the crowded waters of the African Great Lakes, where competition and specialization in the species flock has reached a level of near saturation. To find the right conditions we must journey to the waters of Cuatro Ciénegas, in the state of Coahuila, northern Mexico. There the prescribed species, *Cichlasoma minckleyi*, lives in streams, ponds, and canals. A small, spotted perch-like fish, it coexists with several other fishes of similar size, including one additional cichlid. Its populations contain two radically different feeding types: the papilliform morph, with more slender jaws and slender teeth, and the molariform morph, with thicker jaws and pebble-shaped teeth. They look like entirely different species, but they are not. Both types freely breed together, thus constituting a single species. Both feed readily in the same places on the same wide range of small prey, including insects, crustaceans, and worms. When food grows scarce, however, the molariform cichlasomas—though not their papilliform associates—switch increasingly to snails, which they alone are able to crush with their stronger jaws and thick, flat teeth. By expanding their diet, the molariforms reduce competition with the papilliforms and persist through the hard times. It is easy to imagine such a species as *Cichlasoma minckleyi* invading a new body of water of the Lake Victoria class and then radiating into many niches in short time. The first step would almost certainly be a division into two full, reproductively isolated species, a papilliform *Cichlasoma* concentrating on insects and other soft-bodied prey, and a molariform *Cichlasoma* preying on snails and other mollusks.

Biologists have begun to search more systematically for such species on the threshold of adaptive radiation and hence macroevolution. What may prove to be the most dramatic case discovered so far, exceeding even the Cocos finch and the Mexican cichlasoma, is the arctic char, *Salvelinus alpinus*, a salmon-like fish found in lakes and rivers around the northern polar region. Very few other fish species occur with it, and the char has a relatively wide range of uncontested feeding niches to exploit. Many of the local populations contain several anatomically distinct forms with different food habits and growth rates. In Iceland's Thingvallavatn *(vatn,* lake), there are four such morphs: a large bottom feeder, a small bottom feeder, a predator on other fish, and a vegetarian that grazes algae. Skúli Skúlason and his fellow researchers at the University of Iceland found that these specialists differ one from another genetically but still interbreed freely, forming a single, highly plastic species. Like the finch and cichlasoma, the arctic char seems to be an adaptive radiation waiting to happen, or perhaps certain to happen given more time. The Arctic lakes in which the char lives were created by the retreat of the continental glaciers only a few thousand years ago.

Natural history becomes all the more pleasing and interesting when we look at it through the lens of evolutionary theory and search for the starbursts of adaptive radiation—and all the more foreboding when we learn how quickly these creations can be extinguished. The vast majority of radiated groups stay near the peak of diversity from thousands to millions of years. The cichlid fishes of Lake Victoria, in contrast, are disappearing almost instantaneously by this standard. They are being extinguished en masse by the giant Nile perch, a voracious predator introduced as a game fish by Ugandan officials in the 1920s. This "elephant of the water," reaching 2 meters in length and 180 kilograms, is literally eating its way through the cichlids southward from its northern point of introduction. Where it has become dominant, more than half of the cichlid species have disappeared.

The confinement of groups like the African cichlid fishes and the Hawaiian honeycreepers to a single lake or archipelago renders them extremely vulnerable to environmental change, and they can be obliterated by a swipe of the human hand. In their company are groups of higher taxonomic order and broader geographic distribution that hang on as diminished remnants of a flamboyant past: cycads, crocodilians, lungfishes, rhinoceros, and other so-called living fossils. These too are being pushed by human activity to the brink of extinction, after tenures lasting millions of years. At the opposite extreme are a few select groups that have remained fully radiated for an equivalent period of time. These species display an astonishing array of radically different body forms and life cycles, and are widely distributed and abundant throughout the world. Among these old-money dynasties are ciliated protozoans, spiders, isopod crustaceans, and beetles, as well as one group that in my opinion deserves special attention in any serious talk about diversity and natural history: the sharks.

Sharks, fishes that compose the three superorders Squatinomorphii, Squalomorphii, and Galeomorphii of the class Chondrichthyes, shadows in the sea of our nightmares, lone-wolf predators of frightening quickness, questioners of the Darwinian importance of intelligence, have been on earth for 350 million years. They began as small, stiff-bodied cladodonts in the late Devonian period, then radiated and maintained high diversity in seas throughout the world until the beginning of the Permian period. At that time, 290 million years ago, they declined to a low level of diversity lasting for 100 million years. The survivors recovered, expanded a second time, and somehow passed full-blown through the great extinction spasm at the end of the Age of Dinosaurs. Today they are at least as diverse as they have ever been.

Sharks seen at a distance (fin and back roiling the surface for a heart-stopping moment, then the vaguely torpedo shape slipping into deeper water) may not seem to differ much from one species to the next except in size. In fact the 350 species found in the world vary immensely, so much so as to stretch the very definition of the word *shark*; we must infer their common ancestry from traits of internal anatomy in order to place them all within the same group. The archaic radiation of sharks is marked by differences among species far greater than those in the still youthful radiations of Darwin's finches and Lake Victoria cichlids. It is tempting to think that age has fine-tuned their specializations, pitted them against more competitors, extinguished a higher number over longer periods of time, and produced a generally tougher, more durable group of contemporary species.

If ever there was a prototypical shark in the popular imagination, it is most likely the tiger shark (*Galeocerdo cuvier*), the great fish sometimes called the garbage can of the sea. Reaching 6 meters in length, weighing up to a ton, tiger sharks are often attracted to harbors, where they consume almost anything sizable even hinting of animal protein. From the stomachs of such specimens have been retrieved fish, boots, beer bottles, bags of potatoes, coal, dogs, and parts of human bodies. One dissected giant contained three overcoats, a raincoat, a driver's license, one cow's hoof, the antlers of a

Great white shark

Cookie-cutter shark

The adaptive radiation of sharks has gone to extreme lengths, exemplified by the great white shark *(Carcharodon carcharias)*, which preys extensively on seals and other marine mammals. Another, more bizarre form is the cookie-cutter shark *(Isistius brasiliensis)*, a parasite that carves plugs of flesh from the bodies of marine mammals and large fish without killing them.

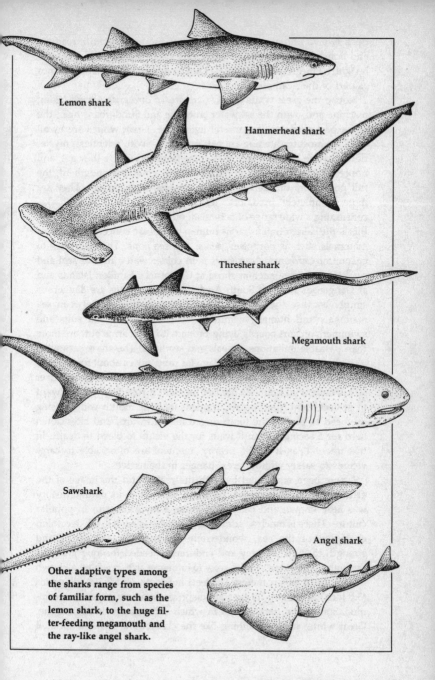

Lemon shark

Hammerhead shark

Thresher shark

Megamouth shark

Sawshark

Angel shark

Other adaptive types among the sharks range from species of familiar form, such as the lemon shark, to the huge filter-feeding megamouth and the ray-like angel shark.

deer, twelve undigested lobsters, and a chicken coop with feathers and bones still inside. Tiger sharks are man-eaters in a casual way, by which I mean taking swimmers not by design but by happenstance as part of their catholic diet.

Not so the great white shark, *Carcharodon carcharias*, famed killing machine and, with the saltwater crocodile and Sundarbans tiger, the last expert predator of man still living free. Great whites are by all odds the most frightening animals on earth—swift, relentless, mysterious (no one knows where they come from and where they go), and unpredictable. They are, in my admittedly emotional judgment, the full possessors of shark *arete*, the essence of sharkness. They are more completely predators, less scavengers, than tiger sharks, consuming a wide range of bony fish, other sharks, sea turtles, and— this is the salient trait as far as human beings are concerned—marine mammals such as porpoises, seals, and sea lions. The best place to encounter *Carcharodon carcharias* is in colder water around seal and sea-lion rookeries, such as those at California's Farallon Islands and at Dangerous Reef off South Australia. Great whites are dangerous simply because they fail to make a clear distinction between sea mammals and human swimmers. Divers in rubberized suits and swimmers on surf boards, lying prone with their arms out, are more than passable imitations of seals and sea lions. The shark sees what it thinks is the silhouette of its familiar prey, noses about for a while, makes up its mind, and sprints toward the swimmer at speeds over 40 kilometers an hour. At the last moment it rolls its eyes backward to protect them from impact. It opens its huge mouth wide, raising the head to project the tooth-ringed maw forward, and bites down hard for a second. Then it waits for the victim to bleed to death. In this interval, as it circles nearby, rescuers are often able to carry victims to safety without great danger to themselves.

I have been enchanted by both the reality and the image of the great white shark for years, back to a time when its natural history was little known and before it became a mythic horror in popular culture. There is much to admire about the species. It is the decathlon champion of the sea, wonderfully designed for the speed and strength to hunt big prey and endurance needed for long journeys through open water. Adults grow to immense size, reaching a known maximum of seven meters (23 feet) in length and 3,300 kilograms (3.6 tons) in weight. Its eyes are disproportionately large, an accommodation to the dark waters in which it hunts much of the time. Great whites have something like the classic tuna shape associated with fast pelagic fishes: body spindle-shaped and rigidly muscular, nose pointed to cut the water like the prow of a submarine. Ridges to the rear along each side of the trunk guide the flow of water evenly past the body. The powerful tail sweeps smoothly from side to side. The mouth, lined with parallel rows of serrated triangular teeth, hangs partly open in a fixed clown's grin, affirming the impression of human divers that the fish is glad to see them. Water flows continuously through the mouth and back over the gills, part of a ramjet system that feeds oxygen efficiently to the large, active body. The great white is warm-blooded, allowing it to cruise the colder waters of most of the world's oceans and to forage from the surface down to at least 1,300 meters.

In 1976 the naturalist Hugh Edwards, who watched for great whites from a shark cage in waters off the Albany whaling station of Western Australia, turned to see a large male suspended two meters away. He later wrote:

> In all our lives there are milestones, important moments we remember long after. This was one of them. For the brief time of his appearance I drank in every detail of the shark—his eyes, black as night; the magnificent body; the long gill slits slightly flaring; the wicked white teeth; the pectoral fins like the wings of a large aeroplane; and above all the poise and balance in the water and the feeling conveyed of strength, power, and intelligence. To see the shark alive was a revelation. He was strong, he was beautiful. No dead shark or second-hand account could convey the vitality and presence of the live creature. A few seconds face to face were worth more than all the years of hearsay, pictures, and slack-jawed corpses.

Working off the image of these classic sharks, I will now argue that, given enough time, evolution can fine-tune and harden adaptive types to create the most extreme radiations. Of greatly different anatomy and biology from the tiger and great white sharks, for example, is the cookie-cutter shark (Isistius brasiliensis). It is not a predator at all but a parasite, of porpoises, whales, bluefin tuna, and even other sharks. Only half a meter in length, shaped like a cigar, the cookie-cutter has a curving row of huge teeth on its lower jaw. It thrusts its maw into the bodies of its victims and twists to slice out 5-centimeter-wide conical plugs of skin and flesh. For many years the circular scars on porpoises and whales were a mystery, attributed variously to bacterial infection or an unknown invertebrate parasite, until the true habits of the small sharks were discovered in 1971.

Cookie-cutters also attack nuclear submarines, taking unnutritious bites from the neoprene coating of sonar domes and hydrophone arrays. The cookie-cutter shark passes what I like to call the test of a complete adaptive radiation: the existence of a species specialized to feed on other members of its own group, other products of the same adaptive radiation.

Equally specialized in a wholly different direction are the filtering sharks, gigantic fishes that cruise placidly near the surface of the open ocean, seining out and swallowing huge quantities of copepod crustaceans and other small planktonic animals in the manner of the baleen whales. The whale shark *(Rhincodon typus)*, reaching 13 meters in length and many tons, may be the largest fish that ever lived. At the opposite extreme is the green lanternshark *(Etmopterus virens)*, which at 23 centimeters—the size of a large goldfish—is the smallest of all sharks.

Other major adaptive types expand the cavalcade of the living sharks:

Horn sharks (example, *Heterodontus japonicus*), inshore bottom dwellers that use their hard, molar-like teeth to feed on mollusks

Frilled sharks (example, *Chlamydoselachus anguineus*), deep-sea dwellers with elongated, eel-shaped bodies and fins, teeth shaped like fishhooks

Angel sharks (example, *Squatina dumerili*), squat bottom dwellers outwardly resembling rays more than sharks, but anatomically sharks

Thresher sharks (example, *Alopias vulpinus*), large pelagic forms that sometimes run in pairs and stun smaller fish by lashing them with long, whiplike tails

Unknown species of sharks almost certainly swim the seas. Some are likely to be very large. I base this conjecture on the megamouth shark *(Megachasma pelagios)*, discovered in 1976. The first specimen was hauled up from deep water off Hawaii by the United States Navy after it became entangled in a cargo parachute used as a sea anchor. It was almost 5 meters long and weighed 750 kilograms. To the surprise of the navy and consulting ichthyologists alike, it was unlike any other shark seen to that time. Four more individuals of the same species have been encountered since. Two were caught in gill nets off California, and one each washed ashore at Japan and Western Australia.

Megamouth is so different in anatomy from all previously known sharks that it has been placed in a taxonomic family of its own, the Megachasmidae. Its most striking feature is an enormous maw, used to draw in water and filter copepods, euphausiid shrimps, and other small planktonic animals on which it feeds. Megamouth is thus in the same ecological guild as the whale shark, as well as the huge basking shark of northern seas. Its body is cylindrical and flabby, its eyes small, its movements stiff and slow. It flees into deep water at small disturbances. Its upper jaw and palate are covered with a silvery, iridescent lining, possibly a deposit of guanine or other reflective waste material. When the Los Angeles specimen became entangled in a gill net, scuba-suited researchers were able to implant transmitters into its body and track it at sea for two days. In that time the shark cruised about 10 to 15 meters below the surface at night and descended to 200 meters during the day. This vertical migration is typical of fishes of the deep scattering layer, the thick concentrations of organisms, detectable by sonar, that travel up and then back down again through each twenty-four hour cycle. Megamouth's deep submergence during the day and shy, aversive behavior overall perhaps explain why the species remained undiscovered for so long.

The metaphor of dynastic succession I chose at the beginning to describe the turnover of adaptively radiated groups implies a balance of nature. One dynasty, this conception holds, cannot tolerate another dynasty of closely similar kind. A limit to organic diversity exists so that when one group radiates into a part of the world, another group must retreat. Because evolution is so dismayingly idiosyncratic, the balance of nature cannot be ranked as a law of biology. But at least it is a rule, a statistical trend: dominant, expanding groups tend to replace those groups found in the same places that are ecologically most similar to them.

The displacement of one group by another is seldom if ever a blitzkrieg. Almost always it is a sitzkrieg, with the newer group pushing gradually into the terrain of the older group, enveloping its rival slowly, and replacing it species by species. Just as often the replacement is favored by decimation of the older dynasty through climatic change or loss of food supply. The rise of the mammals after the fall of the dinosaurs is the textbook case, but examples exist among corals, mollusks, archosaur reptiles, ferns, conifers, and other organisms following the demise of their competitors in one of the major extinction spasms. These temporary winners seized opportunities provided by vacant niches just as the Hawaiian honeycreepers and Lake Victoria haplochromine fishes invaded newly created environments. Their success was global in scope, however, not limited to an archipelago or lake.

We now come to the interesting question implied by the balance of nature: what happens when two full-blown, closely similar dynasties meet head on? If it were possible to play God with geological spans of time to wait and watch, the ideal experiment would be this: allow two isolated parts of the world to fill up with independent adaptive radiations of plants and animals, so that the majority of species in each theater have close ecological equivalents in the other theater; then connect the two regions with a bridge and see what happens. When the organisms intermingle, would those from one theater replace the other, so that a single biota comes to occupy the entire range?

The experiment has in fact been performed once in relatively recent geological time, and we can deduce a great deal of what happened by comparing fossil and living species. Two and a half million years ago the Panama isthmus rose above the sea, allowing the mammals of South America to mix with the mammals of North and Central America. First I should explain that the contemporary mammals of the world are primarily the products of three great adaptive radiations, and three only. The reason is that it takes an entire continent to spawn a mammalian radiation. For insects, a single island is adequate. Beetle species have proliferated luxuriously on St. Helena in the South Atlantic, Rapa in the South Pacific, and Mauritius in the Indian Ocean. Had flightless mammals reached these same specks of land—which they did not and probably could not prior to the coming of man—it is doubtful that the species would have multiplied. Mammals, even rats and mice, are simply too big, active, and wide-ranging. To produce an adaptive radiation of the honeycreeper or cichlid magnitude, their species need a continent.

The first of the three continents on which mammalian radiation did attain full expression is Australia. In biogeographical terms Australia is only an extremely large island, having been isolated from the rest of the world since the breakup of the Gondwanaland super continent more than 200 million years ago. The second body of land spacious enough for mammalian radiation is the "World Continent," comprising Africa, Europe, Asia, and North America as far south as the southern rim of the Mexican plateau. The World Continent has been more or less cohesive throughout the Age of Mammals, during the past 65 million years, because the closeness of its parts has allowed many kinds of plants and animals to emigrate from one to the next. North America, the most isolated of the elements, was joined to Europe across present-day Greenland and Scandinavia during the early Age of Mammals. Alaska and northeastern Siberia have been connected off and on by land bridges, the most recent about 10,000 years ago. The third continental center of mammalian evolution is South America, which was isolated during the breakup of Gondwanaland, drifted north, and was finally linked solidly to North and Central America 2.5 million years ago.

For most people, the mammals of the World Continent are "typical" and "true" mammals, simply because they are the most familiar. These are the animals we were born and raised with. The mammals of Australia and South America are nevertheless highly evolved in their own right.

Today three major groups contribute to the composition of the native, prehuman fauna of Australia. The first are monotremes or egg-laying mammals, remnants of an ancient and largely superseded radiation—faded lines on the palimpsest. They include the duck-billed platypus, an aquatic form that looks as if it were fabricated from the head of a duck and the body of a web-footed muskrat; and the short-beaked echidna, a land dweller best described as a porcupine with the tapered cylindrical snout of an anteater. The second group are placental mammals, which carry their young attached to a placenta in the uterus. Relative newcomers to Australia, yet already composing a third of the species, they comprise a wide array of bats and rodents. Their immediate ancestors island-hopped across Indonesia to reach northern Australia and then spread through parts of the continent.

The third native group of Australia are marsupials, mammals that give birth to their young as tiny fetuses and carry them to an advanced stage in a belly pouch (marsupium). It is this third, relatively ancient and still dominant, group that has converged with the greatest fidelity to the World Continent placental fauna. Here are the chief mammalian analogues across the two regions and the adaptive roles they fill:

Australian marsupial mammals	World Continent placental mammals	Adaptive type
Dibbler (*Parantechinus apicalis*)	Mice	Small, secretive omnivores
Jerboa marsupial mouse (*Antechinomys spenceri*)	Jerboas, kangaroo rats	"Jumping mice" of desert areas; insect feeders in Australia
Bandicoots (*Macrotis lagotis, etc.*)	Rabbits, hares	Long, saltatory hind legs; diet of grass and other plant materials; some omnivorous
Quolls or dasyures (*Dasyurus geoffroii* and *D. viverrinus*)	Small cats	Predators of small mammals, reptiles, and birds
Gliders (*Petaurus sciureus, etc.*)	Flying squirrels	Arboreal gliders, using membranes on sides of body; mostly herbivores
Anteater or numbat (*Myrmecobius fasciatus*)	Anteaters	Feed on termites with long, flexible, sticky tongues
Tree wallabies (*Dendrolagus lumholtzi, etc.*)	Catarrhine monkeys	Arboreal, mostly herbivores
Marsupial mole (*Notoryctes typhlops*)	Moles	Subterranean, feed on insects and worms
Wombats (*Lasiorhinus krefftii, etc.*)	Woodchuck	Secretive, burrowing herbivores
Large kangaroos (*Macropus robustus, etc.*)	Horses, antelopes, other ungulates	Grazers, using chisel-like front teeth and broad grinding molars
Tasmanian devil (*Sarcophilus harrisi*)	Wolverine	Predators of small animals
Tasmanian wolf or thylacine (*Thylacinus cynocephalus*)	Wolves, big cats	Predators of kangaroos, other mammals, and birds

The stage set, we have arrived at the moment of the experiment. The mammalian radiation in South America was as expansive as that in Australia, and its convergence to the World Continent fauna was even closer. Yet the look-alike species are much less familiar—toxodonts, marsupial cats, macrauchenians, glyptodonts—because so few lived to be seen by human beings. They disappeared at about the time the Panama land bridge rose and elements of the World Continent fauna poured into South America. Others that survived failed to diversify at the same rate as the northern invaders. In the exchange North and Central America contributed more to South America than the reverse.

Prior to the overland migrations back and forth, known as the Great American Interchange, the old endemic mammals of South America had been assembled during two waves of radiation and partial extinction. The first began near the end of the Mesozoic era, about 70 million years ago, and climaxed during the succeeding 40 million years. The early stocks of these archaic mammals had risen even earlier in Mesozoic times, in the remnants of Gondwanaland, when South America was still close to Africa and Antarctica and dinosaurs prevailed. Now relieved of the constricting influence of dinosaurs, the mammals expanded to fill the abandoned niches. In the grasslands lived litopterns superficially similar to the "true" horses of the World Continent, members of the family Equidae with which human beings were to evolve in close intimacy. Litopterns possessed fully developed hooves and skulls equipped for grazing long before these specializations evolved in the equids. Other litopterns were more like camels. Toxodonts variously resembled rhinos and hippopotamuses, while some astrapotheres and pyrotheres came passably close to tapirs and elephants. Argyrolagids, good imitations of kangaroo rats but with enormous eyes set far back on the skull, bounced about on springy hind legs. Borhyaenids, whose species resembled shrews, weasels, cats, and dogs, were among the main predators of other mammals. A sabertooth marsupial cat, *Thylacosmilus*, bore an amazing resemblance to the sabertooth tigers of the World Continent fauna.

The herbivores of old South America were mostly placentals, and the carnivores were marsupials. Paleontologists are not sure why this was so, or why in contrast the mammals were primarily marsupials in Australia and placentals in the World Continent. It might have been no more than the luck of the draw: which group penetrated the major adaptive zones first, radiated, and preempted the other. We may never know, for the number of continents with which to test the hypothesis ran out at three. (Being limited to one planet and a small number of continents and archipelagoes is the curse of evolutionary biology.)

About 30 million years ago a long, slow second wave worked its way into South America, this time from the north across island

Mammalian Radiations

AUSTRALIA   AMERICAS

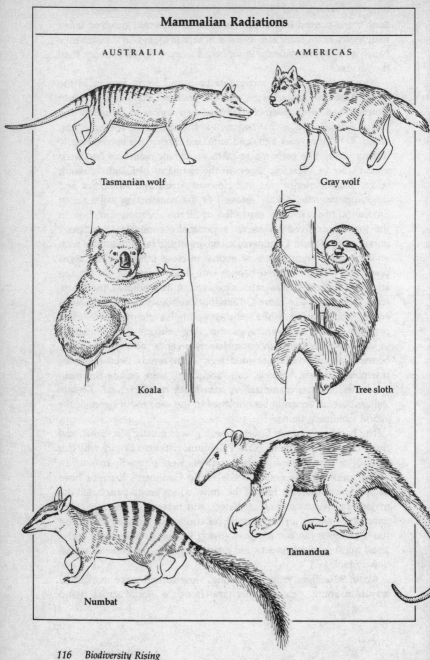

Tasmanian wolf   Gray wolf

Koala   Tree sloth

Numbat   Tamandua

Mammalian Radiations

AUSTRALIA

AMERICAS

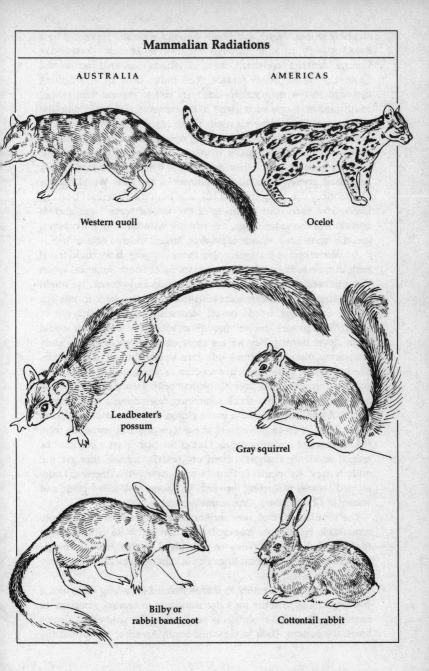

Western quoll

Ocelot

Leadbeater's possum

Gray squirrel

Bilby or rabbit bandicoot

Cottontail rabbit

stepping stones. North and South America were still separated by a broad seaway running through the Bolivar Trough. Present-day Central America consisted then of islands scattered across the Trough, with the newly formed West Indies close by and drifting eastward. A few mammal species were able to expand their ranges southward from one such island to the next and eventually onto the continent of South America itself. These island hoppers included an early monkey-like species of primate, which proceeded to proliferate into howler monkeys, spider monkeys, marmosets, titis, tamarins, capuchins, sakis, and other dwellers of the forest canopy. Many possessed prehensile tails, the hallmark of the New World species (if a monkey can hang from its tail, it is from the American tropics). Even more successful members of the second wave were ancient ancestors of the guinea pigs, the rabbitty viscachas, the porcupines, and the horsefaced aquatic capybaras, largest rodents on earth.

A thousand ages in thy sight are like an evening gone. If we could travel back in time to the mid-Cenozoic savanna of South America, when the continent was still surrounded by straits and oceans, we might think ourselves on safari somewhere in a national park in modern Africa. Everything would be off somewhat, distorted and out of focus, like a picture studied through an astigmatic lens, yet it would seem *almost* normal. Say we are there on the edge of a lake, early one sunny morning, turning our gaze slowly through a full circle. The vegetation looks much like modern savanna. Out in the water a crash of rhinoceros-like animals browse belly deep through a bed of aquatic plants. On the shore something resembling a large weasel drags an odd-looking mouse into a clump of shrubs and disappears into a hole. A creature vaguely like a tapir watches immobile from the shadows of a nearby copse. Out of the high grass a big, cat-like animal suddenly charges a herd of—what?—animals that are not quite horses. Its mouth is thrown open nearly 180 degrees, knife-shaped canines projecting forward. The horse look-alikes panic and scatter in all directions. One stumbles, and . . .

The simulacrum that was ancient South America is all the more remarkable because its mammals had nothing to do with those in the rest of the world. They were evolved in this replicate mega-experiment along different lines out of different stocks, yet roughly to the same effect.

If you limit your scrutiny to faunas isolated for a long time, and if you are willing to settle for loose standards in scoring similarity of anatomy and niche, evolution is predictable. But wild cards always break the pattern. Back to Cenozoic South America, we turn at the sound of tree branches snapping and crashing to the ground, as some large mammal pulls them down to feed. We expect elephants but find ground sloths, immense, ungainly animals covered with thick reddish fur, who gather in the foliage with clawed hands and chew the leaves and tender branches with vaguely horse-like heads. They fill the elephant niche but use different tools. And now a stunning surprise. A *Titanis* appears, a flightless carnivorous bird standing 3 meters (10 feet) tall, its eagle head tipped by a massive hooked beak 38 centimeters (15 inches) long. Loping along on stilt legs like a malevolent ostrich, it flicks its head to left and right in search of prey, which can be as large as a deer. *Titanis* was only the largest of a variety of phororhacoid ground birds; some were as small as geese. Never before or since have mammals faced anything like the phororhacoids, except during their earliest evolution in the Age of Dinosaurs. In South America *Titanis* and its relatives must have been serious rivals to the borhyaenids and other carnivorous marsupials. Since anatomists consider birds as a whole to be direct descendants of dinosaurs, close enough to be called dinosaurs (although that is stretching it), the phororhacoids might be called the final echo of the ruling reptiles.

Phororhacoids, saber-toothed marsupial cats, toxodont rhinoceroids—all that splendid assemblage is gone. We will never ride a litoptern or feed peanuts to a long-trunked pyrothere in the zoo. Although biological history is a flow of events with causes and effects that can in principle be rationally joined, one extraneous accident can change everything. When the Bolivar Trough disappeared and the Panamanian land bridge rose across its center less than 3 million years ago, the final wave of mammals rolled swiftly into South America. Many of the World Continent mammals, which had been blocked for millions of years by the Bolivar straits, now simply walked onto the continent. Most traveled along corridors of grassland, which at the time extended southward along the eastern slopes of the Andes all the way to Argentina.

So successful was this incursion that about half of the most familiar South American mammals today are of geologically recent World Continent origin: jaguars, ocelots, margays, peccaries, tapirs, coatis, kinkajous, bush dogs, giant otters, alpacas, vicunas, llamas, and the recently extinct mastodons. South America's autochthons flowed in the opposite direction. For a while at least, North America was home to giant sloths, armadillos, possums, glyptodonts, porcupines, anteaters, and toxodonts. *Titanis* spread all the way to Florida.

The Great American Interchange resulted in a sharp increase for a time in the mammalian diversity on both continents. Consider first the taxonomic level of the family. Examples of mammalian families are the Felidae, or cats; Canidae, dogs and their relatives; Muridae, the common mice and rats; and of course Hominidae, human beings. The number of mammalian families in South America before the interchange was thirty-two. It rose to thirty-nine soon after the isthmus connection and then subsided gradually to the present-day level of thirty-five. The history of the North American fauna was closely comparable: about thirty before the interchange, rising to thirty-five, and subsiding to thirty-three. The number of families crossing over was about the same from both sides.

When biologists see a number go up following a disturbance and then fall back to the original level, whether body temperature, density of bacteria in a flask, or biological diversity on a continent, they suspect an equilibrium. The restoration of the numbers of mammalian families in both North and South America points to such a balance of nature. In other words, there appears to be a limit to diversity, in the sense that two very similar major groups cannot coexist in their fully radiated condition. A closer examination of the ecological equivalents on both continents, dwellers in the same broad niche, reinforces this conclusion. In South America marsupial big cats and smaller marsupial predators were replaced by their placental equivalents. Toxodonts gave way to tapirs and deer. Still some unusual specialists—the wild cards—were able to persist. Anteaters, tree sloths, and monkeys continue to flourish in South America, while armadillos are not only abundant throughout tropical America but are represented by one species that has expanded its range throughout the southern United States.

In general, where close ecological equivalents met during the interchange, the North American elements prevailed. They also attained a higher degree of diversification, as measured by the number of genera. A genus is a group of related species less well demarcated than those composing a family. The genus *Canis*, for example, comprises domestic dogs, wolves, and coyotes; other genera in the dog family Canidae include *Vulpes* (foxes), *Lycaon* (African wild dogs), and *Speothos* (South American bush dogs). During the interchange, the number of genera rose sharply in both North and South America and remained high thereafter. In South America it began at about 70 and has reached 170 at the present time. The swelling of numbers has come principally from speciation and radiation of the World Continent mammals after they arrived in South America. The old, pre-invasion South American elements were not able to diversify significantly in either North or South America. So the mammals of the western hemisphere as a whole now have a strong World Continent cast. Nearly half of the families and genera of South America belong to stocks that have immigrated from North America during the past 2.5 million years.

Why did the World Continent mammals prevail? No one knows for sure. The answer has been largely concealed by complex events imperfectly preserved in the fossil record—the paleontologist's equivalent of the fog of war. The question remains before us, part of the larger unsolved problem toward which our understanding of dynastic succession is directed. Evolutionary biologists keep coming back to it compulsively, as I did while waiting for the night storm at Fazenda Dimona, in the Brazilian Amazon, surrounded by mammals of World Continent origin. What comprises success and dominance? Before returning a last time to the Great American Interchange, let me try to rephrase these important terms into more useful conceptions.

Success in biology is an evolutionary idea. It is best defined as the longevity of a species with all its descendants. The longevity of the Hawaiian honeycreepers will eventually be measured from the time the ancestral finch-like species split off from other species, through its dispersal to Hawaii, and finally to that time when the last honeycreeper species ceases to exist.

Dominance, in contrast, is both an ecological and an evolutionary concept. It is best measured by the relative abundance of the species group in comparison with other, related groups and by the relative impact it has on the life around it. In general, dominant groups are likely to enjoy greater longevity. Their populations, simply by being larger, are less prone to sink all the way to extinction in any given locality. With greater numbers, they are also better able to colonize more localities, increasing the number of populations and making it less likely that every population will suffer extinction at the same time. Dominant groups often are able to preempt the colonization of potential competitors, reducing still further the risk of extinction.

Because dominant groups spread farther across land and sea, their populations tend to divide into multiple species that adopt different ways of life: dominant groups are prone to experience adaptive radiations. Conversely, dominant groups that have diversified to this degree, such as the Hawaiian honeycreepers and placental mammals, are on average better off than those composed of only a single species; as a purely incidental effect, highly diversified groups have better balanced their investments and will probably persist longer into the future. If one species comes to an end, another occupying a different niche is likely to carry on.

The mammals of North American origin proved dominant as a whole over the South American mammals, and in the end they remained the more diverse. Over two million years into the interchange, their dynasty prevails. To explain this imbalance, paleontologists have forged a widely held theory, an evolutionary-biologist kind of theory, in other words a rough consensus that violates the minimal number of facts. The fauna of North America, they note, was not insular and discrete like that of South America. It was and remains part of the World Continent fauna, which extends beyond the New World to Asia, Europe, and even Africa. The World Continent is by far the larger of the two land masses. It has tested more evolutionary lines, built tougher competitors, and perfected more defenses against predators and disease. This advantage has allowed its species to win by confrontation. They have also won by insinuation: many were able to penetrate sparsely occupied niches more decisively, radiating and filling them quickly. With both confrontation and insinuation, the World Continent mammals gained the edge.

The testing of this theory has just begun. Right or wrong, whether decisive in empirical support or not, its pursuit alone promises to link paleontology in interesting new ways to ecology and genetics. That synthesis will continue as the study of biological diversity expands in widening circles of inquiry to other disciplines, to other levels of biological organization and farther reaches of time.

CHAPTER EIGHT

The Unexplored Biosphere

I N 1983 a previously unknown creature, *Nanaloricus mysticus*, which vaguely resembles an ambulatory pineapple, was described as a new species, new genus, new family, new order, and new phylum of animals. Barrel-shaped, a quarter of a millimeter long (one-hundredth of an inch), sheathed in neat rows of scales and spines, it possesses a snout up front and, when young, a pair of flippers shaped like penguin wings at the rear. *Nanaloricus mysticus* lives in the gravel and coarse sand 10 to 500 meters deep on ocean bottoms around the world. Almost nothing is known about its ecology and behavior, but we can guess from the body shape and armament that it burrows like a mole in search of microscopic prey.

To place a species in its own phylum, the decision taken in this case by the Danish zoologist Reinhardt Kristensen, is a bold step. He said, and other zoologists agreed, that *Nanaloricus mysticus* is anatomically distinct enough to deserve placement alongside major groups such as the phylum Mollusca, comprising all the snails and other mollusks, and phylum Chordata, consisting of all the vertebrates and their close relatives. It is like ranking Liechtenstein with Germany, Bhutan with China. Kristensen named the new phylum Loricifera, from the Latin *lorica* (corset) and *ferre* (to bear). The corset in this case is the cuticular sheath that encases most of the body.

The loriciferans—now a larger group, since about thirty other species have been discovered in the past decade—live among a host of other tiny bizarre animals found in spaces between grains of sand and gravel on the ocean bottom. They include gnathostomulids (raised to phylum status in 1969), rotifers, kinorhynchs, and cephalocarid crustaceans. This Lilliputian fauna is so poorly known that most of the species lack a scientific name. They are nevertheless cosmopolitan and extremely abundant. And they are also almost certainly vital to the healthy functioning of the ocean's environment.

The existence of loriciferans and their submicroscopic associates is emblematic of how little we know of the living world, even that part necessary for our own existence. We dwell on a largely unexplored planet. Consider that our earth is a planet of a certain size, its continents and seas are arrayed in such and such a way, and all its life is based on a single nucleic-acid code, in the same sense that all of written English is based on twenty-six letters. In the universe there must exist a vast array of life-bearing planets of other sizes and geographies, and perhaps different codes as well, each combination fixing a particular level of natural biodiversity. Several lines of evidence, including the history of adaptive radiations, suggest that earth is at or close to its own particular capacity. But what exactly is that capacity? No one has the faintest idea; it is one of the great unsolved problems of science.

In the realm of physical measurement, evolutionary biology is far behind the rest of the natural sciences. Certain numbers are crucial to our ordinary understanding of the universe. What is the mean diameter of the earth? It is 12,742 kilometers (7,913 miles). How many stars are there in the Milky Way, an ordinary spiral galaxy? Approximately 10^{11}, 100 billion. How many genes are there in a small virus? There are 10 (in ϕX174 phage). What is the mass of an electron? It is 9.1×10^{-28} grams. And how many species of organisms are there on earth? We don't know, not even to the nearest order of magnitude. The number could be close to 10 million or as high as 100 million. Large numbers of new species continue to turn up every year. And of those already discovered, over 99 percent are known only by a scientific name, a handful of specimens in a museum, and a few scraps of anatomical description in scientific journals. It is a myth that scientists break out champagne when a new species is discovered. Our museums are glutted with new species. We don't have time to describe more than a small fraction of those pouring in each year.

With the help of other systematists, I recently estimated the number of known species of organisms, including all plants, animals, and microorganisms, to be 1.4 million. This figure could easily be off by a hundred thousand, so poorly defined are species in some groups of organisms and so chaotically organized is the literature on diversity in general. More to the point, evolutionary biologists are generally agreed that this estimate is less than a tenth of the number that actually live on earth.

To see why the biodiversity audit is so far short of reality, consider the phylum Arthropoda, which includes all the insects, spiders, crustaceans, centipedes, and related organisms with jointed, chitinous exoskeletons. About 875,000 arthropod species have been described, or more than half the total for all organisms. Insects in particular, with 750,000 species known, compose the unchallenged dynasty of animals in the small to small-medium range on the land, and they have been thus installed since late Carboniferous times, more than 300 million years ago. Their terrestrial corulers in the plant kingdom for the past 150 million years have been the angiosperms, or flowering plants, constituting about a quarter million species, 18 percent of the total known for all organisms.

The immense diversity of the insects and flowering plants combined is no accident. The two empires are united by intricate symbioses. The insects consume every anatomical part of the plants, while dwelling on them in every nook and cranny. A large fraction of the plant species depend on insects for pollination and reproduction. Ultimately they owe them their very lives, because insects turn the soil around their roots and decompose dead tissue into the nutrients required for continued growth.

So important are insects and other land-dwelling arthropods that if all were to disappear, humanity probably could not last more than a few months. Most of the amphibians, reptiles, birds, and mammals would crash to extinction about the same time. Next would go the bulk of the flowering plants and with them the physical structure of most forests and other terrestrial habitats of the world. The land surface would literally rot. As dead vegetation piled up and dried out, closing the channels of the nutrient cycles, other complex forms of vegetation would die off, and with them all but a few remnants of the land vertebrates. The free-living fungi, after enjoying a population explosion of stupendous proportions, would decline precipitously, and most species would perish. The land would return to approximately its condition in early Paleozoic times, covered by mats of recumbent wind-pollinated vegetation, sprinkled with clumps of small trees and bushes here and there, largely devoid of animal life.

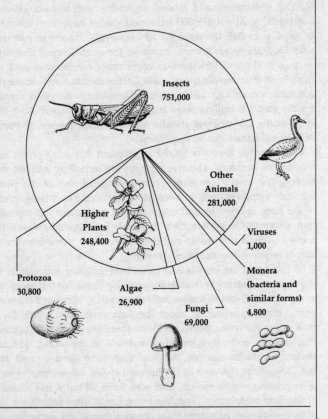

Number of Living Species of All Kinds of Organisms Currently Known
(According to Major Group)

ALL ORGANISMS: TOTAL SPECIES, 1,413,000

Insects
751,000

Other
Animals
281,000

Higher
Plants
248,400

Viruses
1,000

Protozoa
30,800

Algae
26,900

Fungi
69,000

Monera
(bacteria and
similar forms)
4,800

Insects and higher plants dominate the diversity of living organisms
known to date, but vast arrays of species remain to be discovered in the
bacteria, fungi, and other poorly studied groups. The grand total for all
life falls somewhere between 10 and 100 million species.

Number of Living Species of Higher Plants Currently Known
(According to Major Group)

HIGHER PLANTS: TOTAL SPECIES, 248,000

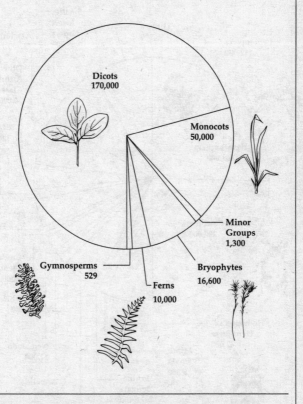

Dicots
170,000

Monocots
50,000

Minor
Groups
1,300

Gymnosperms
529

Ferns
10,000

Bryophytes
16,600

The plant diversity of the world consists primarily of angiosperms (flowering plants), which in turn make up grasses and other monocots and a huge variety of dicots, from magnolias to asters and roses. Most flowering plants live on the land; algae (26,900 known species) prevail in the sea.

Number of Living Animal Species Currently Known

(According to Major Group)

ANIMALS: TOTAL SPECIES, 1,032,000

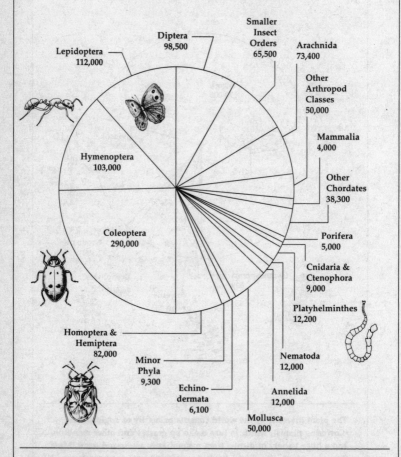

Diptera 98,500

Smaller Insect Orders 65,500

Lepidoptera 112,000

Arachnida 73,400

Other Arthropod Classes 50,000

Mammalia 4,000

Hymenoptera 103,000

Other Chordates 38,300

Coleoptera 290,000

Porifera 5,000

Cnidaria & Ctenophora 9,000

Platyhelminthes 12,200

Homoptera & Hemiptera 82,000

Nematoda 12,000

Minor Phyla 9,300

Annelida 12,000

Echino-dermata 6,100

Mollusca 50,000

Among animals known to science, the insects are overwhelming in number. Because of this imbalance, most animal species live on the land; but most phyla (Echinodermata, etc.), the highest units of classification, are found in the sea.

Arthropods are thus all around us, life-giving, and we have never taken their measure. There are far more species than the 875,000 given a scientific name to date. In 1952 Curtis Sabrosky, working at the U.S. Department of Agriculture, conjectured on the basis of the flood of new species pouring continuously into museums that there are about 10 million kinds of insects among an unknown diversity of other arthropods. In 1982 Terry Erwin of the National Museum of Natural History raised the ante threefold, estimating that there are 30 million species of arthropods in the tropical forest alone, of which the great majority are insects. Most of the variety, he said, is concentrated in the crowns of rain-forest trees. This layer of leaves and branches, which conducts most of the photosynthesis for the forest, was already known to be rich in animal diversity. Yet it has been inaccessible because of the height of the trees, 30–40 meters, the slick surface of the trunks, and the swarms of stinging ants and wasps awaiting human climbers at all levels.

To overcome these difficulties, entomologists developed the "bug bomb," a method for blowing fogs of a rapidly acting insecticide from the ground up into the treetops, enveloping the arthropods and chasing them out of their hiding places even as it kills them. The specimens are then collected as they fall dying to the ground. The particular fogging procedure used by Erwin and his research team in South and Central America is conducted mostly at night. Walking into the rain forest in the evening, they select a tree for sampling and lay out a grid of 1-meter-wide funnels beneath it. The funnels feed into bottles partly filled with 70 percent alcohol, the specimen preservative of choice. In the predawn hours next morning, when the wind in the treetops dies down to a minimum, the crew forces the insecticide upward into the canopy from a motor-driven "cannon." They continue the treatment for several minutes. Then they stand by for five hours while the dead and dying arthropods rain down in the thousands, many falling into the funnels. Finally, the collected specimens are then sorted, roughly classified to major taxonomic group (such as ants, leaf beetles, or jumping spiders), and sent to specialists for further study.

Erwin himself studied the beetles of the canopy. He made some counts in a small sample in a Panamanian rain forest, then proceeded by an arithmetical progression to arrive at a guess of the total number of arthropod species in tropical forests worldwide. Erwin first estimated that there are 163 species of beetles living exclusively in the canopies of one tree species, the tiliaceous *Luehea seemannii*. There

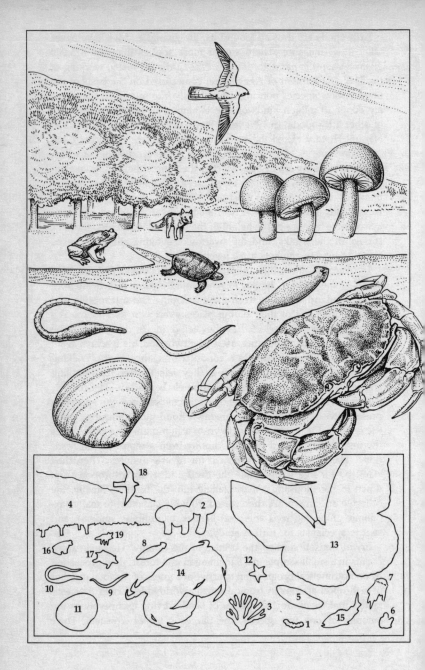

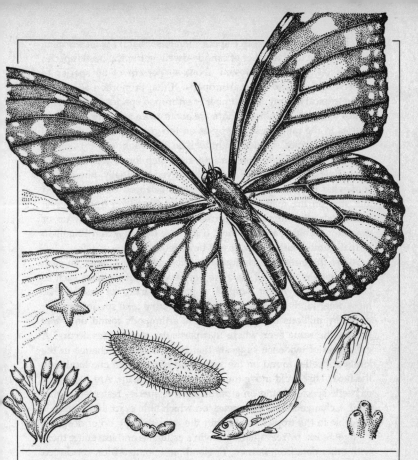

The species-scape. The size of the representative organism in each group has been made to be roughly proportional to the number of species currently known to science. The code and number of species are given below. Viruses and some minor invertebrate groups have been omitted.

1. Monera (bacteria, cyanobacteria), 4,800
2. Fungi, 69,000
3. Algae, 26,900
4. Higher plants, 248,400
5. Protozoa, 30,800
6. Porifera (sponges), 5,000
7. Cnidaria and Ctenophora (corals, jellyfish, comb jellies, and relatives), 9,000
8. Platyhelminthes (flatworms), 12,200
9. Nematoda (roundworms), 12,000
10. Annelida (earthworms and relatives), 12,000
11. Mollusca (mollusks), 50,000
12. Echinodermata (starfish and relatives), 6,100
13. Insecta, 751,000
14. Noninsectan arthropods (crustaceans, spiders, etc.), 123,400
15. Fishes and lower chordates, 18,800
16. Amphibians, 4,200
17. Reptiles, 6,300
18. Birds, 9,000
19. Mammals, 4,000

are about 50,000 tropical tree species in all, so that if *Luehea seemannii* is typical, the total number of canopy-dwelling tropical beetle species is 8,150,000. Beetles represent about 40 percent of all species of insects, spiders, and other arthropods. If that proportion also exists in the tropical canopy, the number of arthropod species in the habitat comes to about 20 million. There are about twice as many arthropod species in the rain-forest canopy as on the ground, so that the total number of tropical species might well be 30 million.

Erwin's calculations were an important step forward in the study of biodiversity. The explicit figure he arrived at initially, however, is somewhat like an upside-down pyramid balanced on its point. At any step on the road to the final total of 30 million tropical-forest arthropods, the number of species can be shifted drastically up or down by changing assumptions. If the true total is within 10 million of that number either way, it will be sheer luck.

So are there really such a great number of beetle species on each tree species worldwide? Data are very sparse, but legumes such as *Luehea seemannii* might support a greater variety of insects than most other kinds of trees. This could move the total species number down by millions of species. Are the arthropods found on a tree species the same everywhere that particular tree species occurs? A great deal of evidence suggests that there is often a change in the kinds of beetles found on the same kind of tree from one locality to the next. This could move the total number back up. Are 10 percent of beetle species found on a particular tree species restricted to that tree? A change in this parameter, on which little exact information is available in the tropics, could shift the total strongly up or down.

Nigel Stork, by reevaluating Erwin's estimates and leavening them with other data from Borneo, England, and South Africa, concluded that the total number of tropical arthropod species is indeed very large, but likely to be lower than projected by Erwin, perhaps anywhere from 5 to 10 million. Kevin Gaston interviewed specialists on different kinds of insects and found them to be conservative, also pointing to a total of 5 to 10 million species. Illumination by these and other studies has been only partial. In a sense we are back to square one: the number of species of organisms on earth is immense but still cannot be placed to the nearest order of magnitude.

The great naturalist-explorer William Beebe said of the rain-forest canopy in 1917: "Yet another continent of life remains to be discovered, not upon the earth, but one to two hundred feet above it." The subsequent decades have revealed a second unexplored continent

1,000 meters and more beneath the surface of the oceans, on the floor of the deep sea. This vast domain, 300 million square kilometers in extent, is with the possible exception of Antarctica's valleys the least hospitable habitat on earth, bitterly cold, crushed by pressure from the water columns above, and pitch-dark except for rare pin-points of light from passing luminescent organisms. Biologists of the early nineteenth century thought the deep sea to be lifeless. They were proved wrong by dredging operations conducted during the Challenger expedition of 1872–1876, whose mud samples disclosed a wide array of previously unknown organisms. Thus was the abyssal benthos discovered, the community of organisms living on or close to the ocean floor. In the 1960s a major advance was made by the introduction of the epibenthic sled, which rakes the top layer of the floor with fine mesh nets and traps the residue with a closing door to prevent the winnowing and loss of smaller organisms. The new sample yielded animal life diverse to a degree beyond even the boldest imaginings of the biologists. From these collections, and from photographs and more recent selective sampling made from deep-sea vehicles, we know that the abyssal benthos contains swarms of polychaete worms, peracarid crustaceans, mollusks, and other animals found nowhere else on earth. Many of the invertebrate animals are minuscule and subsist at low metabolic rates for life spans that may last decades. Bacteria are present that can grow and divide only in cold water under very high pressure. The abyssal benthos is a subdued miniaturized world. There is no way of guessing the full number of species present, but it certainly ranges into the hundreds of thousands and probably beyond. J. Frederick Grassle, after reviewing data on all the samples taken up to 1991, conjectured that the number of animal species could range in the tens of millions. The diversity of bacteria and other microorganisms cannot even be guessed to order of magnitude.

In an ecological sense, the animals of rain forests and the abyssal benthos occupy opposite ends of the earth; one could say that they dwell on two planets. Their environments are as physically different as possible, and their biotas share not a single species of plant or animal. Yet all the diversity they contain may be dwarfed by that of the bacteria, organisms that saturate the two extreme environments and every other place on earth. It is a common misperception among both biologists and nonbiologists that bacteria are relatively well known because they are so important in medicine, ecology, and molecular genetics. The truth is that the vast majority of bacterial types remain completely unknown, with no name and no hint of the means needed to detect them. Take a gram of ordinary soil, a pinch held between two fingers, and place it in the palm of your hand. You are holding a clump of quartz grains laced with decaying organic matter and free nutrients, and about 10 billion bacteria. How many bacterial species are present? Take one-millionth of the pinch of soil and spread it evenly over nutrients poured into standard culture dishes. If each and every one of the bacteria in this near-invisible soil sample could multiply, we would expect to see over 10,000 little colonies growing on the nutrient surface, one from each bacterium. But they cannot, and we do not. We will get only between 10 and 100 colonies.

Some of the bacterial cells that failed to respond were dead at the moment of implantation, but most simply did not find the conditions of the culture medium congenial for division and colony formation. Such species are incommunicado—they refuse to respond to the microbiologists who use standard techniques. They await the right temperature, acidity, air pressure, and combination of sugars, fats, proteins, and minerals appropriate to their genetically dictated metabolic needs. Moreover, each of these silent species may be represented in the pinch of soil by as few as one or two individuals per million. To find them, microbiologists must offer one culture medium and ambient environment after another until they hit on exactly the right combination. Then a colony proliferates and enough of the bacteria become available to be sorted and analyzed by standard microscopy and biochemical techniques.

Microbiologists rarely try to find silent bacteria. They are interested only in the select group of species already proved to be of some scientific or practical interest. One of the most famous species in the world, the colon bacterium *Escherichia coli*, is the key experimental organism of molecular biology. All beginning biology textbooks celebrate the knowledge won from its short life cycle and ease of culture. But from an evolutionary biologist's point of view, *E. coli* is only a somewhat peculiar symbiont of the large intestine of mammals, one that helps to convert exhausted food into feces. The bacterial proletariat, that vast majority of other species representing three billion years of adaptive radiation, remains unstudied and unheralded.

How many species of bacteria are there in the world? *Bergey's Manual of Systematic Bacteriology*, the official guide updated to 1989, lists about 4,000. There has always been a feeling among microbiologists that the true number, including the undiagnosed species, is much greater, but no one could even guess by how much. Ten times more? A hundred? Recent research suggests that the answer might be at least a thousand times greater, with the total number ranging into the millions.

Jostein Goksøyr and Vigdis Torsvik went in search of silent bacterial species in a natural environment. They chose to cut the Gordian knot of selective culturing by separating and comparing the DNA of the bacteria directly. They took small quantities of soil from a Norwegian beech forest near their laboratory. Using a succession of steps in extraction and centrifugation, they separated the bacteria from the soil and removed and purified the DNA of these organisms in a single common batch. They employed extreme high pressure to shear the double-stranded DNA molecules into fragments of uniform size. When heated, DNA molecules separate into their constituent single strands.

To divide DNA into single strands means that the letters of the DNA code, the base pairs, have been split apart. In most DNA, the base pairs are adenine-thymine and cytosine-guanine, AT and CG for short. As you proceed down the DNA helix, any of the four bases can be on the right or left in each nucleotide, so that in reading off the genetic code four permutations are possible: AT, TA, CG, GC. A typical sequence might then be TA-CG-CG-AT-GC, and so on for thousands or millions of such letters per cell. When the DNA helix is split, the two complementary single strands will read, for the segment just cited, T-C-C-A-G and A-G-G-T-C respectively.

When cooled to about 25°C below the melting temperature, the DNA strands are easily brought together again to form a double helix; they are "annealed," as the molecular biologists say. The higher the concentration of complementary strands in a solution, the faster the annealing will occur. If there are mixes of different species and strains of species, like those in the Norwegian soil bacteria, the concentration of complementary single strands will be much lower than is the case when DNA from only one species is present; the annealing process will be slowed to a corresponding degree. The rate at which annealing occurs can be measured exactly and calibrated against standards consisting of DNA single strands from a bacterium (the familiar *E. coli*) with known quantities of DNA. By this means, it is possible indirectly to estimate the overall percentage of matches among the diverse single strands of DNA in an entire bacterial community—that is, in all the bacteria living in a pinch of soil.

The percentage of DNA matching can be used as an indirect means to calculate the number of bacterial species. In so doing, microbiologists cannot use the biological-species concept directly. It is beyond their power to observe which bacterial cells exchange DNA, as if these organisms were so many birds and oak trees in the Norwegian forest. They are forced to rely on the similarity in DNA of one cell to the next. The arbitrary standard proposed by classifiers of bacteria is the following: a bacterial species consists of all those cells whose nucleotides are at least 70 percent identical, and hence are at least 30 percent different from the nucleotides of other species. This cutoff point is actually conservative; many higher plant and animal species are separated by far less than a 30 percent difference.

I have provided so much technical detail in order to reveal the difficulties faced by microbiologists and emphasize why it has taken so long to make inroads into bacterial diversity. Here is the result of the Norwegian research group: between 4,000 and 5,000 bacterial species were found in a single gram of beech-forest soil. A similar number of species, with little or no overlap, was found in a gram of sediment from shallow water off the Norway coast.

"It is obvious," Jostein Goksøyr has written, "that microbiologists will not run out of work for a couple of centuries." If over 10,000 microbial types exist in two pinches of substrate from two localities in Norway, how many more await discovery in other, radically different habitats? Again, no one has the faintest idea. It seems inevitable that entire new complexes of bacteria await discovery on the floor of the deep sea, in the axils of rain-forest orchids, amid the algal scum of mountain lakes, and so on around the world in thousands of sites that ordinarily escape our notice. Recent drilling of deep aquifers in South Carolina revealed large numbers of peculiar bacteria down to at least 500 meters below ground. The species changed from one stratum to the next. More than 3,000 forms, all new to science, were found in the early probes.

Still another uncharted world of bacteria and other microorganisms exists in and on the bodies of larger organisms. Some of the species are neutral guests, neither harming nor helping their hosts. Others have been characterized, however, that assist their hosts in digestion, excretion, and even the production of light by luminescent chemical reactions within their tiny bodies. So useful—indeed vital—are many that their hosts maintain specialized cells and tissues to carry them, while resorting to elaborate steps in physiology and behavior to pass the symbionts between the sexes and from parent to offspring. The phenomenon is well illustrated in the transmission of bacteria and yeasts by the scale insect *Rastrococcus iceryoides*. The microorganisms move to the offspring by an elegantly choreographed *pas de deux* within the developing egg. It has been described by the great authority on symbiosis, Paul Buchner:

> Both types of symbionts infect at the same place, thus forming in the mature egg a roundish ball at the upper pole. When the germ band approaches them, the two partners, which were united until now, separate and it is interesting to see how differently the host treats each of them. First, it shows interest only in the yeasts. While the yolk nuclei migrate toward them and soon penetrate them on all sides, the bacteria, which in the meantime have markedly increased, glide toward the periphery of the embryo in irregular groups, yet without allying themselves with nuclei. Yeasts and bacteria are soon separated. By the time the extremities sprout, cell limits have been formed around the yeasts, whereas the bacteria groups, unchanged, are situated here and there in the plasma.

Although hundreds of such peculiar partnerships have been discovered, they are described in the literature in only fragmentary manner. Very few of the bacterial species have been given a scientific name or described beyond adjectives such as "rod-shaped" or "vesicle-shaped."

To plumb the depth of our ignorance, consider that there are millions of insect species still unstudied, most or all of which harbor specialized bacteria. There are millions of other invertebrate species, from corals to crustaceans to starfish, in similar state. Consider that each bacterial type, each species if we employ the DNA-matching rule, can utilize at most a hundred carbon sources, such as different sugars or fatty acids. Most can in fact metabolize only one to several such compounds. Consider further that bacteria can evolve rapidly to exploit these sources. Different strains and even species readily exchange genes, especially during periods of food shortage and other forms of environmental stress. Their generations are extremely short, allowing natural selection to act on new assortments of genes within days or even hours, shifting the heredity this way or that, perhaps creating new species.

Consider, finally, a centimeter-wide patch of soil, the area of a fingernail, at a randomly chosen spot on the forest floor. A decaying splinter of wood lying on the surface contains one set of bacterial forms, leached sand grains a millimeter away another flora, and specks of humus a centimeter down yet another. All told there are thousands of species. Now assemble all such microfloras across an

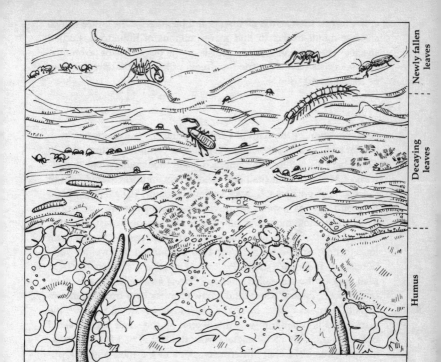

Newly fallen leaves

Decaying leaves

Humus

The teeming life of a North American deciduous forest appears to exist on a two-dimensional plane when viewed in a typically human perspective from above *(left)*. In this assemblage, the lithobiid centipede in the center is surrounded, from the top clockwise, by a greenbottle fly, social wasp, long-snouted acorn weevil, bess beetle, termite, wood cockroach, carpenter ant, sowbug, ground beetle, tick, ichneumon wasp, aphid, earwig, and harvestman (daddy-longlegs). When the litter and soil are cut vertically and viewed from the side *(right)*, a different, three-dimensional world is revealed. The dead leaves piled loosely at the top provide a dry, airy living space, inhabited in this example by small globular springtails, tiny turtle-shaped oribatid mites, harvestman (feeding on a snail), jumping spider, centipede, and ground beetle. A few centimeters deeper, in denser and moister litter amid piles of arthropod and earthworm fecal matter, are scattered more springtails and mites, pseudoscorpion (claws but no stinger), and two slug-shaped crane fly larvae. Deeper still in the now tightly packed humus and soil, two earthworms rest in their burrows.

entire forest, then across all forests and habitats for the entire world, and we might expect to find many millions of hitherto unstudied species. The bacteria await biologists as the black hole of taxonomy. Few scientists have even tried to dream of how all that diversity can be assayed and used.

As exploration of the natural world continues, new species of even the largest, most conspicuous organisms continue to turn up. In Colombia's Chocó Region, embracing the mountainous rain forests west of Medellín, half the plant species remain unregistered, and of these a large portion still lacks a scientific name. An average of two new bird species a year are discovered somewhere in the world, usually in remote mountain valleys and recesses of the last surviving tropical forests. Even new kinds of mammals are discovered from time to time. In the banner year 1988 the following novelties were published: Tattersall's sifaka (*Propithecus tattersalli*), a new lemur from Madagascar; the sun-tailed guenon (*Cercopithecus solatus*), a monkey from Gabon, central Africa; and a new muntjak deer from the mountains of western China. In 1990 a previously unknown primate, the black-faced lion tamarin, was discovered on the small coastal island of Superaqui, only 65 kilometers from the city of São Paulo. It was, in the words of Russell Mittermeier, "one of the most amazing primatological discoveries of this century." And caught in the nick of time, I might add, because the species is represented by only several dozen individuals. One hunter could have extinguished the species in a matter of days.

Not even the order Cetacea, containing the largest animals on earth, the whales and porpoises, is fully known. It is true that the species of the very largest, the baleen whales, including the blue, right, and humpback whales, had all been described by 1878. In contrast, the toothed whales, which include the giant sperm whale, killer whales, and their smaller relatives, the beaked whales and porpoises, have continued to yield new species at an average rate of one a decade during the twentieth century. Here are the eleven discovered since 1908, composing 13 percent or more of all the living cetaceans known:

Andrews' beaked whale, *Mesoplodon bowdonini* Andrews, 1908

Spectacled porpoise, *Australophocaena dioptrica* (Lahille), 1912

True's beaked whale, *Mesoplodon mirus* True, 1913

Chinese river porpoise (baiji), *Lipotes vexillifer* Miller, 1918

Longman's beaked whale, *Mesoplodon pacificus* Longman, 1926

Tasman beaked whale, *Tasmacetus shepherdi* Oliver, 1937

Fraser's porpoise, *Lagenodelphis hosei* Fraser, 1956

Vaquita (Gulf of California harbor porpoise), *Phocoena sinus* Norris and McFarland, 1958

Ginkgo-toothed beaked whale, *Mesoplodon ginkgodens* Nishiwaki and Kamiya, 1958

Hubbs' beaked whale, *Mesoplodon carlhubbsi* Moore, 1963

Pygmy beaked whale, *Mesoplodon peruvianus* Reyes, Mead, and Van Waerebeek, 1991

Many of the smaller whales and porpoises have been recorded solely from scattered carcasses or body parts washed ashore in remote parts of the world, and their natural history remains a mystery. Of the Tasman beaked whale, the cetacean expert Willem Mörzer Bruyns wrote in 1971: "A total 6 specimens washed up on beaches of Stewart Island, Bank's Peninsula and Cook's strait, east coast New Zealand." Of Hector's beaked whale, discovered in 1871: "Originally described from three skulls of very immature, perhaps neonatal calves, found in New Zealand waters . . . In 1967 the skull of an adult female was found in Tasmania." And of Longman's beaked whale: "Described from a skull found near Mackay, Queensland, Australia. In March 1968 Dr. Maria Louise Azzaroli described a second skull found in 1955 near Mogadiscio (Somali) which establishes the identity of a separate species." The rarity and elusiveness of these species suggest that other ocean giants await discovery. Individuals of at least one distinctive new species of beaked whale have in fact been sighted several times in the waters of the eastern tropical Pacific, but none has been captured.

A large part of species diversity stares us in the face but goes unrecognized. Previously I defined sibling species as two or more populations reproductively isolated from one another and yet so similar in outward appearance as to be lumped together even by expert taxonomists. Only a careful study of fine details in anatomy, cell structure, biochemistry, and behavior brings the differences to light and allows systematists to define the species with certainty. Early in my career as a classifier of ants, I combined all the slave-making ants of eastern North America into two species, thinking that there were only two reproductively isolated populations. I was wrong. A second entomologist, William Buren, took a closer look and broke the slave makers into five species, on the basis of small differences in their hair patterns, the shapes and color of their bodies, and the other species of ants they kidnap as slaves. There is little doubt that all are in fact reproductively isolated populations, each with a unique genetic constitution.

Groups exist, such as the protozoans and fungi, that are rife with sibling species for a purely technical reason: they have few outward traits by which species can be separated even with refined microscopical techniques. Owing to the frailties of the human sensory apparatus, their species are hidden. The recorded diversity of such groups can be expected to rise sharply as the DNA sequences and physiological requirements of more and more species are teased apart. It is further true that closer examination will turn many more subspecies into species. When the exact geographic limits of populations are mapped, many of those previously thought to be widely distributed species are revealed to comprise multiple species with exclusive ranges.

Most biological diversity, however, in the old-fashioned way, awaits discovery by foot, net, and scuba gear. To confront diversity, biologists continue to go out of the laboratory and into the world. They count species in three ways, according to the breadth of the geographic area surveyed. Alpha diversity is the number of species at one habitat in one locality. Two of my coworkers, Stefan Cover and John Tobin, and I recently set out to break the world record for alpha diversity in ants. We got it: 275 species collected in 8 hectares of rain forest near Puerto Maldonado, Peru. Beta diversity, the second measure, is the rate at which the species number increases as nearby habitats are added. If the Puerto Maldonado study were extended to swamp forest, river banks, and grassland patches, our catalog would almost certainly increase to over 350. Finally, gamma diversity is the totality of species in all habitats across a broad area. To survey thoroughly all the ants of Peru, valley by valley and across all the tributaries of the Amazon, might easily yield 2,000 species. It is gamma diversity, of course, that biologists have assayed with least precision. Knowing this, they press on into unexplored mountain ridges, river headwaters, and coral reefs. For most countries in the world, especially the tropics, the plumb line is still being let out; we have no idea where it will all end. The rewards of physical adventure, the excitement of grimy, sweat-soaked exploration into remote corners of the earth, still beckon in science.

But imagine for a moment that all the diversity of the world were finally revealed and then described, say one page to a species. The description would contain the scientific name, a photograph or drawing, a brief diagnosis, and information on where the species is found. If published in conventional book form, with pages bound into ordinary thousand-page volumes 17 centimeters wide inside cloth covers, this Great Encyclopedia of Life would occupy 60 meters of library shelf per million species. If there are 100 million species of organisms on earth, they would extend through 6 kilometers of shelving, the size of a medium-sized public library. Of course biodiversity studies will never come to that. Long before all the species are discovered, long before we put away our butterfly nets and plant presses, descriptions will be electronically recorded, so that the Great Encyclopedia can reside on disks placed in a box on one side of an office desk. Far more information will be added for each species as it becomes available, from its genetic code to its role in ecosystems, and these data will be readily available through networks arrayed around international and regional biodiversity centers.

The Great Encyclopedia of Life will record additional measures of diversity in current use by biologists. One is equitability, or the evenness of the abundance of species. Up to this point I have spoken of the measure of diversity only as the number of species: so many bacteria in a pinch of soil, so many ants in a stretch of rain forest. What also matters very much is the relative abundance of species. Suppose that we encounter a fauna of butterflies consisting of 1 million individuals divided into 100 species. Say one of the species is extremely abundant, represented by 990,000 individuals, and each of the other species therefore comprises an average of about 100 individuals. One hundred species are present but, as we walk along the forest paths and across the fields, we encounter the abundant butterfly almost all the time and each of the other species only rarely. This is a fauna of low equitability. Then in a nearby locality we encounter a second butterfly fauna, comprising the same 100 species, but this time all are equally abundant, represented by 10,000 individuals each. This is a fauna of high equitability, in fact the highest possible. Intuitively we feel that the high-equitability fauna is the more diverse of the two, since each butterfly encountered in turn is less predictable and therefore gives us more information on average, just as each word in a rich vocabulary fully employed gives more information. To study a highly diverse fauna is to be continually informed—and therefore pleased in an ultimate aesthetic sense. Diversity in this dimension also has practical importance in ecology. A fauna of high equitability is likely to have a very different impact on an ecosystem than one of low equitability, sustaining a larger variety of plants and other animals dependent on it.

Biologists measure the diversity of life not only by species, but by genera, families, and other higher categories of classification up to and including phyla and kingdoms. Each such higher unit is a cluster of species that resemble one another and are thought to share a common ancestry. In particular, a genus is a group of species placed together in the classification because they are very similar and of more or less immediate common ancestry. A family is a group of similar, related genera (its species overall are more distantly related than those within a genus); an order is a group of similar, related families; and so on through the hierarchy of classification all the way up to kingdoms, which embrace plants as a whole, animals as a whole, and so on. Here in briefest form is the complete taxonomic placement of the domestic cat, *Felis domestica:*

Species: *domestica*
  Genus: *Felis*
    Family: Felidae
      Order: Carnivora
        Class: Mammalia
          Phylum: Chordata
            Kingdom: Animalia

The basic principles of classification follow a transparent logic that can be stated in a few words. First principle: the species is the pivotal unit. Second principle: two definitions are used in setting up the hierarchical classification, as follows. A category is an abstract level of classification used universally in classification. The categories are the species, the genus, the family, and so on. A taxon, in contrast, is a concrete group of organisms, a particular set of populations given the rank of one or the other of the categories. Examples of taxa include the species *Felis domestica* and the family Felidae. Categories are the abstraction, taxa the reality. Third principle: a higher taxon such as the genus *Felis* is a group of species that are all descended from a single ancestral species. The species of a different taxon of equal rank, such as the great cats of the genus *Panthera*, are all descended from *another* ancestral species. When two genera are placed together to form a family, however, in this case the Felidae, they are considered to be descended from an even older ancestral species; this earlier progenitor gave rise to the two more recent ancestral species, which in turn gave rise to the species composing the two respective genera. Fourth principle: as these last examples make clear, the higher categories are a mental construct invented for convenience. They are based on a conception of species splitting into new species through time, and reflect the branching pattern that the successive splittings produce. The construction of branching patterns to map evolutionary change is called *cladistics*, and the devising of higher classifications (genus and up) to conform with the results of cladistics is called *phylogenetic systematics*. Classifications are expected to be consistent with phylogeny: in other words, the family trees of species.

Fifth and final principle: the exact limits of the higher taxa are arbitrary. The species themselves, the atomic units, are natural—more or less. So are their phylogenetic (ancestral) trees, if we have deduced them correctly. But the *limits* of genera, families, and still higher taxa, those lines drawn around clusters of species, are arbitrary. This statement may seem paradoxical because I just finished saying that the whole point of cladistics is to create a natural classification at the level of the genus and above. That much is true. Cladistics does allow us to judge which species are most likely to share a common ancestor, validating their placement in the same genus or family or higher taxon. What is arbitrary are the limits of each higher taxon. Should *Felis* and *Panthera* be kept as separate genera, or should they be lumped together in the single genus *Felis*? Either classification is correct by the standards of cladistics. Again: should the Felidae be allowed to stand as the only cat family, or should it be divided into two families, say the Felidae, the "true" cats, and Acinonycidae, the cheetahs? Cladistics is silent on this matter.

Systematists look at the reconstructed evolutionary trees. They see which species are descended from which common ancestors and can be bound into higher taxa, that is, into clusters of related species. They use criteria—common sense for the most part—to decide how to divide the clusters into smaller clusters. If all the species are very similar, it is sensible to place them in a single genus. If one species is very different from the others, even if it shares the same ancestor, the best procedure is to erect a new genus, thus drawing attention to its unusual properties. To recognize one genus or two, in the large number of borderline cases, is a judgment call. Systematics is mostly science but also a bit of art.

This fuzzy resolution is the right compromise to reach. The subjective nature of higher taxonomic categories reflects the chaotic nature of organic evolution. Like stars in an expanding universe, species are always evolving away from all other species, until they become extinct—or, in a few cases, break down the reproductive barriers between them and hybridize. This principle of evolution stems in turn from the immense variety made possible by rearrangements in the sequence of nucleotide letters of the genetic code. The code contains about a million nucleotide pairs in bacteria and between 1 to 10 billion nucleotide pairs in higher plants and animals. Evolution proceeds mostly by the accidental substitution of one or more of the letters, followed by the winnowing of these mutations and their combinations through natural selection. Because mutations occur at random, and because natural selection is affected by idiosyncratic changes in the environment that differ from one place and time to the next, no two species ever follow exactly the same path for more than a couple of steps. The real world, then, consists of species that differ from one another in infinitely varying directions and distances. So far as we know, no way exists to lump or to split them into groups except by what the human mind finds practical and aesthetically pleasing.

There is another consequence of evolution-as-expanding-universe that affects the taxonomic rank of species and their perceived value to humanity. Every species born, given enough time to evolve and proliferate into multiple species, is a potential genus or taxon of still higher rank. The longer this assemblage survives and evolves, the more it comes to differ genetically from the remainder of life. Because extinction is all but inevitable, the assemblage ordinarily dwindles until only a few relict species remain. These survivors are old, unique, and precious. Think of a species that has lived a very long time. Either its sister species have been stripped away by extinction, or else it is the sole occupant of an ancient line that never proliferated into multiple species in the first place. It stands alone now, accorded the rank of genus, family, or still higher category. It deserves special consideration by the human race for the story it tells. Such is the status of the giant panda, sole member of the genus *Ailuropoda;* the coelacanth fish *Latimeria chalumnae,* most celebrated of living fossils; and the tuatara, *Sphenodon punctatus,* a lizard-like reptile limited to small islands off New Zealand and only one of two members of the order Rhynchocephalia to survive from the Mesozoic era.

The known diversity of life has expanded within each level of the taxonomic hierarchy: so many species per genus, so many genera per family, and on up. At the very top, a total of 89 living phyla are distributed among the kingdoms of life. According to one widely used but highly subjective classification, the kingdoms are five in number:

Plantae: multicellular plants, from algae to flowering plants

Fungi: mushrooms, molds, and other fungi

Animalia: multicellular animals from sponges and jellyfish to vertebrates

Protista: single-celled eukaryotic organisms (protozoans and other single-celled organisms)

Monera: single-celled prokaryotic organisms (such as bacteria and cyanobacteria)

To describe diversity by organizing species into clusters according to how closely they resemble one another was a fundamental advance of eighteenth-century biology. A later and equally important way of describing diversity is by level of biological organization. The organizational levels of importance to biological diversity are arrayed in this hierarchy:

Ecosystem
 Community
  Guild
   Species
    Organism
     Gene

The idea is best conveyed by a concrete example that can be followed down through the full stack. The example I choose: *A goshawk* (Accipiter gentilis) *hunts for songbirds through the Black Forest of Germany, flying fast and low among the fir trees, shifting direction abruptly. It sees a wood warbler* (Phylloscopus sibilatrix) *resting on a pine branch. With a few short wingbeats and a long silent glide, it closes on its prey.*

The goshawk lives in a particular ecosystem, the upland fir forest of the Black Forest. The land is covered by granitic soil weathered from round low hills, laced with streams that form the headwaters of the Danube and Neckar rivers. The ecosystem consists of this physical base plus all the organisms living in the habitats of the forest, in the woods, the glades, and small bodies of fresh water. The combined physical and biotic elements, from rock and streams to trees, hawks, and warblers, are tightly bound to one another. Energy is carried as in a leaky bucket from one species to another through the food webs of organisms. Nutrients flow through the organisms, soil, water, and air, and back in unending biogeochemical cycles. The character of the soil cover and water drainage depends intimately on the organisms living in the forest. The Black Forest ecosystem is unique in its particular combination of physical environment and tenant organisms. We look up and across southern Germany, at Europe as a whole, and finally around the entire world to measure the diversity of ecosystems in existence. We find that the possibilities are astronomical in number—the combinations of millions of species that can live in all of the distinguishable physical settings is a number beyond practical calculation. This incapacity is interesting but not important. The real number of ecosystems is what counts. Each ecosystem has intrinsic value. Just as a country treasures its finite historical episodes, classic books, works of art, and other measures of national greatness, it should learn to treasure its unique and finite ecosystems, resonant to a sense of time and place.

Within the Black Forest ecosystem, the goshawk belongs to a particular community of organisms, defined as all the species connected in the food web and by any other activity that influences the life cycles of the species. The fir tree is in the goshawk's food web because it nourishes the moth larvae that feed the songbirds that feed the hawk. The common buzzard, a European buteo hawk, is a member of the same community by virtue of competition and accidental symbiosis. The occasional small bird it kills reduces the goshawk's larder. The nest it abandons and makes available to the goshawk improves the breeding chances of that less discriminating bird. The diversity of communities is measured within a particular ecosystem. More accurately, it is subjectively assayed, since the limits of a community can seldom be drawn with exactitude.

Inside the community, the goshawk is a member of a guild, a set of species that live in the same place and harvest the same food by similar means. Strictly speaking, the goshawk shares its guild with only one species in the Black Forest community, the sparrow hawk *(Accipiter nisus)*. They are both accipiters, possessed of short rounded wings and long tails. They hunt small birds by rapid, twisting flight through the forest, occasionally ascending to brief soars above the trees. Other guilds of the Black Forest include the aster-eating flower insects, the warblers, and the forest-dwelling shrews and small mice. Because guilds tell us something about ecology, they are as valuable a measure of diversity within an ecosystem as the number of species.

We are approaching the lowest levels of biological diversity. The goshawk is a species, a nexus of ill-defined local populations that altogether range from continental Europe across Asia to Canada and the northern and western United States. The individual birds composing it are the repositories of genetic diversity, the differences that exist among the chromosomes and genes and the level below species diversity. This is a level most quickly grasped by using familiar examples in human heredity. A single gene difference determines whether the ear has a free-hanging lobe or is attached all the way along its base. Possession of an earlobe is a dominant trait. If either one of the two genes in each cell is an earlobe gene, the person acquires a fully developed earlobe. Only if both genes are for the recessive, lobeless condition does the person develop that alternative trait. The earlobe genes, the particular pair of genes determining the presence or absence of earlobes, occur at just one of the 200,000 or more sites arrayed in a row along the 46 chromosomes. Other examples of human variation based on single genes are the blood types, the ability to roll the tongue into a tube, the presence of a widow's peak in the hairline, whether the last segment of the thumb angles back sharply when the thumb is stretched out (the condition called "hitchhiker's thumb"), and a myriad of hereditary diseases from sickle-cell anemia to albinism, hemophilia, and Huntington's chorea. Many other traits, such as height, skin color, and predisposition to diabetes, are affected by combinations of genes at many chromosome sites that work together, the so-called polygenes.

By counting such variations in outward traits known to arise from mutations in single genes and polygenes, it is possible to arrive at a figure of total genetic diversity. The estimate, however, would fall short by orders of magnitude. The reason is that differences among alternative genes at the same chromosome position cause differences that are often invisible; they prescribe variations in proteins detectable only by chemical analysis. In the 1960s an advance was made in resolving power with the introduction of gel electrophoresis, a technique that allows the rapid purification and identification of enzymes. When molecules are placed in a charged field on material through which they can move, such as a porous gel, they migrate at a rate proportionate to their own electric charge. As a result they space out like runners with varying abilities. Enzymes are protein molecules whose design, including their electric charge, is prescribed by genes.

Even small differences in the genes caused by mutations translate to variations in the enzymes, often (but not always) translating into differences in electric charge, which cause the enzymes to travel at different velocities and separate on the electrified gel plates. Geneticists follow a straightforward procedure to take advantage of this chain of events. They crush tissues from the organisms to be studied, extract materials containing enzymes, and place the extracts at one end of a gel plate. They let the enzymes run in the electrical field for a while, then stain them with dyes to reveal their positions. They count the stained enzymes that have separated on the plate. They are then able to determine the number and identity of the enzymes and hence infer the number and identity of the prescribing genes. By sampling many individuals in a species and by proceeding from one kind of enzyme to another—thus from one set of genes to another—they can estimate the amount of overall genetic diversity in the species.

Surveys using electrophoresis have ranged widely over many kinds of organisms, from flowering plants and insects to fishes, birds, and mammals. Of all the discoveries made, one stands out: the amount of genetic diversity revealed is very large, much greater than had been expected in pre-electrophoresis days when researchers relied principally on such visible traits as earlobes and skin color. In order to express the diversity as a number, geneticists use the concept of polymorphism. A gene is said to be polymorphic when it occurs in multiple forms, or multiple alleles as they are more technically called. The rarer alleles are not counted unless they exceed some arbitrarily selected frequency, usually 1 percent of the total for that particular gene. In other words, only if earlobe alleles were more than 1 percent in the human population would they be included in the count (in fact, they are 45 percent), and only then would the controlling gene be called polymorphic (it is in this case). The electrophoretic studies have shown that in the great majority of species, somewhere between 10 and 50 percent of genes are polymorphic. A typical figure is roughly 25 percent.

High levels of polymorphism per gene through the population also produce high levels of polymorphism within the bodies of individual organisms. In each individual on the average, again according to species, between 3 and 20 percent of the genes are polymorphic. This means that each organism is heterozygous for that number of genes; in human beings it means possessing one gene for earlobes and one for the lobeless state in each cell, or one for A blood type and one for B, and so on through the 200,000 or more genes that make up the total hereditary composition of a human being.

Yet even the unexpectedly high electrophoretic numbers are minimal estimates and almost certainly too low. Some enzyme variants have no special electrical charge or molecular arrangement by which they can be separated, so they stay silent on the electrophoretic field. In order to take an exact and final measure of genetic diversity, it is necessary to go past the proteins and straight to the genes themselves and learn the nucleotide sequence, the letters of the genetic code. The true, ultimate measure of genetic diversity is nucleotide diversity. It must be determined base pair by base pair through a large part of the chromosomes, and in many individuals belonging to the same species.

During the 1980s rapid advances were made in DNA sequencing. The human genome project was born, aiming at nothing less than a complete nucleotide map of one human being. A similar project is planned for a species of fruit fly. When sequencing becomes cheap enough, and reading genetic codes is as routine as counting feathers and molar teeth, we will be technically prepared in full to address the question of how much biodiversity exists on earth.

Meanwhile I will venture a guess at the final outcome: to the nearest order of magnitude, or powers of ten, 10^8 (100 million) species, multiplied by 10^9 (1 billion) nucleotide pairs on average per species; hence a total of 10^{17} (100 quadrillion) nucleotide pairs specifying the full genetic diversity among species. Nucleotide diversity, it should be noted in passing, is limited to a maximum of four kinds of nucleotide per site and hence does not add as much as an order of magnitude.

That figure, 10^{17}, is in one sense the entire diversity of life. Yet it still does not take into account the differences among individuals belonging to the same species. When that dimension is added, the potential grows still more. Consider that, within a typical sexually reproducing species, two nucleotides occurring at the same site on different chromosomes can generate three combinations; the letters AT and CG, for example, can generate the combinations (AT) (AT), (AT) (CG), and (CG) (CG). If only one out of a thousand sites had two such variants somewhere in the species, then with 10^6 positions (in other words, one-thousandth of 10^9 positions in the whole genetic makeup of the species) there would be 10^{18} possible combinations for each species. This immense figure is still an underestimate. Whatever the true figure, it represents the potential of biodiversity at the level of the organism, the great field of possible genetic combinations through which each species travels with the raw materials it possesses, guided by natural selection and now, increasingly, the ignorant hands of humanity.

The Creation
of Ecosystems

THE BALD EAGLE, one species, flies above the Chippewa National Forest of Minnesota. A thousand species of plants compose the vegetation below. Why does this particular combination obtain rather than a thousand eagles and one plant? Or a thousand eagles and a thousand plants? It is natural to ask whether the numbers that do exist are governed by mathematical laws. If there are such laws, it follows that we can someday predict diversity in other places, in other groups of organisms. To master complexity by such an economical means would be the crowning achievement of ecology.

There are no laws unfortunately, at least none that biologists have hit upon yet, not in the sense ordained by physicists and chemists. But, as in any study of evolution, there are principles that can be written in the form of rules or statistical trends. The discipline formulating these weaker statements, community ecology, is still youthful and rapidly growing, which is a polite way of saying that it is a long way behind the physical sciences—but there is progress, and ambition.

Before us now is the overwhelmingly important problem of how biodiversity is assembled by the creation of ecosystems. We can address it by recognizing two extreme possibilities. One is that a community of organisms, like that occupying the Chippewa National Forest, is in total disorder. The species come and go as free spirits. Their colonization and extinction are not determined by the presence or absence of other species. Consequently, according to this extreme model, the amount of biodiversity is a random process, and the habitats in which the various species live fail to coincide except by accident. The second extreme possibility is perfect order. The species are so closely interdependent, the food webs so rigid, the symbioses so tightly bound that the community is virtually one great organism, a superorganism. This means that if only one of the species were named, say the Acadian flycatcher, marbled salamander, or goblin fern, the thousands of other species could be ticked off without further information about that particular community.

Ecologists dismiss the possibility of either extreme. They envision an intermediate form of community organization, something like this: whether a particular species occurs in a given suitable habitat is largely due to chance, but for most organisms the chance is strongly affected—the dice are loaded—by the identity of the species already present.

In such loosely organized communities there are little players and big players, and the biggest players of all are the keystone species. As the name implies, the removal of a keystone species causes a substantial part of the community to change drastically. Many other species decline to near or total extinction or else rise to unprecedented abundance. Sometimes other species previously excluded from the community by competition and lack of opportunity now invade it, altering its structure still more. Put the keystone species back in and the community typically, but not invariably, returns to something resembling its original state.

The most potent keystone species known in the world may be the sea otter (Enhydra lutris). This wonderful animal, large and supple in body, cousin to the weasels, whiskered like a cat, staring with a languorously deadpan expression, once thrived among the kelp beds close to shore from Alaska to southern California. It was hunted by European explorers and settlers for its fur, so that by the end of the nineteenth century it was close to extinction. In places where sea otters disappeared completely, an unexpected sequence of events unfolded. Sea urchins, normally among the major prey of the otters, exploded in numbers and proceeded to consume large portions of the kelp and other inshore seaweeds. In otter times, the heavy kelp growth, anchored on the sea bottom and reaching to the surface, was a veritable forest. Now it was mostly gone, literally eaten away. Large stretches of the shallow ocean floor were reduced to a desert-like terrain, called sea-urchin barrens.

With strong public support, conservationists were able to restore the sea otter and with it the original habitat and biodiversity. A small number of the animals had managed to survive at far opposite ends of the range, in the outer Aleutian Islands to the north and a few localities along the southern California coast. Some of these were now transported to scattered intermediate sites in the United States and Canada, and strict measures were taken to protect the species throughout its range. The otters waxed and the sea urchins waned. The kelp forests grew back to their original luxuriance. A host of lesser algal species moved in, along with crustaceans, squid, fishes, and other organisms. Gray whales migrated closer to shore to park their young in breaks along the kelp edge while feeding on the dense concentrations of animal plankton.

Ecologists, like the organisms they study, cannot make nature conform to their perfect liking. They search for openings and seize opportunity, exploiting the occasional discovery of keystone species like sea otters to gain insight into the organization of communities in different environments. Other examples have been found. In the undisturbed forests of Central and South America—more precisely, in the dismayingly few such forests remaining—jaguars and pumas prey on a wide variety of small animals encountered on the ground. They are "searchers," taking whatever animals they meet, as opposed to "pursuers" like cheetahs and wild dogs, which select only a few kinds of animals and then chase them down. The big cats are especially fond of coatis, members of the raccoon family with elongated bodies and tapered noses, and agoutis and pacas, outsized rodents variously resembling jackrabbits and small deer. When jaguars and pumas disappeared from Barro Colorado Island in Panama, because the forest was no longer extensive enough to support them, the prey species soon increased tenfold. Effects from this shift in balance now appear to be rippling downward through the food chain. Coatis, agoutis, and pacas feed on large seeds that fall from the rain-forest canopy. When they become superabundant, as on Barro Colorado Island, they reduce the reproductive ability of the particular tree species that produce these seeds. Other tree species whose seeds are too small to be of interest to the animals benefit by the lessened competition. Their seeds set and their seedlings flourish, and a larger number of the young trees reach full height and reproductive age. Over a period of years, the composition of the forest shifts in their favor. It seems inevitable that the animal species specialized to feed on them also prosper, the predators that attack these animals increase, the fungi and bacteria that parasitize the small-seed trees and associated animals spread, the microscopic animals feeding on the fungi and bacteria grow denser, the predators of these creatures increase, and so on outward across the food web and back again as the ecosystem reverberates from the removal of the keystone species.

In a very different way, elephants, rhinoceros, and other big herbivores rank as keystone species in the savannas and dry woodlands of Africa. When allowed to reach natural high densities, they control the entire physical structure of these habitats. "Modern African elephants," Norman Owen-Smith has written, "push over, break, or uproot trees,"

> altering vegetation physiognomy and hence habitat conditions for other animal species. Trees killed by elephants are replaced by regenerating shrubs or grasses that offer more accessible foliage for consumption by smaller herbivores. The leaves of rapidly growing woody plants are less strongly defended chemically than those of the slower-growing trees they replace. Rates of nutrient cycling are also accelerated. Grazing pressure from white rhinoceroses and hippopotamuses transforms medium-tall grasslands into a mosaic of short and tall grass patches. Short, creeping grasses are generally less fibrous and more nutrient rich than taller grasses. As a result of such vegetation changes, food quality is improved for smaller, more selective grazers. Animal species dependent upon a dense cover of woody vegetation or tall grasses for predator evasion may persist in areas of low impact.

For millions of years the great herbivores of sub-Saharan Africa ranged freely across the vast parklands, creating a mosaic of habitats, a swath of short grassland here, an acacia copse or remnant of riverine forest there, reed-lined pools grown from mud wallows scattered widely about. The total effect was a huge enrichment of biological diversity.

Focusing now from the kilometer reach of elephants down to grassroot level, we find a wholly different class of keystone species. Where big mammals control the vegetation structure, a colony of driver ants at their feet captures millions of victims each day and alters the nature of the community of small animals. Viewed a few meters away, a driver-ant raiding column seems a living thing, a giant pseudopodium reaching out to engulf its prey. The victims are snared with hook-shaped jaws, stung to death, and carried to the bivouac, a labyrinth of underground tunnels and chambers housing the queen and immature forms. Each expeditionary force comprises several million workers who flow out of this retreat. The hungry legions emerging from the bivouac are like an expanding sheet that lengthens into a treelike formation. The trunk grows from the nest, the crown expands as an advancing front, and numerous branches pour back and forth between the two. The swarm is shaped but leaderless. Excited workers rush back and forth throughout its length at an average speed of a centimeter per second. Those in the van press forward for a short distance and then fall back to yield their front position to other runners. The feeder columns resemble thick black ropes laid along the ground, slowly writhing from side to side. The front, advancing at 20 meters an hour, blankets all the ground and low vegetation in its path. The columns expand into it like a river entering a delta, where the workers race back and forth in a feeding frenzy, consuming most of the insects, spiders, and other invertebrates in their path, attacking snakes and other large animals unable to move away.

Day after day the driver ants scythe through the animal life around their bivouac. They reduce its biomass and change the proportions of species. The most active flying insects escape. So do invertebrate animals too small to be noticed by the ants, particularly roundworms, mites, and springtails. Other insects and invertebrates are hit hard. One driver-ant colony, comprising as many as 20 million workers— all daughters of a single mother queen—is a heavy burden for the ecosystem to bear. Even the insectivorous birds must fly to a different spot to find enough food.

It has become clear that an elite group of species exercises an influence on biological diversity out of all proportion to its numbers. Scientists are drawn to such strong cases, not just in ecology but in other fields as disparate as astrophysics and neurobiology, because they yield quick information and an entry into systems that are otherwise intractable. But they can be misleading if overgeneralized. There comes a time, in all science, when it is profitable to move away from the bold and obvious and circle around a bit, inventing more subtle approaches to search for concealed phenomena. In the study of communities, this strategy requires greater attention to context, history, and chance.

One successful recent approach has been to deduce the assembly rules of faunas and floras. Although the attempt to identify keystone species takes a community pretty much as it is and figures out what happens when the candidate species is removed, the assembly rules reconstruct the sequence in which species were added when the

A keystone species at the grassroots level: a swarm of driver ants marches across a savanna in Kenya. The ant armies drastically alter the abundance of insects and other small animals in the habitats through which they pass.

community came into being. It does more: it postulates the sequences that are possible and those that are not. Let me express the idea with an imaginary example chosen for clarity. A certain plant species establishes itself, say on a mountainous island. Its presence allows the colonization of the habitat by a beetle species that feeds only on it. A wasp species that parasitizes the beetle is added next. In another dimension, entailing competition, a second assembly rule is manifested. A woodpecker species arrives; call it A. It multiplies to such abundance and dominates so much of the food supply that when two more woodpecker species arrive, B and C, either one of them but not both can squeeze into the community. Now we have a woodpecker fauna consisting of either AB or AC, depending on which of the latecomers arrived first. Finally, woodpecker species D

Assembly rules determine which species can coexist in a community of organisms (such as the bird species occupying a forest patch). The rules also determine the sequence in which species are able to colonize the habitat. A set of imaginary rules is represented here as pieces of a jigsaw puzzle that can be fitted together in one of two combinations, ABD or AC.

appears. Occupying a distinctive niche of its own—say foraging on large conifers—it can squeeze in with the other species if the preexisting combination is AB but not if the combination is AC. So the first stable woodpecker fauna in the community is either ABD or AC.

Ecologists deduce assembly rules by observing which species actually live together in nature. One approach, used by Jared Diamond in pioneering work on the birds of New Guinea, compares communities in many different localities to see which combinations of species occur and which ones rarely or never occur. The preliminary conclusions reached in this way can then be tested further by detailed studies of habitat preferences of the individual species. Suppose, in the woodpecker case just cited, B and C almost never coexist in the same localities because they compete until one or the other is extinguished. Suppose further that additional studies refine the pattern: B and C occur on some mountains together but almost always at different elevations, so in fact they are seldom members of the same community. On mountains where both occur, B ranges from 200 to 1,000 meters and C from 1,000 to 2,000 meters. Where only one of the species occurs on a mountain, it spreads all the way from 200 to 2,000 meters. This expansion in the absence of a competitor is the same phenomenon we have already met, ecological release. The constriction in the presence of the competitor is called *ecological displacement*. The existence of ecological release and displacement is considered strong presumptive evidence that even when B and C occupy the same geographical range they cannot live closely together in the same habitat and community. They withdraw to elevations where each in turn is the superior competitor, in this case B in the lowlands and C higher up.

It will be interesting now to return to Krakatau and to recall the example it offers of the assembly of species. A community does not arrive on the shores of such an island as a finished product. Instead, it is stacked like a house of cards, one species on another, loosely obedient to assembly rules. Most propagules, whether plant seeds or wandering bird flocks, are doomed to failure. For them the soil is wrong, the forest glades are still too small, the prey species have not yet arrived, or formidable competitors wait at the shore. Even many of the species established earlier cannot hold on as conditions inevitably change: grassy swales are closed out by forest growth, disease strikes, a stronger competitor invades, chance fluctuations in members bounce the population to zero. The community shifts continuously, and by an unconscious trial and error, through innumerable fits and starts, its biodiversity slowly rises. Species excluded earlier at last find room, symbiotic pairs and trios are fitted together, the forest grows deeper and richer, new niches are prepared. The community thus approaches a mature state, actually a dynamic equilibrium with species forever arriving and disappearing and the total species numbers bobbing up and down inside narrow limits.

Throughout the process of colonization, accommodations are made. Species in collision sometimes compromise through ecological displacement. They yield part of the environment to their competitors and survive. Fire ants, for example, are among the most aggressive territorial animals known, and it is unusual to find more than two or three species coexisting in the same community. Their colonies, made up of a mother queen and thousands of stinging, biting workers, engage each other in organized combat. They seek out and destroy smaller colonies and settle territorial boundaries with larger ones by continuous *combat d'usure*, pushing at one another until a balance of power is attained. Sometime during the 1930s a South American species, the imported fire ant *Solenopsis invicta*, was accidentally introduced into the port of Mobile, Alabama. It was successful from the start, needing just forty years to spread across the southern United States from the Carolinas to Texas. Throughout most of that range it confronted a native fire ant, *Solenopsis geminata*, which up to that time had been a dominant ant in both woodland and open habitats. The native fire ant is still abundant, but it has been largely forced back into scattered woodland sites. The habitats most favored by fire ants generally, pastureland, lawns, and roadsides, are now owned by the newcomers. If the imported fire ant could somehow be removed (an event fervently but hopelessly desired by southerners), the native fire ant would almost certainly reoccupy its old haunts.

The case of the fire ants illustrates the well-documented principle that closely similar species can fit together when their requirements are elastic. Elasticity is the hallmark of Darwin's finches of the Galápagos, for the simple reason that their long-term survival depends on it. They live on volcanic desert islands with harsh, variable environments, changing in the quality of life they offer from month to month and year to year. During the wet season, when most plant growth occurs and food is relatively abundant, the birds enjoy a broad diet. Species that live on the same island and are anatomically similar to one another feed to substantial degree on the same items. In the dry season, food grows scarce and the species come to differ in the items they select. Some become specialists, while others broaden their diets.

The tiny island of Daphne Major is home to two resident species, the medium ground finch *Geospiza fortis* and the cactus finch *Geospiza scandens*. Both live in dense stands of opuntia cactus. In the wet season, when the cactus is in full bloom, the two species consume much the same food. They take the nectar and pollen of the flowers, and they also feast on various kinds of seeds and insects. In the dry season, as the food supply drops, *scandens* narrows its diet to concentrate on edible parts of the opuntia plant. *Fortis* broadens its diet to include an even broader range of items than before, wherever it can find them.

Imagine a case in which two such species have been squeezed together in the same communities long enough for evolution to occur. When they first came into contact, they were elastic and could diverge in their habits enough to lessen competition. The differences were phenotypic, the result of environment and not genes. The compression occurred in traits that were relatively easy to change, most likely by a retreat from parts of the habitat and diet by one or both of the species. As the generations passed, genetic differences arose and hardened the distinction between the two species. Individual birds found it advantageous to excel in those portions of the niche to which they had been driven. The success of those genetically predisposed to do so caused the population as a whole to specialize— to consume a certain food or to build nests in one or another habitat. The differences between the two species next extended to anatomy and physiology. Then the two species competed less with one another, most likely at the price of some of their original elasticity. They experienced the evolutionary change called *character displacement*.

The classic example of character displacement is the change in bill size and food habits of Darwin's finches. Adaptive radiation among the thirteen Galápagos species was based to a large degree on variations in thickness of the bill, and this trait has been engineered in part by character displacement. The selection pressure behind the evolution is improved efficiency during specialization. The deeper the bill at the plane of its attachment to the head, the more power it can exert along the cutting edges and at the tip. Finches with thick bills are well equipped to rip open tougher fruits and to crush bigger and more brittle seeds. Finches with thin bills are limited to softer fare, but they are compensated by an ability to probe narrow crevices and manipulate small objects. A rough analogy from human technology is the adaptive radiation of pliers. To turn a bolt or twist a thick wire with dispatch calls for either linesman's pliers or parrot-head gripping pliers. To manipulate fine bolts and wires you need nose pliers, which are thinner and proportionately longer.

The shape of the bill is not the whole story of displacement and radiation in Darwin's finches. The size of the jaw muscles, the stereotyped movement of the birds during feeding, and perhaps even the chemistry of the digestive traits have been altered as part of dietary specialization among the species. But bill depth remains the most obvious and easily measured among all the traits. It is a convenient proxy by which the larger syndrome of specialized changes can be studied.

The surest test of character displacement as an engine of adaptive radiation is the demonstration of a certain two-part geographic pattern: species have evolved away from each other in places where they are in contact, but they have failed to do so, or have even converged, where they live alone. In the special case of Darwin's finches, we look for enhanced differences among the species on islands where they live together, and particularly in those traits such as bill shape that allow them to specialize and reduce competition. And we need a control: on other islands harboring only a single species, the competitors should more closely resemble each other, again in those traits believed to be most subject to competition. If this dual pattern is strong and convincing, we may reasonably conclude that where the species has been forced to compete, it evolved away from its opponent to fill a special niche, and where it lacked competition it stayed put—or else evolved in the direction of the opponent to fill both niches.

In testing for character displacement in Darwin's finches, Peter Grant put to use the fact that some of the species occur on many islands in the Galápagos. He looked at thirteen cases where pairs of closely related species occur together on various islands. In eleven such instances he found them to differ more in beak depth than when they occur alone, on islands of their own. The evidence was nevertheless short of decisive. Grant recognized that there is another way in which such a pattern can arise in the absence of competition. Character displacement could also occur by reproductive reinforcement of the differences that isolate species as distinct gene pools. If two species hybridize to some extent when they meet, and the hybrids are inferior or sterile, it is to the advantage of both species to avoid interbreeding altogether. One device might be to evolve traits such as distinctive bill shapes that allow individuals to select members of their own species with greater accuracy. Using stuffed female birds, which in spite of their immobility are courted by unsuspecting males, Grant discovered that the males prefer females with the right bill shape on islands where similar species live together. They are much less selective, however, where the same species lives alone. In other words, bill shape *is* used as a cue by male finches to choose females of their own species, and reproductive reinforcement does occur as an evolutionary process. By closely weighing the factors, however, Grant showed that character displacement occurs primarily through competition, and reproductive reinforcement is hooked onto it as a secondary effect. This means that once bills evolve apart as a consequence of competition, related species of Darwin's finches also use the differences to avoid hybridization.

Character displacement has been persuasively documented in a few other groups of organisms, including frogs, fruit flies, ants, and snails, but it is far from a universal biological process. It allows a bit of compression here and there, and enables a few more species to squeeze into local communities. It represents one process by which communities can be organized to some degree, mediating a rise in general biological diversity.

To the forces that increase biodiversity, add predators. In a celebrated experiment on the seacoast of Washington state, Robert Paine discovered that carnivores, far from destroying their prey species, can protect them from extinction and thereby salvage diversity. The starfish *Pisaster ochraceus* is a keystone predator of mollusks living in rock-bound tidal waters, including mussels, limpets, and chitons. It also attacks barnacles, which look like mollusks but are actually shell-encased crustaceans that remain rooted to one spot. Where the *Pisaster* starfish occurred in Paine's study area, fifteen species of the mollusk and barnacle species coexisted. When Paine removed the starfish by hand, the number of species declined to eight. What occurred was unexpected but in hindsight logical. Free of the depredations of *Pisaster*, mussels and barnacles increased to abnormally high densities and crowded out seven of the other species. In other words, the predator in this case was less dangerous than the competitors. The assembly rule is this: insert a certain predator, and more species of sedentary animals can invade the community later.

Still another dimension of complexity is added by symbiosis, defined broadly as the intimate association of two or more species. Biologists recognize three classes of symbiosis. In parasitism, the first, the symbiont is dependent on the host and harms but does not kill it. Put another way, parasitism is predation in which the predator eats the prey in units of less than one. Being eaten one small piece at a time and surviving, often well, a host organism is able to support an entire population of another species. It can also sustain many species simultaneously. A single unfortunate and unmedicated human being might, theoretically at least, support head lice *(Pediculus humanus capitis)*, body lice *(Pediculus humanus humanus)*, crab lice *(Pthirus pubis)*, human fleas *(Pulex irritans)*, human bot flies *(Dermatobia hominis)*, and a multitude of roundworms, tapeworms, flukes, protozoans, fungi, and bacteria, all metabolically adapted for life on the human body. Each species of organism, especially each kind of larger plant or animal, is host to such a customized fauna and flora of parasites. The gorilla, for example, has its own crab louse, *Pthirus gorillae*, which closely resembles the one on *Homo sapiens*. A mite has been found that lives entirely on the blood it sucks from the hind feet of the soldier caste of one kind of South American army ant. Tiny wasps are known whose larvae parasitize the larvae of still other kinds of wasps that live inside the bodies of the caterpillars of certain species of moths that feed on certain kinds of plants that live on other plants.

Raising diversity still more are the commensals, symbiotic organisms that live on the bodies of other species or in their nests but neither harm nor help them. Without any awareness of the fact, most human beings carry around on their foreheads two kinds of mites, slender creatures with wormlike bodies and spidery heads so small as to be almost invisible to the naked eye. One *(Demodex folliculorum)* dwells in the hair follicles, the other *(Demodex brevis)* in the sebaceous glands. You can get to know your own forehead mites the following way: stretch the skin tight with one hand, carefully scrape a spatula or butter knife over the skin in the opposite direction, squeezing out traces of oily material from the sebum glands. (Avoid using too sharp an object, such as a glass edge or sharpened knife.) Next scrape the extracted material off the spatula with a cover slip and lower the slip face down onto a drop of immersion oil previously placed on a glass microscope slide. Then examine the material with an ordinary compound microscope. You will see the creatures that literally make your skin crawl.

People would never notice their forehead mites in any other way. These acarines and other commensals slip the thin wedge in, sip small amounts of nutrients and energy virtually useless to their hosts, and live secure lives of flawless modesty. Their biomass is small to microscopic, their diversity immense. They are everywhere, but it takes a special eye to find them. On the leaves of trees in the tropical rain forests grow flat, centimeter-wide gardens of lichens, mosses, and liverworts. Among the epiphylls—plants that live on the leaves—thrive a host of tiny mites, springtails, and barklice. Some of the animals browse on the epiphylls, others prey on the epiphyll browsers. Thus a single leaf of a tree, often composing less than one part in 10,000 of that single large organism, is home to an entire miniaturized fauna and flora.

The tightest bond of all among species, the one that gives the word *community* more than metaphorical meaning, is mutualism. This third kind of relationship, often considered the true symbiosis and employed that way in less formal prose, is an intimate coexistence of two species benefiting both. A large part of dead wood is decomposed by termites—not by the termites really, but by protozoans and bacteria that live in the hind guts of the termites. And not entirely by these microorganisms either, since they need the termites to provide them a home and a steady stream of wood chewed into digestible pulp. So the right way to put the original phrase is: a large part of wood is decomposed by the termite-microorganism symbiosis. The termites harvest the wood but cannot digest it; the microorganisms digest the wood but cannot harvest it. It might be said that over millions of years the termites domesticated the microorganisms to serve their special needs. That, however, would be big-organism chauvinism. It is equally correct to say that termites have been harnessed to the needs of the microorganisms. Such is the nature of mutualistic symbiosis: to attain the highest level of intimacy, the partners are melded into a single organism.

Mutualistic symbioses are more than simply curiosities for the delectation of biologists. Most life on land depends ultimately on one such relationship: the mycorrhiza (literally from the Greek for fungus-root), the intimate and mutually dependent coexistence of fungi and the root systems of plants. Most kinds of plants, from ferns to conifers and flowering plants, harbor fungi that are specialized to absorb phosphorus and other chemically simple nutrients from the soil. The mycorrhizal fungi give up part of these vital materials to their plant hosts, and the plants repay them with shelter and a supply of carbohydrates. Plants deprived of their fungi grow slowly; many die.

According to species, the fungi either enter the outer root cells of their plant hosts or envelop the entire roots to form dense webs. A plant pulled up almost anywhere in the world reveals a tangle of delicate fibers clutching masses of soil particles. Some of the extensions are likely to be rootlets of the plant, but others are the moldlike hyphae of the symbiotic fungi. In many kinds of plants, fungal hyphae have completely replaced the rootlets during evolution.

Without the plant-fungus partnership, the very colonization of the land by higher plants and animals, 450 to 400 million years ago, probably could not have been accomplished. The barren, rain-lashed soil of that time was not hospitable to organisms more complex than bacteria, simple algae, and mosses. The earliest vascular plants were leafless, seedless forms that superficially resembled modern-day horsetails and quillworts. By allying themselves with fungi, they took hold of the land. Some of the pioneers evolved into the lycophyte trees and seed ferns of the great Paleozoic coal forests. They also gave rise to the ancestors of modern conifers and flowering plants, whose vegetation came in the fullness of time to harbor the largest array of animal life that has ever existed. Today the tropical rain forests, which may contain more than half the species of plants and animals on earth, grow on a mat of mycorrhizal fungi.

Coral reefs, the marine equivalents of rain forests, are also built on a platform of mutualistic symbiosis. The living coral organisms, which cover the carbonaceous bulk of the reef, are the polyps, close relatives of the jellyfish. Like the jellyfish and other coelenterates, they use feathery tentacles to capture crustaceans and other small animals. They also depend on the energy provided by single-celled algae, which they shelter within their tissues and to which they donate some of the nutrients extracted from their prey. In most coral species, each individual polyp lays down a skeletal container of calcium carbonate that surrounds and protects its soft body. Coral colonies grow by the budding of individual polyps, with the skeletal cups being added one on another in a set geometric pattern particular to each species. The result is a lovely, bewildering array of skeletal forms that mass together to make the whole reef—a tangled field of horn corals, brain corals, staghorn corals, organ pipes, sea fans, and sea whips. As the colony grows, the older polyps die, leaving their calcareous shells intact beneath; and in time the living members form a layer on top of a growing reef of skeletal remains. These massive remains, many of which are thousands of years old, play a major role in the formation of tropical islands, in particular the fringing reefs of volcanic islands and the atolls left behind when the volcanoes erode away. They create the physical basis and photosynthetic energy for tightly packed communities of thousands of species, from sea hornets and mantis shrimps to carpet sharks.

What, in summarizing to this point, do we understand of the assembly of communities? Obviously we know that there is a large amount of organization in the connections among species. But how much? The answer is unknown for any kind of community—all the organisms in a patch of hardwood forest, for example, or in a coral reef or desert spring. We know some keystone species, some assembly rules, some processes of competition and symbiosis that serve as a weak gravitational force.

We know how some species fit together in twos and threes, but not how the whole community fits together. There are a few hints of what is to come as research grows more sophisticated. Think of the community as a food web, a connection of species that prey on other species. Consider what might happen when a species is extinguished, simply plucked out of the food web as were the sea otters. What is the effect? With field studies and mathematical models, ecologists have pieced together a few of the most general properties of food webs that bear on the result of such an experiment. They have learned that the food chains making up the web are very short. If you track who eats whom in different parts of the web, you will usually find the number of links in the chain to be five or fewer. For example: in a marshy glade of the north central states, reedgrass is eaten by short-horned grasshoppers, the grasshoppers are eaten by orb-weaver spiders, the spiders are eaten by palm warblers, and the warblers are eaten by marsh hawks. Because the grass eats no one and the hawks are eaten *by* no one (except by bacteria and other decomposers when they die), these two species form the ends of the chain. A second rule is that the number of links in the food web does not increase as the size of the community increases. No matter how many species manage to persist in the community, the average number of links from a given plant species to a given top predator does not increase.

I cite these two generalizations to illustrate the more solid principles of community ecology. But I cite them also to show how incomplete and tenuous those principles are. Imagine that you excise the palm warblers from the marsh food web. That food chain is broken, but the ecosystem remains intact, more or less. The reason is that each species in the chain is linked to additional chains. Other species of birds still present in the marsh will eat more spiders, and the marsh hawks will turn, almost imperceptibly, to a larger number of birds, rodents, snakes, and other creatures. Feather mites, bird lice, and other symbionts found only on palm warblers, part of yet other chains, disappear with their host, but their loss has a negligible effect on the community at large.

Expand the thought experiment to extirpate two warbler species, then all warbler species, and finally all the songbirds in the community. As the knife cuts deeper, its effects will spread with increasing severity through a large but indeterminate part of the community. Take out the ants, the principal predators and scavengers of insects and other small animals, and the effects will intensify—yet the details are even less predictable. Most species of birds, ants, and other plants and animals are linked to multiple chains in the food web. It is very difficult to assess which survivors will fill in for the extinguished species and how competently they will perform in that role. Physicists can chart the behavior of a single particle; they can predict with confidence the interaction of two particles; they begin to lose it at three and above. Keep in mind that ecology is a far more complex subject than physics.

The reverse of the extinction process is species packing. Ecologists are unable for the most part to predict which species can still invade the community and add to its diversity. Select a habitat at random. How tightly packed are the species? What is the upper limit of stable diversity, the highest number of species that can be maintained without human intervention? It is easy to enhance local diversity by the artificial introduction of more and more species—orchids affixed to tree trunks, zoo-bred tigers released into the jungle—but most would eventually perish. Without constant and intrusive manipulation, most overloaded communities will revert to a lower state of diversity, perhaps resembling the original, perhaps not.

The indeterminacy of community structure is increased by the existence of connections between species lying beyond the conventional food webs, and for which few reliable laws or rules exist. Competition—especially that resulting in the exclusion of one species by another—is especially difficult to call. So are the effects of removing scavengers and symbionts. Most difficult of all to assess is the impact of species that alter the physical environment over many years. Dominant tree species overgrow and change the temperature and humidity regimes in which other plants and animals must live. Mound-building termites turn and enrich the soil; they alter the composition of chemical elements and determine the species of plants that can grow near their underground tunnels. Populations of mites and springtails bloom, and fungus spores and humus correspondingly decline—all to indeterminate degree.

The unpredictability of ecosystems is a consequence of the particularity of the species that compose them. Each species is an entity with a unique evolutionary history and set of genes, and so each species responds to the rest of the community in a special way. I will finish with my own favorite example of law-destroying idiosyncrasy. Tree holes often fill with rainwater, creating small aquatic habitats for animals and microorganisms. On the west coast of the United States live larvae of a tree-hole mosquito species, *Aedes sierrensis*. They feed on microscopic ciliated protozoans, *Lambornella clarki*, which resemble the familiar paramecia used in biology courses. The protozoans in turn feed on bacteria and other microorganisms breeding in the tree-hole water. After the protozoans have been exposed to the odor of the mosquito larvae for one to three days, they turn the tables on their tormentors. Some of them metamorphose into parasitic forms that invade the bodies of the larvae and start to feed on their tissues and blood. Thus a segment of the food chain is flipped upside down, creating a food cycle where each species is simultaneously predator and prey of the other.

The mosquito-protozoan cycle of predation and counterpredation is emblematic of the direction that community ecology must take: analyze ecosystems in detail, from the bottom up. Biologists are returning to natural history with a new sense of mission. They cannot expect to learn much more from the top down, from the properties of whole ecosystems (energy flow, nutrient cycles, biomass) interpolated to the properties of communities and species. Only with a detailed knowledge of the life cycles and biology of large numbers of constituent species will it be possible to create principles and methods that can precisely chart the future of ecosystems in the face of the human onslaught.

Then there might be an answer to the question I am asked most frequently about the diversity of life: if enough species are extinguished, will the ecosystems collapse, and will the extinction of most other species follow soon afterward? The only answer anyone can give is: possibly. By the time we find out, however, it might be too late. One planet, one experiment.

Biodiversity Reaches
the Peak

THREE BILLION years ago the land was virtually devoid of life. More than that, it was uninhabitable. No ozone layer existed in the stratosphere, and the progenitor molecules of oxygen in the air below were too thin to create it. Short-length ultraviolet radiation traveled unimpeded to the earth and beat down on the dry basaltic rocks. It assaulted organisms venturing there out of the sea, shutting down their enzymatic synthesis, opening their membranes to ambient poisons, and rupturing their cells. But in the water, safe from the lethal rays, microscopic organisms swarmed. They were close to modern cyanobacteria (sometimes called blue-green algae) and a mélange of bacteria and bacteria-like species. Most were single-celled and prokaryotic, and a few were composed of cells strung together in filaments. These simple organisms were devoid of nuclear membranes, mitochondria, chloroplasts, and the other organelles that give structural complexity to the cells of higher plants and animals.

A large portion of the early life forms were concentrated in thin scummy sheets called microbial mats. Under the mats they built up distinctive rock formations called stromatolites, resembling stacks of mattresses (*stroma*, mattress) strewn about the shallow sea bottoms like packages on a warehouse floor. Modern versions of these organism-topped rocks still grow in subtidal marine waters in a few scattered localities such as Baja California and northwestern Australia. Some are soft enough to be cut by a hunting knife. Others have been infiltrated by enough calcium carbonate to make them as hard as the fossil stromatolites. The formations grow by accretion. The living organisms on them are periodically covered by silt and debris carried by the tide and storms. They multiply upward, pressing through the fouling cover to touch once again clear water and sunlight, and by this means they add height to the stromatolite foundations year by year.

Not all modern microbial mats have thick columns under them. Many form thin, unsupported sheets in marginal habitats where physical conditions are severe and predators and competing organisms scarce, such as hot springs, salty lagoons, Antarctic lakes, deep-sea sediments, and damp rock surfaces on the land. They are scarce and scattered in comparison with most ecosystems. But three billion years ago all available space in the shallow seas was probably covered by a variety of such microbial formations, each kind specialized for a particular niche of light, temperature, and acidity.

Since the beginning of life, the denizens of microbial mats have gathered into communities of considerable complexity. The plain appearance of the outer coat viewed with the naked eye is misleading. When a mat is sliced vertically and examined under the microscope, it is seen to be packed with photosynthetic organisms from the surface to a depth of a millimeter. Across that short distance, half the height of a capital letter on this page, sunlight attenuates to 1 percent of the intensity it has in the water above. That is about the same amount of energy lost by sunlight in traveling from tree crown to floor in a dense forest. And the analogy runs deeper: the mat community is even organized somewhat like a forest. Cyanobacteria, which capture solar energy, are distributed in succession from top to bottom like different kinds of trees, with least shade-tolerant species near the surface and most shade-tolerant species toward the bottom. They use the energy to combine water and carbon dioxide into organic molecules, giving off oxygen in the process. Farther down, in the miniature equivalent of the dark forest interior (or deep sea, below the upper lighted waters), live sulfur-oxidizing bacteria. These archaic organisms, of a kind that may have preceded cyanobacteria in evolution, are not photosynthesizers. They do not decompose water into hydrogen and oxygen with the aid of solar energy, but instead split the weaker sulfide bonds unaided by sunlight.

Swimming and drifting in open water around the ancient microbial mats were almost certainly populations of cyanobacteria and other prokaryotic forms different from the mat organisms. Some lived by photosynthesis, others by preying on prokaryotes or scavenging their

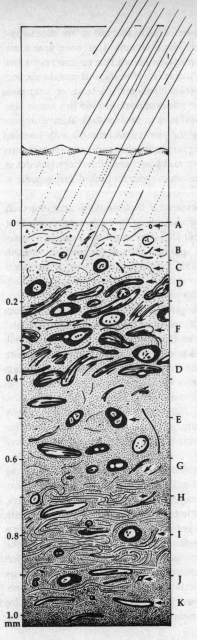

Among the most primitive ecosystems on earth are the microbial mats, thin assemblages of microscopic organisms that date back in geological time almost to the beginning of life. In this living mat from shallow marine water, organisms are arrayed by species according to their depth across 1 millimeter of the mat and hence by the amount of light and kind of nutrients reaching them.

A   diatoms (microscopic algae)
B   *Spirulina* (cyanobacteria or blue-green algae)
C   *Oscillatoria* (cyanobacteria)
D   *Microcoleus* (cyanobacteria)
E   nonphotosynthesizing bacteria
F   mixed single-celled cyanobacteria
G   bacterial mucilage
H   *Chloroflexus* (green bacteria)
I   *Beggiatoa* (sulfide-oxidizing bacteria)
J   unidentified grazing organisms
K   discarded sheaths of cyanobacteria

dead cells. Life must have been already diverse at the microscopic level, appropriating relatively large quantities of energy and nutrients. Yet the early organisms were not all that diverse, not when compared with present-day biotas. No forests and grasslands succoring millions of animal species blanketed the land; no kelp beds choked the ocean margins; no flocks of terns hunted fish across blue waters. If you and I could travel back in time to wade along the shore of the ancient sea, searching for plants and animals with unaided vision, we would find nothing certifiably alive, only unprepossessing pond-scum smudges of brown and green and slimy rock surfaces of uncertain provenance. Visible organisms and high diversity were to come much later.

Biological diversity has increased a thousandfold since the early days of the microbial mats, pulled along by evolutionary progress, measured in turn by four great steps that mark the passage of eons:

• The origin of life itself, spontaneously from prebiotic organic molecules, about 3.9 to 3.8 billion years ago. The first organisms were single-celled and hence microscopic. Stromatolite ecosystems appeared no later than 3.5 billion years ago.

• The origin of eukaryotic organisms—"higher organisms"—about 1.8 billion years ago. Their DNA was enveloped in membranes, and the remainder of the cell contained mitochondria and other well-formed organelles. At first eukaryotes were single-celled, in the manner of modern protozoans and the simpler forms of algae, but soon they gave rise to more complex organisms composed of many eukaryotic cells organized into tissues and organs.

• The Cambrian explosion, 540 to 500 million years ago. Newly abundant macroscopic animals, large enough to be seen with the naked eye, evolved in a radiative pattern to create the major adaptive types that exist today.

• The origin of the human mind, in later stages of evolution in the genus *Homo*, probably from a million to 100,000 years ago.

Some biologists and philosophers have trouble with that term, "evolutionary progress." The expression is inexact and loaded with humanistic nuance, granted, but I use it just the same to identify a paradox pivotal to the understanding of biological diversity. In the strict sense, the concept of progress implies a goal, and evolution has no goal. Goals are not inherent in DNA. They are not implied by the impersonal forces of natural selection. Rather, goals are a specialized form of behavior, part of the outer phenotype that also includes bones, digestive enzymes, and the onset of puberty. Once assembled by natural selection, human beings and other sentient organisms formulate goals as part of their survival strategies. Because goals are the ex-post-facto responses of organisms to the necessities imposed by the environment, life is ruled by the immediate past and the present, not by the future. In short, evolution by natural selection has nothing to do with goals, and so it would seem to have nothing to do with progress.

And yet there is another meaning of "progress" that does have considerable relevance to evolution. Biological diversity embraces a vast number of conditions that range from the simple to the complex, with the simple appearing first in evolution and the more complex later. Many reversals have occurred along the way, but the overall average across the history of life has moved from the simple and few to the more complex and numerous. During the past billion years, animals as a whole evolved upward in body size, feeding and defensive techniques, brain and behavioral complexity, social organization, and precision of environmental control—in each case farther from the nonliving state than their simpler antecedents did. More precisely, the overall averages of these traits and their upper extremes went up. Progress, then, is a property of the evolution of life as a whole by almost any conceivable intuitive standard, including the acquisition of goals and intentions in the behavior of animals. It makes little sense to judge it irrelevant. Attentive to the adjuration of C. S. Peirce, let us not pretend to deny in our philosophy what we know in our hearts to be true.

An undeniable trend of progressive evolution has been the growth of biodiversity by increasing command of earth's environment. New methods to detect microscopic fossils in billion-year-old sedimentary rocks, chemical analyses of ancient environments, and statistical estimates of the relative abundances of extinct species have allowed geochemists and paleontologists during the past decade to bring this history into sharper focus.

By two billion years before the present, a large fraction of earth's organisms were generating oxygen through photosynthesis. But this element, so vital to life as we know it today, did not accumulate in the water and atmosphere. It was captured by ferrous iron, which dissolves in water and was abundant enough to saturate the early seas. The two elements combined to form ferric oxides, insoluble in water, which settled to the ocean floor. As J. William Schopf neatly summarized the situation, the world rusted.

Denied oxygen by the ferrous sink, the organisms were forced to remain anaerobic. The aerobic pathways of metabolism, which are highly efficient means to obtain and deploy free energy, could at most have evolved as an auxiliary adaptation. By 2.8 billion years ago, the sink had partially filled, and a few local habitats sustained low levels of molecular oxygen. Aerobic organisms, still single-celled prokaryotes, appeared about this time. During the next billion years, oxygen levels rose worldwide to constitute about 1 percent of the atmosphere. By 1.8 billion years ago, the first eukaryotic organisms appeared: alga-like forms, forerunners of the dominant photosynthesizers of the modern seas. By no later than 600 million years ago, near the end of the Proterozoic era, the first animals evolved. Members of this Ediacaran fauna, named after the Ediacara Hills of South Australia where many of the first specimens were found, were soft-bodied and typically flat. They vaguely resembled jellyfish, annelid worms, and arthropods, and some may have been members of those surviving groups.

Approximately 540 million years ago, near the beginning of the Cambrian period, earliest of the time segments of the Phanerozoic eon in which we now live, a seminal event occurred in the history of life. Animals increased in size and diversified explosively. The supply of free oxygen in the atmosphere was by this time near the 21 percent level of today. The two trends are probably linked, for the simple reason that large, active animals need aerobic respiration and a rich supply of oxygen. Within a few million years, the fossil record held almost every modern phylum of invertebrate animals a millimeter or more in length and possessed of skeletal structures, hence easily preserved and detectable later. A large portion of present-day classes and orders had also come on stage. Thus occurred the Cambrian explosion, the big bang of animal evolution. Bacteria and single-celled organisms had long since attained comparable levels of biochemical sophistication. Now, in a dramatic new radiation, they augmented their niches to include life on the bodies and waste materials of the newly evolved animals. They created a new, microscopic suzerainty of pathogens, symbionts, and decomposers. In broad outline at least, life in the sea attained an essentially modern aspect no later than 500 million years ago.

By this time a strong ozone layer existed as well, screening out lethal short-wave radiation. The intertidal reaches and dry land were

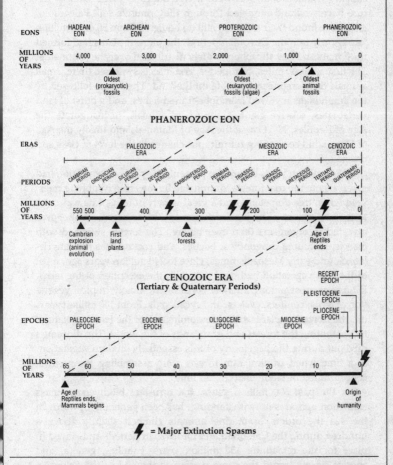

Full History of Life

EONS	HADEAN EON	ARCHEAN EON	PROTEROZOIC EON	PHANEROZOIC EON

MILLIONS OF YEARS: 4,000 — 3,000 — 2,000 — 1,000 — 0

- Oldest (prokaryotic) fossils
- Oldest (eukaryotic) fossils (algae)
- Oldest animal fossils

PHANEROZOIC EON

ERAS	PALEOZOIC ERA	MESOZOIC ERA	CENOZOIC ERA

PERIODS: CAMBRIAN PERIOD, ORDOVICIAN PERIOD, SILURIAN PERIOD, DEVONIAN PERIOD, CARBONIFEROUS PERIOD, PERMIAN PERIOD, TRIASSIC PERIOD, JURASSIC PERIOD, CRETACEOUS PERIOD, TERTIARY PERIOD, QUATERNARY PERIOD

MILLIONS OF YEARS: 550 500 — 400 — 300 — 200 — 100 — 0

- Cambrian explosion (animal evolution)
- First land plants
- Coal forests
- Age of Reptiles ends

CENOZOIC ERA
(Tertiary & Quaternary Periods)

RECENT EPOCH
PLEISTOCENE EPOCH
PLIOCENE EPOCH

EPOCHS	PALEOCENE EPOCH	EOCENE EPOCH	OLIGOCENE EPOCH	MIOCENE EPOCH

MILLIONS OF YEARS: 65 60 — 50 — 40 — 30 — 20 — 10 — 0

- Age of Reptiles ends, Mammals begins
- Origin of humanity

⚡ = Major Extinction Spasms

The full geological history of life goes back more than 3.5 billion years, when the first single-cell organisms appeared. Key episodes in evolution are placed within the divisions of geological time: eons divided into eras, eras into periods, and periods into epochs. Biodiversity was sharply reduced by the great extinction spasms, indicated here by lightning flashes.

safe for life. By the late Ordovician period, 450 million years ago, the first plants, probably derived from multicellular algae, invaded the land. The terrain was generally flat, lacking mountains, and mild in climate. Animals soon followed: invertebrates of still unknown nature burrowed and tunneled through the primitive soil. Paleontologists have found their trails but still no bodies. Within 50 or 60 million years, early into the Devonian period, the pioneer plants had formed thick mats and low shrubbery widely distributed over the continents. The first spiders, mites, centipedes, and insects swarmed there, small animals truly engineered for life on the land. They were followed by the amphibians, evolved from lobe-finned fishes, and a burst of land vertebrates, relative giants among land animals, to inaugurate the Age of Reptiles. Next came the Age of Mammals and finally the Age of Man, amid continuing tumultuous change at the level of class and order.

By 340 million years before the present, the pioneer vegetation had given way to the coal forests, dominated by towering lycophyte trees, seed ferns, tree horsetails, and a great variety of ferns. Life was close to the attainment of its maximal biomass. More organic matter was invested in organisms than ever before. The forests swarmed with insects, including dragonflies, beetles, and cockroaches. By late Paleozoic and early Mesozoic times, close to 240 million years ago, most of the coal vegetation had died out, with the exception of the ferns. Dinosaurs arose among a newly constituted, mostly tropical vegetation of ferns, conifers, cycads, and cycadeoids. From 100 million years on, the flowering plants swept to domination of the land vegetation, reconstituting the forests and grasslands of the world. The dinosaurs died out during the hegemony of this essentially modern vegetation, at a time when tropical rain forests were assembling the greatest concentration of biodiversity of all time.

For the past 600 million years, the thrust of biodiversity, mass extinction episodes notwithstanding, has been generally upward. In the sea the orders of marine animals climbed slightly above a hundred during the Cambrian and Ordovician periods and stayed in place for the remaining 450 million years. Families, genera, and species closely traced the same pattern to the end of the Paleozoic era, 245 million years ago. They were knocked down sharply by the extinction catastrophe that closed the Paleozoic era, followed in only 50 million years by a smaller spasm in the Triassic period. Thereafter they climbed steeply, with a dip at the end of the Mesozoic, reaching unprecedented levels during the past several million years. Plant and

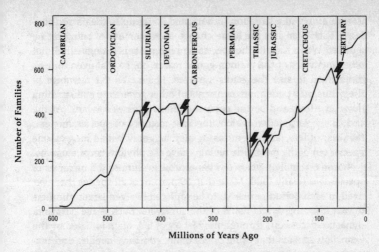

Biological diversity has increased slowly over geological time, with occasional setbacks through mass global extinctions. There have been five such extinctions so far, indicated here by lightning flashes. The data given are for families (groups of related species) of marine organisms. A sixth major decline is now underway as a result of human activity.

animal diversity on the land, after a delay of 100 million years during which colonization took place, followed the same trajectory to our time.

Each extinction spasm reduced the numbers of species most and the numbers of classes and phyla least. The lower the taxonomic category, the more it was diminished. At the end of the Paleozoic era, as many as 96 percent of the species of marine animals and foraminiferans vanished, compared with 78 to 84 percent of the genera and 54 percent of the families. Apparently no phyla came to an end.

This descending vulnerability by taxonomic rank is an artifact, a straightforward consequence of the hierarchical manner in which biologists classify organisms. But it is an interesting and technically useful artifact, for reasons most clearly expressed by what has been called the Field of Bullets Scenario. Imagine advance infantrymen walking forward in eighteenth-century manner, posting arms in one serried rank into a field of fire. Each man represents a species, belonging to a platoon (genus), which is a unit of a company (family), in turn a unit of a battalion (order), and so on up to the corps (phylum). Each man has the same chance each moment of being hit by a bullet. When he falls the species he represents is extinguished, but other members of the platoon-genus march on, so that even though diminished in size the genus survives. In time all the members of the genus may perish, yet remnants of other genera are still standing down the line, and so the company-family presses forward. At the end of the long lethal march, the vast majority of species, genera, families, orders, and even classes may be gone, but so long as one species out of the multitude remains alive the phylum-corps survives.

Across 600 million years of Phanerozoic evolution, the turnover of species was nearly total. More than 99 percent of all species that ever lived in each period perished, to be replaced by even larger numbers drawn from the descendants of the survivors. Such is the nature of dynastic successions through the history of life, often initiated by the extinction spasms that bring down entire company-families and battalion-orders. Ninety-nine percent is not a surprising figure. Imagine a group such as the archaic amphibians of the Paleozoic. A thousand species die, and one survives to produce the primitive reptiles. A thousand of those also die, but one survivor carries on to become the ancestor of the Mesozoic dinosaurs. The survival rate of species in this sequence is 1 in 2,000. In other words, only a single line persists out of the 2,000 created, and yet life flourishes as diversely as before.

This brings us to the notable possibility that no phylum has ever been extinguished. Let me phrase this conjecture in a more operational manner: no major group that has gone extinct can be definitely assigned a taxonomic rank as high as the phylum. A great many battalions and regiments have disappeared in full, but we cannot be sure that any one of them constitutes an entire corps. If any true phylum has vanished, it is most likely to have been one that originated during the Cambrian explosion of animal diversity. Something happened to the environment then, most likely the availability of atmospheric oxygen, that opened the sea to large animals. The waters of the world became a new continent on which animals above a centimeter long could evolve and adaptively radiate; and so they did, creating most or all of the remaining phyla we know today.

It has been widely believed that the Cambrian explosion was a period of wild experimentation during which basic body plans never seen before or afterward were invented and discarded. If this view is correct, some of the most extreme of the short-lived species must qualify as phyla that went extinct. It would also follow that at the phylum level biodiversity reached its peak during the Cambrian explosion and subsided soon afterward to modern levels. This interpretation is supported by the existence of well-preserved fossils in the Burgess Shale of British Columbia, early to middle Cambrian in age, which appear to fit no established phylum. Other Burgess-type fossils, equally bizarre, have been found in Europe, China, and Australia.

These fossils taken together leave little doubt that many unusual animal types arose during Cambrian times and disappeared after brief tenures. In taxonomic language, orders and classes originated that lasted only a few million years. But the fossils do not yet tell us definitively that radical new body types—innovations that confer phylum rank—were created and discarded. In 1989 Simon Conway Morris, a leading authority on Burgess Shale faunas, recognized eleven modern phyla present in those ancient assemblages, together with "19 distinct body plans that for the most part are as different from each other as any of the remaining phyla in the fauna." The swift radiation, Conway Morris continued, is reflected in the still-living phylum Arthropoda, in which the Burgess Shale "cavalcade of morphologies seems to be almost inexhaustible. The overall impression is of an enormous mosaic, individual species being assembled according to differences in number and types of jointed appendages, number of segments and extent of tergite fusion, and overall proportions of the body."

Still, the totality of known arthropod diversity in the Cambrian fossils does not exceed that in living arthropods, and it probably falls far short. The array of classes and orders in the sea remains vast. The insects had not yet arisen to shower the land and fresh water with burrowers, swimmers, and flying machines of fantastical design. I cannot help thinking that if we were to take just four living species of this single class (the Insecta)—say the maggot of a blackfly, the giant bottlenosed lantern fly, a female coccid scale, and the water penny—and preserve them with the same fidelity as the Burgess rock fossils, the remains would be classified erroneously as four distinct phyla. They display what appear superficially to be wholly different body designs.

The paleontologists working on the Burgess faunas have approached their subject with admirable caution. They refer to the fossils with body types not assignable to modern phyla as "problematica." When better-preserved fossils turned up, as in the Chengjiang beds of southern China, and improved methods were developed to study the old specimens, the problematica were whittled down. The armored creatures of the genera *Hallucigenia, Microdictyon,* and *Xenusion* have recently been placed among the Onychophora, the living phylum of caterpillar-like animals thought to be intermediate between arthropods and annelid worms. The essence of weirdness, *Wiwaxia corrugata,* thought to resemble a scaly slug with long spikes sprouting from its back, has proved to be fragments of a polychaete worm, member of an extant phylum, the Annelida.

To summarize the grand parade to date: the number of living animal phyla, all of which have representatives in the sea, is about thirty-three. Of these, approximately twenty comprise animals large and abundant enough to leave fossils of the kind preserved in beds of the Burgess Shale type. The number of Cambrian animal phyla identified with confidence remains at eleven. To the best of our knowledge, no phylum has yet gone extinct. The level of diversity in the sea inched upward after the Cambrian explosion, and it also increased on the land after the assembly of the coal forests and their insect and amphibian inhabitants. The greatest rise in general has occurred during the past 100 million years.

We may now ask why, in spite of major and minor temporary declines along the way, in spite of the nearly complete turnover of species, genera, and families on repeated occasions, the trend in biodiversity has been consistently upward. Part of the answer is that the continental land masses have changed in a way that enhances species formation. During late Paleozoic times, the earth's land surface was composed of the single supercontinent Pangaea. By the early Mesozoic, Pangaea had split into two great fragments, Laurasia to the north and Gondwana to the south; India had broken off as a smaller fragment and was crawling northward toward its rendezvous with the Himalayan arc. Around 100 million years ago, the modern continents were in place, with the waters between them progressively widening. Major faunas and floras evolved in a state of deepening isolation. The length of coastline increased everywhere, along with the area available for inshore bottom-dwelling organisms. Shallow seas spread and retreated back and forth over the land, creating and abolishing new habitats and assemblages of organisms adapted to them. These inhabited worlds, the "faunistic and floristic prov-

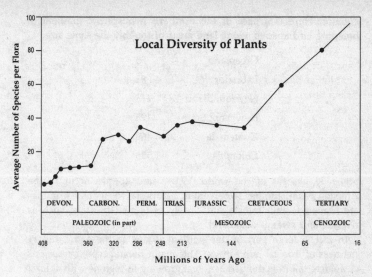

Local Diversity of Plants

Y-axis: Average Number of Species per Flora

DEVON.	CARBON.	PERM.	TRIAS.	JURASSIC	CRETACEOUS	TERTIARY
PALEOZOIC (in part)				MESOZOIC		CENOZOIC

408 — 360 — 320 — 286 — 248 — 213 — 144 — 65 — 16

Millions of Years Ago

The average number of plant species found in local floras has risen steadily since the invasion of the land by plants 400 million years ago. The increase reflects a growing complexity in terrestrial ecosystems around the world.

inces" as they are called, have waxed and waned, but what we live with today is a peak of enrichment.

Global biodiversity was brought to its Cenozoic pinnacle first by the creation of the aerobic environment and second by the fragmentation of the land masses. But that is far from the whole story. The number of species living together within particular habitats, shallow bays and tropical forests for example, have also increased by fits and starts. The number at least doubled in marine organisms and more than tripled in land plants during the past 100 million years. The trends imply that more species are now packed into local communities—that species either originate more quickly or die more slowly.

To see why local communities have grown richer in diversity through geological time, we must return to living faunas and floras, where the biology of individual species can be examined in finer detail. First we can compare living communities poor in species with those that are rich in species. The major clue before us is the latitudinal diversity gradient, the increase in species (or any other taxonomic category) encountered while traveling from the poles to the equator. Here is a slice of the northern hemispheric gradient in breeding bird species, using land areas of roughly the same size:

Greenland	56
Labrador	81
Newfoundland	118
New York State	195
Guatemala	469
Colombia	1,525

Some 30 percent of the world's 9,040 bird species occur in the Amazon Basin, another 16 percent in Indonesia. Most of these faunas are limited to rain forests and closely associated habitats, such as riverine and swamp woodland.

In fact, a large part of the gradient is due to the extraordinary richness of tropical rain forests. This major habitat type, or *biome* as ecologists call it, is defined as forest growing in regions with at least 200 centimeters of annual rainfall spread evenly enough through the year to allow a heavy growth of broad-leaved evergreen trees. The forest is arrayed in multiple layers, from upper canopy 30 meters or more in height, broken by scattered emergent trees soaring over 40 meters, down through ragged middle levels to chest-high understory shrubs. Lianas, strangler vines, and creepers coil around the tree trunks and dangle from the high limbs straight to the ground. Gardens of orchids and other epiphytes festoon the thicker branches. Palms are common in the lower and middle layers of many rain forests, adding lush beauty and a deceptive feel of benignity for the visitor on foot. So effective are the staggered canopy layers at intercepting sunlight that the energy-starved lower vegetation is as sparse as in a juniper woodland. You can walk through it easily, pushing the crisscrossing fronds and branches aside, stepping around the trunks of the great trees, bending to pass under lianas and low tree limbs. There is almost never a need to slash a path with a machete through tangled vegetation, in accord with the popular image of jungles. Machetes are for second growth and forest borders, the true jungles. Rain forests are green cathedrals. They are like the gentle temperate forests familiar to most, except that they tower high and have somehow stayed mysterious and wild.

In the deep rain forest, sunlight dapples the ground, which is covered in spots by a thin layer of leaves and humus and is completely bare in others. Away from the sunlit patches, the ground is so dark that a flashlight is needed to study it closely, to see the insects, spiders, sowbugs, millipedes, daddy-longlegs, and other small creatures swarming there, together forming the cemetery squads of scavengers and the predators that hunt them in turn.

Tropical rain forests, though occupying only 6 percent of the earth's land surface, are believed to contain more than half the species of organisms on earth. I say "believed" because no exact estimates of diversity have been made either for the world as a whole or for rain forests in particular. The more-than-half figure has simply emerged as a consensus from technical reports and conversations among experts, bolstered by educated guesses and logical extensions by the theoreticians of biological diversity. It is largely based, I admit, on anecdotes and piecemeal analysis. Yet in the aggregate this circumstantial evidence grows more persuasive with time.

Here are the elements of reasoning behind the more-than-half argument. The latitudinal diversity gradient, which I illustrated with birds, is a true general principle of biology: the largest numbers of species occur in the equatorial regions of South America, Africa, and Asia. Another explicit example is offered by the vascular plants, which include the flowering plants, ferns, and a mix of lesser groups such as the club mosses, horsetails, and quillworts. These groups together compose more than 99 percent of land vegetation. Of the approximately 250,000 species known, 170,000 (68 percent) occur in the tropics and subtropics, especially in the rain forests. The peak of global plant diversity is the combined flora of the three Andean countries of Colombia, Ecuador, and Peru. There over 40,000 species occur on just 2 percent of the world's land surface. The world record for tree diversity at one site was set by Alwyn Gentry in the rain forest near Iquitos, Peru. He found about 300 species in each of two 1-hectare (2.5-acre) plots. Peter Ashton discovered over 1,000 species in a combined census of ten selected hectare plots in Borneo. These numbers are to be compared with 700 native species found in all of the United States and Canada, in every major habitat from the mangrove swamps of Florida to the coniferous forests of Labrador.

Butterflies are even more disproportionately rich in the rain forests. The largest documented faunas in the world occur in the Río Madre de Dios drainage in southeastern Peru. To date Gerardo Lamas and his coworkers have recorded 1,209 butterfly species within the 55 square kilometers of the Tambopata Reserve. In a close race, Thomas Emmel and George Austin have identified 800 species in a forest patch several square kilometers in extent at Fazenda Rancho Grande, near the center of the state of Rondonia, western Brazil. By adding probable species numbers from still poorly researched groups, they estimate the total inventory to fall between 1,500 and 1,600 species. Nearby, at Jaru, on October 5, 1975, one entomologist sighted an astonishing 429 butterfly species within twelve hours (this site has since been cleared for agriculture and almost all its butterflies are gone). In contrast, there are only about 440 species in all of eastern North America and 380 in Europe and the Mediterranean coast of North Africa combined.

The ants rival the butterflies in the steepness of their latitudinal gradient. At the Tambopata Reserve, Terry Erwin used a bug bomb to collect all the insects from a single leguminous tree in the rain forest. I identified the ants in his sample and found 43 species in 26 genera, approximately equal to the entire ant fauna of the British Isles. The ants are in turn dwarfed by the beetles. Erwin estimated that over 18,000 species occur in 1 hectare of Panamanian rain forest, with most previously unknown to science—in other words, still lacking a scientific name. To date, only 24,000 beetle species are known from all of the United States and Canada, 290,000 from the entire world.

Thus it goes, biodiversity pyramiding southward, group by group on the land. A few kinds of plants and animals, including conifers, aphids, and salamanders, are more diverse in the temperate zones. These are the exceptions and in any case are not unusually diverse. There are, for example, fewer than 400 species of salamanders known in the entire world. Other groups of plants and animals are mostly tropical but specialized to live in deserts, grassland, and dry forest. They too are generally less diverse than the inhabitants of nearby rain forests.

The organisms of the shallow-water marine environments follow the same latitudinal trend: plankton and bottom dwellers increase in diversity toward the tropics, and the densest concentrations of all are found in the coral reefs, the marine equivalent of rain forests. They abound in absolute diversity, most of it unexplored. Hundreds of species of crustaceans, annelid worms, and other invertebrates can be found in a single coral head, the equivalent of a rain-forest tree.

To summarize the present global pattern, latitudinal diversity gradients rising toward the tropics are an indisputable general feature of life. And on the land biodiversity is heavily concentrated in the tropical rain forests. So immense are the insect faunas alone in these forests, comprising possibly tens of millions of species, overwhelming even the opulence of the coral reefs, that on this basis alone it is reasonable to suppose that over half of all species are found there.

The cause of tropical preeminence poses one of the great theoretical problems of evolutionary biology. Biologists have focused variously on climate, solar energy, amount of habitable terrain, variety of habitats available, amount and frequency of environmental disturbance, degree of isolation of the faunas and floras, and the mostly intangible idiosyncrasies of history. Many have called the problem intractable, supposing its solution to be lost somewhere in an incomprehensible web of causes or else dependent on past geological events that have faded beyond recall. Yet a light glimmers. Enough solid analyses and theory have locked together to suggest a relatively simple solution, or at least one that can be easily understood: the Energy-Stability-Area Theory of Biodiversity, or ESA theory for short. In a nutshell, the more solar energy, the greater the diversity; the more stable the climate, both from season to season and from year to year, the greater the diversity; finally, the larger the area, the greater the diversity.

The evidence for this theory has come from several directions and tells us a great deal not only about biological diversity but also about the importance of physical environment in the organization of ecosystems. David Currie, for example, studied the effects of a wide range of environmental variables on the number of tree and vertebrate species in different parts of North America. This continent offers an excellent laboratory for such a multifactorial analysis. It lies entirely within the temperate zone and hence possesses the same well-marked seasons everywhere, yet varies greatly in precipitation and topography from east to west. Under these conditions—no tropics to worry about for the moment—the overwhelming factor is the amount of solar energy and humidity available to organisms throughout the year. The measurement that captures both variables is evapotranspiration, the quantity of water evaporating from a saturated surface. This amount depends in turn on the energy available to evaporate water, and that energy comes from the heat of the sunlight combined with the temperature of the ambient air and the movement of drying air currents. It depends to a lesser degree on humidity. In North America at least, warm and humid environments support more species of trees. The diversity of land vertebrates, including mammals, birds, reptiles, and amphibians, rises with solar energy but is less dependent on humidity. Put as briefly as possible, dry spots are bad news for trees, much less so for vertebrates. For both kinds of organisms, however, more solar energy means more diversity.

The parts of the world with the highest year-round temperatures are the equatorial tropics, and the habitats with the greatest combined heat and humidity are the tropical rain forests. Given an equal amount of nutrients, the hottest, most humid places are also the most productive in terms of the quantity of plant and animal tissue grown each year. It would seem to follow that the higher the production of living matter, the greater the number of species that can coexist in the same community. Put another way, the larger the pie, the greater the number of possible slices big enough to sustain the lives of individual species.

But energy and biomass production alone cannot explain the tropical dominance of biological diversity. What is to prevent one superbly adapted species in each broad category—one flowering plant, one frog, one wood-boring beetle—from taking over the entire habitat? Something like that has in fact occurred in red mangrove swamps and cordgrass marshes, two of the most productive wetlands in the world. In each of these habitats one species of plant composes more than 90 percent of the vegetation. But simple ecosystems are the rare exceptions worldwide, and diverse ecosystems are the rule. A fuller explanation of latitudinal gradients requires shifting the analysis to look at the role of the seasons. In the temperate and polar regions, organisms experience wild swings in temperature through the year. They must adapt to a wide range of physical and biological environments as part of their life cycles. In winter they variously hibernate, die off after setting seed, shed leaves, move down the mountainside, descend trees to the ground, burrow deeper in the soil, change their diet to cold-hardy prey, change activity from nocturnal to diurnal peaks, or, in the case of migratory birds and monarch butterflies, flee the country altogether. In the spring animals enjoy a flush of fresh vegetation, which tapers off with the dryness of late summer and forces a shift to new foods and habitats.

Because animal and plant species of cold climates are therefore adapted to a greater breadth of local environments, they also occupy larger geographical ranges. In particular, they are distributed across a wider range of latitudes. If a butterfly can thrive in the cool, wet spring of New England, it can endure the winter of Florida. This trend has been called Rapoport's rule, after the Argentine ecologist Eduardo Rapoport, who first suggested it in 1975. It means that as you travel southward down North America or northward up temperate South America, the ranges of individual species shrink steadily the closer you come to the equator. Of equal importance, the altitudinal range of species along the sides of mountains also contract. Thus more species are packed into the same amount of space in the tropics than in the colder temperate zones.

Higher energy, greater biomass production, the narrowing of geographical ranges within a less varying environment—all these properties elevate the level of biodiversity in the tropics over long stretches of evolutionary time. But there is still more to the engine of tropical exuberance. Stable climates with muted seasons allow more kinds of organisms to specialize on narrower pieces of the environment, to outcompete the generalists around them, and so to persist for longer periods of time. Species are packed more tightly. No niche, it seems, goes unfilled. Specialization is likely to be pushed to bizarre, beautiful extremes. In the sunlit clearings of Central American rain forests there are giant helicopter damselflies, drifting back and forth in the still air, with bands on their beating transparent wings that seem to spin around their bodies. Their nymphs are not to be found in ponds and streams, the conventional haunts of damselflies elsewhere, but in the water-filled axils of epiphytes high in the canopy. The adults feed by plucking spiders from their webs. On the hindfeet of army-ant soldiers nearby are fastened parasitic mites found nowhere else in the world. While sucking the blood of the ants, they allow themselves to be used as artificial feet; the ants walk on the bodies of the parasites with no sign of discomfort on either side. The mite covers the claws of the ant by which the ant hangs while nesting and renders it useless, but no matter: the mite has curved hindlegs the size of the claws, and the ant uses them instead. On the vegetation of mountain rain forests of Papua New Guinea live weevils half the size of a human thumb, sluggish and long-lived, their backs covered with algae, lichens, and mosses. In this miniature traveling garden dwell distinctive species of tiny mites and nematodes. I could go on with this bestiary, moving from one country to the next—the literature of tropical biology never runs out of surprises. Where conventional niches have already been filled, it seems that enterprising species invent new ones.

Walk the floor of a tropical rain forest, searching for specimens of almost any group, whether orchids, frogs, or butterflies. You will find that the species change subtly every hundred or thousand meters. A certain kind is common at one spot, then fades and disappears, to be replaced by closely similar species not encountered farther back. Then a stroke of luck: there appears a single individual of a species never before encountered in the entire area. Collect or at least photograph it carefully because you will never see it again. In Central American rain forests, the nymphaline butterfly *Dynamine hoppi* is a pretty species distinguished by large white spots on the forewing and metallic blue fringes on the hindwing bands. It has been encountered three times in history. One female was collected by the lepidopterist Philip DeVries in a clearing at Finca La Selva, Costa Rica, in July. It was the only individual he found during a six-month study of the butterflies of this particular forest. A second female was taken the following year in the same spot, also in July. Then no more. If you return to the forest day after day and from one year to the next, quartering it on foot with net and binoculars, your list of orchids or frogs or butterflies will grow—and grow.

At first the picture of diversity in the forest seems hopelessly confusing, but in time a general pattern emerges: there are a few common species, most distributed in scattered patches, and a great many rare species, including some like *Dynamine hoppi* that are extremely rare. How has this skewed statistical curve come into existence?

Some of the rare species are on the edge of extinction, especially where the forest has been disturbed or cut back, but there is another, more likely explanation. Most species are specialized for a particular set of conditions within the forest. Trees of a certain kind grow best when they are touched by direct sunlight for more hours of the day (or fewer hours), when the slope in which they are rooted is well drained (or poorly drained), and when symbiotic root fungi of the needed species are available (or not). If any of these three conditions changes, the trees are likely to yield to other species. Particular ensembles of insect species flourish when logs at a certain stage of decomposition are available (for example, wood firm but soft enough to break apart by hand, with the bark still in place), and vanish when the logs rot still further (wood crumbling, the bark sloughing off from its own weight). The condition of particular rotting shifts from spot to spot.

The rain forest may look uniform through the windows of an airplane passing over, but on foot it is seen to be endlessly heterogeneous, a daunting labyrinth of transient local physical environments and overlapping species distributions. Individual species are most abundant in places that are just right for them. Their populations breed luxuriantly there, grow in size, and send out colonists in all directions. Such locations are called *source areas* for the favored species. The colonists often land in less-suitable sites, where they may survive and even reproduce for a while, but not well enough to be self-sustaining. These places are called *sink areas*. In the source-sink model of ecology, the successful populations subsidize the failing populations. If you rope off a sample area at random—one hectare, a hundred hectares, whatever—you will have within it sources for some species, which are relatively common, and sinks for others, which are relatively rare. Sources and sinks occur in all kinds of habitats, but they contribute far more to biodiversity in the tropical forests, where the environmental requirements of species have been rigidly set in the course of evolution.

A balance between source and sink was one of the key features discovered by Stephen Hubbell and Robin Foster during their superb study of tree diversity in a 50-hectare plot on Barro Colorado Island, Panama. The investigators and their diligent assistants tracked more than 238,000 trees and shrubs belonging to 303 species for a period of years. From their data Hubbell and Foster concluded:

> Many (at least one third) of the rare species (fewer than 50 total individuals) do not appear to have self-maintaining populations in the plot. Their presence appears to be the result of immigration from population centers outside the plot, and their numbers are probably kept low by a combination of unfavorable regeneration conditions, lack of appropriate habitat, or both, in the plot.

There is yet another way that abundant solar energy and the evenness of climate have contributed to the rise of biodiversity: the piggybacking of species. Benign, less variable environments permit the existence of larger life forms that cannot survive in harsher climates. Other, smaller species live on these large organisms, often in great variety. In the tropical forests, but not in the temperate deciduous and coniferous forests, woody lianas abound. They sprout on the forest floor as herbs, then send long shoots up the trunks of nearby trees and any other available vegetation. At maturity all traces of their conventional origins are gone. They are like heavy anchored ropes, reaching all the way from their roots in the ground to their branches and leaves intertwined high above with those of the supporting trees. They create a supplemental form of vegetation, a source of food and hiding places for animals that could not otherwise survive. Alongside them grows another class of vines, the stem climbers, which fasten themselves onto tree trunks by suckerlike roots. One prominent group is the arums, including the philoden-drons and monsteras. They have large heart-shaped leaves and a tolerance for deep shade, qualities that make them favorite house plants. In the rain forests stem climbers are plastered in dense clus-ters over the surfaces of the tree trunks. Their stems and roots accumulate layers of soil and decaying organic matter, home to yet another unique assemblage of tiny plants, insects, scorpions, sowbugs, and less common invertebrates. These serial forms are adapted to a way of life virtually absent in the temperate zones.

The greatest multipliers of tropical diversity, however, are the epiphytes, plants growing on trees that do not extract water or nutrients from them. Orchids make up a majority of the species of epiphytes, but they are accompanied by vast arrays of ferns, cacti, gesneriads, arums, members of the pepper family, and others, together comprising 28,000 known species in 84 families, or a little less than 10 percent of all the higher plants. These arboricolous plants turn tree limbs into Babylonian hanging gardens. Each one is a small habitat, complete with soil captured from the dust of the air and home to animals that can range from mites and roundworms to snakes and small mammals. The tank bromeliads of the American tropics hold as much as a liter of water in their stiff, upturned leaves. In these catchments live aquatic animals found nowhere else, includ-ing tadpoles of tree-dwelling frogs and the specialized larvae of mosquitoes and damselflies.

At the Monteverde Cloud Forest Reserve of Costa Rica, Nalini Nadkarni and other botanists have encountered what may be the ultimate piggyback phenomenon and the physically most complex arboreal ecosystem in the world. The epiphyte gardens on some of the heavier horizontal limbs are so profuse and tangled as to resemble a thicket of miniature woodland. Even small trees, of a kind usually found only on the ground, sprout from the dense clusters. The ecological house of cards has become a tower, an emblem of the prodigiousness of life on earth: big trees supporting orchids and other epiphytes, epiphytes supporting smaller trees with their root masses, lichens and other tiny plants growing on the leaves of the smaller trees, mites and small insects browsing among these leaf-top plants, and protozoans and bacteria living in the tissues of the insects.

Area counts in the buildup of diversity: the larger the forest or desert or ocean or any other definable habitat, the greater the number of species. As a rule of thumb, a tenfold increase in area results in a doubling of the number of species. If a forest-covered island 1,000 square kilometers in area has 50 species of butterflies, a nearby forest-covered island of 10,000 square kilometers can be expected to have about double that number, or 100 butterfly species. The reasons for this logarithmic increase are complicated, but two factors stand out. The larger island can support a greater overall population, say of butterflies, and so more rare species can be squeezed into the same forest. And the larger island is more likely to have additional habitats in which species can find refuge. It may have a central mountain, offering the beginnings of a closed zone—higher rainfall and lower temperatures for butterflies specialized for life in that climatic regime. The tropics contain vast areas of both land and shallow water that serve as theaters for the evolution of extreme diversity.

And time counts, by which I mean evolutionary time, enough time for the large piggyback organisms to evolve, symbiotic bargains to be struck, competition to be moderated, extinction rates to drop, and species thereby to assemble in respectable numbers. We have come back to climatic stability as a factor, but on a larger scale. The tropical rain forests, unlike large portions of the temperate forests and grasslands, were not obliterated by the continental glaciers of the Ice Age. They were never overridden by ice sheets or forced into new lands hundreds of kilometers from their original ranges. In the prolonged droughts that accompanied glacial cycles at higher latitudes, the lowland rain forests did retreat and were replaced by grasslands and in a few places by semideserts. The change was especially drastic in equatorial Africa. There were nevertheless abundant refuges into which species assemblages could persist more or less intact, along river courses, in regional pockets of persisting moderate rainfall, and in remnant tracts partway up the sides of cloud-enveloped mountains. Each time year-round rains came back to the equatorial river basins, the tropical rain forests expanded to carpet the earth again. The historical circumstance of interest is that the forests have persisted over broad parts of the continents since their origins as strongholds of the flowering plants 150 million years ago. Just prior to the coming of man they occupied more than 10 percent of the land surface, about 20 million square kilometers, and in earlier times much more than that. During the Eocene epoch, 60–50 million years ago, the marginal land mass that is now the British Isles was covered by forests similar in broad features to those of present-day Vietnam.

Let us put this notion of the importance of climatic stability to a test. If stability over a large area through evolutionary time is a prerequisite for high biological diversity, we should expect to find the most diversity anywhere in the world where stability prevails, not only in tropical forests. The ideal testing ground would be a very stable environment with small amounts of energy. Energy can then be discounted and the role of stability more confidently identified. The floor of the deep sea has the right geographical and historical qualifications. It covers an area of over 200 million square kilometers, has lain relatively undisturbed in most places for millions of years (no winters, no dry seasons), and is devoid of all energy except for widely scattered volcanic vents and a thin rain of organic debris from the lighted zones above. The animals are for the most part small annelid worms, starfish and other echinoderms, and clams and other bivalve mollusks. Compared with similar forms inhabiting shallow, lighted sea bottoms, they are sparse in numbers, sluggish, and long-lived. But in accord with the stable-environment hypothesis, they are extremely diverse. Their species number into the hundreds of thousands, perhaps millions. The stability part of the general biodiversity theory is thus upheld, to surprising degree.

Where are the niches on the deep-sea floor into which species can be packed? It has no forests or rivers. Large stretches of the terrain appear flat and outwardly barren, like a desert. Yet the floor is far from uniform in a biological sense. If examined millimeter by millimeter, the scale at which small animals and microorganisms live, it is seen to offer finely divided niches on which deep-sea life can specialize. Sediments pile up in small mounds, while the burrows of digging worms and bivalves create ridges and depressions. The concentration of food varies immensely from spot to spot. Almost all the energy comes drifting down in the form of dead animals and plant matter. Each piece—fish head, waterlogged fragment of driftwood, strand of seaweed—is a bonanza on which animals gather to feed and bacteria and other microscopic organisms proliferate. Their predators also assemble, and in time a tiny local community is created, often different in composition from other communities existing a few meters away. The bonanzas vary not just on a local scale but also regionally across thousands of square kilometers of ocean floor. Sections near the mouths of great rivers receive logs and tree branches carried downstream and out to sea, as well as more nutrient-rich sediments washed from the land by rain. In the horse latitudes of the North Atlantic, the benthic deeps are the cemetery of the sargassum beds, receiving dead vegetation and animals directly from that unique clearwater ecosystem high above.

In every habitat on land or in the sea, whether richly diverse or impoverished, the size of an organism exercises an important influence on the number of species in its group. Very small plants and animals are more diverse by far than very large organisms. Herbs and epiphytes exceed trees, and insects exceed vertebrates. The rule also holds within finer taxonomic divisions: among the 4,000 species of mammals found throughout the world, a thousandfold decrease in weight means (very roughly) a tenfold increase in the number of species. This translates to about ten times as many species the size of mice as species the size of deer.

The reason for the pyramid of diversity stacked by size is that small organisms can divide the environment into smaller niches than large organisms can. In 1959 the ecologists G. Evelyn Hutchinson and Robert MacArthur suggested that the number of species increases directly with a decrease in the area of the body of animals, or the square of the decrease in their weight. The reason for this rule, they proposed, is that animals living on surfaces require spaces that are the square of their body lengths. In other words, the animals move around not along a straight line, and not up and down in the three-dimensional space of midair, but over a surface, so that with each millimeter increase in their length they need a square millimeter more to find new roles, open new niches, and split into new species. Hence the more millimeters in an animal's length, the fewer species by the square of that length.

Though this mathematical exercise is beguiling, it is not very accurate. Nature is always too devious to obey simple formulations in any but a slovenly manner. To see why this is so, and to come closer to the truth, picture in your mind a large beetle 50 millimeters long, living on the side of a tree. As it walks around the trunk, browsing on lichens and fungi, it measures a trunk circumference of 5 meters.

But it cannot take account of the much smaller world at its feet. The beetle is scarcely aware of the many dips and hollows in the bark only a millimeter across. In that irregularity live other species of beetles small enough to make it home. They exist in an entirely different scale of space. To them, irregularities are not trivial. As they crawl down the sides of the crevices and up again, the circumference of the tree trunk is about ten times what it is for the giant beetle, which knows nothing of tiny crevices. The surface of the trunk is a hundred times greater for the small beetles, the square of the difference in the circumference perceived by them and that perceived by the big beetle. The disparity translates into more niches. Different crevices contain their own regimes of humidity and temperature and a variety of combinations of algae and fungi on which insects can feed. Hence small beetles have many more dwelling places and foods on which they can specialize, and as a result a correspondingly larger number of species can evolve.

Let us descend deeper into the microscopic. At the feet of the small beetles are still smaller crevices and patches of algae and fungi too narrow for them to enter. Living there, however, are the smallest of all insects together with armored oribatid mites, measuring under a millimeter in length. A close scan of the surface geometry reveals that the species of this diminutive fauna live as if the surface of the tree trunk were a hundred times or more greater than the surface embraced by beetles the next size up, and thousands of times greater than the titan beetle looming over the whole ensemble. Finally, the tiny insects and mites stand on grains of sand lodged in algal films and the rhizoids of mosses, and on a single grain of sand may grow colonies of ten or more species of bacteria.

I have dwelled on this tree-trunk microcosm to stress that in the real world, where species multiply until halted, space is not measured in ordinary Euclidean dimensions but in fractal dimensions. Size depends on the span of the measuring stick or, more

---

In the evolution of biodiversity, smaller size means more species. Within particular groups of animals, such as the insects, the smallest organisms are able to exploit more niches and thus pack more species into local communities. In the mountain rain forests of Papua–New Guinea, the large weevil (*Gymnopholus lichenifer*) carries a garden of lichens on its back, a microhabitat that supports several species of mites and springtails. At its feet, in a world of their own, are miniature anobiid beetles of an unknown species.

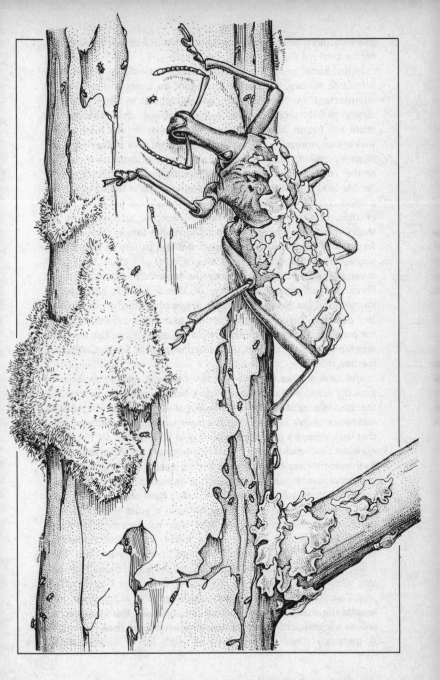

precisely, on the size and foraging ambit of the organisms dwelling on the tree.

In the fractal world, an entire ecosystem can exist in the plumage of a bird. Among the prominent organisms living in that peculiar environment are feather mites, spidery organisms apparently subsisting on oily secretions and cellular detritus. Individuals are so small and territorial that they can spend most of their lives on one part of one feather. Each species is specialized on a feather type and feather position, such as the outer quill of a primary wing feather, or the vane of a body contour feather, or the interior of a downy feather, and so on through what to feather mites is the equivalent of a forest of trees and shrubs. A single parrot species, the green conure of Mexico, is host to as many as thirty species, each with four life stages, making a total of over a hundred life forms. Each of these forms in turn has its own preferred site and pattern of behavior. A single conure harbors fifteen or more species of feather mites, with seven occupying different sites on the same individual feather. Tila Pérez of the National University of Mexico recently collected six species from the plumage of museum specimens of the extinct Carolina parakeet. If this near-microscopic fauna was indeed unique to the parakeet, which seems likely, then the mite species also vanished when the last bird died in the Santee Swamp of South Carolina in the late 1930s.

Statistical studies have shown that the most diverse animals are not only small in size but also highly mobile, giving them access to the most bountiful variety of foods and other resources. The ultimate exemplars of this principle are the insects, so diverse and abundant that they project a popular image of near invincibility. *(In the nuclear aftermath a cockroach surveys the scorched landscape atop a blasted beer can.)* Entomologists are often asked whether insects will take over if the human race extinguishes itself. This is an example of a wrong question inviting an irrelevant answer: insects have already taken over. They originated on the land nearly 400 million years ago. By Carboniferous times, 100 million years later, they had radiated into forms nearly as diverse as those existing today. They have dominated terrestrial and freshwater habitats around the world ever since. They easily survived the great extinction spasm at the end of the Paleozoic era, when life survived more than the equivalent of a total nuclear war. Today about a billion billion insects are alive at any given time around the world. At nearest order of magnitude, this amounts to a trillion kilograms of living matter, somewhat more than the weight of humanity. Their species, most of which lack a scientific name, number into the millions. The human race is a newcomer dwelling among the six-legged masses, less than two million years old, with a tenuous grip on the planet. Insects can thrive without us, but we and most other land organisms would perish without them.

Richard Southwood has explained the preeminence and hyperdiversity of insects with three words: size, metamorphosis, and wings. Size for the small niches to be defined and the many species therein generated. Metamorphosis for the transition from one life stage to another—from larva or nymph to adult—that allows the penetration of more than one habitat and the fabrication of still more niches. And wings, for dispersal to far corners of the land environment, across lakes and desert corridors, to outermost leaf tips and distant sanctuaries, putting insects within easy reach of additional food sources and places to mate and to escape from enemies. To this may be added preemption: because insects were the first to expand into all of the terrestrial niches, including the air, they were no doubt too well entrenched to be evicted by newcomers.

The human species came into the world as a late product of the radiations that, 550 million years into the Phanerozoic, lifted global biodiversity to its all-time high. In a more than biblical sense, humanity was born in the Garden of Eden and Africa was the cradle. During most of its recent geological history, from the Mesozoic era to approximately 15 million years ago, that continent was cut off from Europe to the north and Asia to the east by the Tethys Sea, a body of shallow tropical water connecting the Atlantic and Indian Oceans. As the Tethys dwindled to its remnant as the Mediterranean Sea, Africa was joined to Europe and Asia and became part of the World Continent, the loosely united biogeographical realm through which major groups of plants and animals were able to spread. Before that time Africa was an island continent similar in bulk and isolation to Australia and South America. Like those severed land masses, it developed a distinctive mammalian fauna: elephants, hyracoids, giraffes, barytheres, elephant shrews, and, not least, man-apes and the earliest true humans. Some of the groups were indigenous to Africa. Others, including the big cats and the primates, flourished all across Europe and Asia and periodically invaded Africa, where occasional lines then branched into multiple species during secondary bursts of evolution. The man-apes and early men were one of the final products of the post-Tethyean secondary radiation of the primates. They walked upright onto the stage, bearing Promethean fire—self-awareness and knowledge taken from the gods—and everything changed.

The Human Impact

The Life and Death of Species

E VERY SPECIES lives a life unique to itself, and every species dies a different way. The New Zealand mistletoe *Trilepidea adamsii* was a pretty plant with pale green glabrous leaves, red tubular flowers tinged with yellowish green, and bright-red ellipsoidal fruits. It disappeared from its last stronghold on North Island in 1954. The species grew as a parasite on shrubs and small trees in the understory of native forest. Never common, it was limited at the time of the first European botanical explorations to a few localities in the northern peninsula, around Auckland.

Trilepidea adamsii came to an end by a combination of circumstances that no one could have foreseen a hundred years ago. Its habitat was reduced by deforestation, first by the original Maoris during a thousand years of occupation and then at an accelerating rate by British settlers in the late nineteenth century. Placed at risk, the population was reduced still further by collectors eager to secure specimens of what was recognized as a rare and desirable plant. The dispersal of the mistletoe dropped still more as bird populations declined in the area, depressed by clearing of their forest habitats and predation from introduced mammals. Birds are necessary for the transport of the seed from one host tree or shrub to another. By the early 1950s *Trilepidea adamsii* was close to extinction. The nature of its final days are unknown. The last few plants may have been eaten by brush-tailed possums, a species of arboreal browsing mammal deliberately introduced from Australia during the

1860s to establish a fur trade. The possums were never abundant enough to destroy the mistletoe while it flourished, but they could have tipped it into extinction at the very end.

Consider this familiar paradox of biological diversity: almost all the species that ever lived are extinct, and yet more are alive today than at any time in the past. The solution of the paradox is simple. The life and death of species have been spread across more than three billion years. If most species last an average of, say, a million years, then it follows that most have expired across that vast stretch of geological time, in the same sense that all the people who ever lived during the past 10,000 years are dead though the human population is larger than it has ever been. The turnover would have been even greater if the grand pattern were dynastic, with one species giving rise to many species, most or all of which yielded to later ascendant groups.

Evolution is indeed dynastic, and million-year longevities are close to the mark for many kinds of organisms. The precise measure of interest is not the longevity of species but of the clade, composed of the species and all its descendants, taken from the time the ancestral species first splits off from other species to the moment the last organism belonging to that species and all of its descendants disappears. Chronospecies extinction, or pseudo-extinction as it is also called, does not count. If a population of organisms evolves so much that biologists declare it to be a new species, or a chronospecies, the species did not go extinct; it just changed a great deal. The life of the clade goes on, and that particular lineage of genes endures.

Each major group of organisms appears to have a characteristic clade longevity. Because of the relative richness of fossils in shallow marine deposits, the duration of fish and invertebrate clades living there can often be determined with a modest degree of confidence. During Paleozoic and Mesozoic times, the average persistence of most lay between 1 and 10 million years—for example, 6 million years for starfish and other echinoderms, 1.9 million years for graptolites (colonial animals distantly related to the vertebrates), and 1.2 to 2 million years for ammonites (shelled mollusks resembling modern nautiluses). On the land, the longevities of clades of flowering plants during Cenozoic times also appear to fall within the range of 1 to 10 million years. Those of mammals vary from .5 million to 5 million years, depending on the geological epoch.

The probability of extinction of species within clades is more or less constant through time. As a result the frequency of species in a

The extinct New Zealand mistletoe (*Trilepidea adamsii*).

clade surviving to a greater and greater age falls off as an exponential decay function. If, to use an oversimplified example, one half the species are alive at the end of a million years, about one half of those (or one quarter of the original) persist 2 million years, one half of those again (or one eighth of the original) last 3 million years, and so on. The progression is often accelerated by shifts in climate that cause waves of extinction and subsequent rebirth—not only the great catastrophes that ended the Paleozoic and Mesozoic eras but smaller, more frequent, and more local events. Clades of buffalos and antelopes in Africa south of the Sahara have persisted from 100,000 to several million years. But about 2.5 million years ago many came to an end, and others first appeared almost simultaneously. The controlling event was apparently a period of cooling and diminished rainfall that caused grasslands to spread over a large part of the African continent.

Local climatic instability is only one of the reasons not to generalize too quickly about the lifespan of species from the fossil record. Sibling species, so similar in anatomical detail as not to be traceable in fossils, could come and go in rapid succession without being detected. Small local species might also turn over at a high rate in places where fossilization seldom occurs, such as desert valleys and the interiors of small islands, leaving no evidence whatever of their existence.

We know that contemporary species formation in the northern Andean cloud forests is both profuse and resistant to the formation of fossils. In the mountain habitats of Colombia, Ecuador, and Peru, populations of plants and animals are prone to fast evolution and early extinction by geographical location alone. The ridges on which they live are isolated and differ from one another in temperature, rainfall, and the species composing local communities. The populations are small. Alwyn Gentry and Calaway Dodson estimate that in these places some orchid species could multiply in only fifteen years. By implication, the longevity of species might be short as well, measured in decades or centuries. Orchids are by far the most diverse of living plants, comprising at least 17,000 species or 8 percent of all flowering plants. Many are rare and local like the Andean endemics, and they could originate and die at a high rate without leaving a trace. The general biology of orchids is also such as to erase their history. They live mostly in the tropics, where the fossil record is poor. Most grow as epiphytes in the crowns of forest trees, a habitat not conducive to the fossilization of plant parts. And unlike the vast majority of other flowering plants, they do not scatter their pollen as simple grains, letting much of it fall into lakes and streams where it can form easily studied microfossils. Instead, orchid plants bind pollen together in solid bodies, the pollinia, which are carried from flower to flower by insects. These two traits taken together, rapid speciation and the difficulty of fossilization, mean that orchid floras leave almost no record by which we can hope to measure the longevity of species.

The orchids are not alone. They merely instruct us that, in addition to the species whose fossils suggest a longevity of 1 to 10 million years, there is a large hidden group of species that appear and disappear at a far higher rate. New species occupy small ranges on the average and are often started by small numbers of pioneers that land on island shores or distant mountain ridges. If extinction of such young, vulnerable populations were high, the equivalent of infant mortality among organisms struggling in a harsh environment, a large percentage of species would die young with no record of their existence. The birth and death of most species may therefore lie behind a veil of artifact. Only the more widespread populations in or near bodies of water are fossilized with consistency for direct measurement. Behind the veil lie vast numbers of species that once lived in restricted habitats and are forever beyond direct access.

To pull the veil back, to visualize how rare species live and die, we must take a less direct approach by returning to the principles of ecology and to natural history, which is ecology expressed in the details of the biology of individual species that still live or have recently perished. Consider first the laws of ecology. They are written in the equations of demography. The number of plants or animals belonging to a particular species is exactly determined by the rate at which new individuals are born, the age at which they reproduce, and the age at which they die. The distribution of the population by age (how many newborn, how many juvenile, how many young and old adults) is set by these schedules of birth and death. The schedules themselves are influenced by the size of the population or, more precisely, by its density. The number of birds crowded into a woodlot or the number of algal cells living on a wet stone affects food supply, how heavily predators and disease pathogens strike, to what degree reproduction is delayed, how long individuals live, which competitors can force themselves into the same community. All this has an important consequence: if ecology is ultimately a matter of demography, then demography eventually must turn into natural history, with parameters expressed as a function of particular time and place. The equations of demography are specified by context.

So it is when we move on up to the life and death of particular species. The laws of biological diversity are written in the equations of speciation and extinction. Ecologists and paleontologists have begun to pursue these laws, aware of the importance of data on the birth rates of species and the longevity of the clades they spawn. The equations are beginning to resemble those of ecology, and they too are being fleshed out by the details of natural history.

Consider a newly formed island in the sea, devoid of life, say Krakatau in 1883, Surtsey off Iceland in 1963, or Kauai 5 million years ago. Plants and animals soon arrive, showering down as aerial plankton or blown ashore by storms. Next focus on a particular group, say landbirds, reptiles, or grasses. At first the rate of arrival of new species in the group is relatively high, but it inevitably drops because the strong dispersers become established early. On islands and continents nearby, there are other species that can cross the sea, but they compose a less able pool of potential colonists. As the island fills up with more and more species of the group of interest—birds, reptiles, grasses—the rate of arrival of species not already established keeps dropping. It might start with an average of one new species a year and decline during the next century to one every ten years. At the same time, the rate of extinction rises as more and more species contend for the available space and resources.

In time the rate of extinction of species already on the island, say in species per year, will just about equal the rate of immigration of new species onto the island, again in species per year. The number of species is in a dynamic equilibrium. New species are arriving, old species are disappearing, the composition of the fauna and flora is constantly changing, but the number of species present at any moment in time stays the same.

This very simple model of a balance between immigration and extinction is the basis of the theory of island biogeography that Robert MacArthur and I developed in 1963. We had noticed that faunas and floras of islands around the world show a consistent relation between the area of the islands and the number of species living on them. The larger the area, the more the species. Cuba has many more kinds of birds, reptiles, plants, and other organisms than does Jamaica, which has a larger fauna and flora in turn than Antigua. The relation showed up almost everywhere, from the British Isles to the West Indies, Galápagos, Hawaii, and archipelagoes of Indonesia and the western Pacific, and it followed a consistent arithmetical rule: the number of species (birds, reptiles, grasses) approximately doubles with every tenfold increase in area. Take an actual case, the landbirds of the world. There is an average of about 50 species on islands of 1,000 square kilometers, and about twice that many, 100 species, on islands of 10,000 square kilometers. In more exact language, the number of species increases by the area-species equation $S = CA^z$, where A is the area and S is the number of species. C is a constant and z is a second, biologically interesting constant that depends on the group of organisms (birds, reptiles, grasses). The value of z also depends on whether the archipelago is close to source areas, as in the case of the Indonesian islands, or very remote, as with Hawaii and other archipelagoes of the eastern Pacific. In short, z is a parameter. It holds constant for a given group of organisms and set of islands, such as the birds of the West Indies, but can change when we proceed to other organisms on other islands, such as the grasses of Indonesia. It ranges among faunas and floras around the world from about 0.15 to 0.35. To give the rule of thumb that a tenfold increase in area doubles the fauna and flora is the same as saying that $z = 0.30$, or $\log_{10} 2$. Notice that we can if we wish, and this is very important for conservation, state the rule in reverse: a tenfold *decrease* in area cuts the number of species in half.

The rise of biodiversity with the size of islands is called the *area effect*, and it follows from the equilibrium model in a straightforward way. Think of a row of newly emerged islands along the edge of a continent, all located an equal distance from the shore of that larger body of land but varying in size. As they fill up with species, the fringing islands will all have about the same immigration rate—the same number of new species arriving each year—since they are all equally far from the continent. On the other hand, the extinction rates will rise more slowly on the larger islands. The reason is that more area means more space, more space means larger populations for each species, and finally larger populations mean an expectation of longer life for the species. You are less likely to go completely broke if you are rich at the start, and more people can be crowded onto large tracts of land before they become poor. So the overall

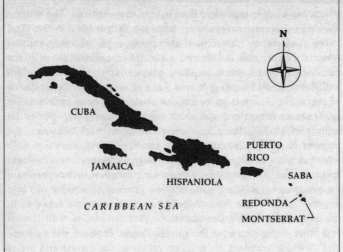

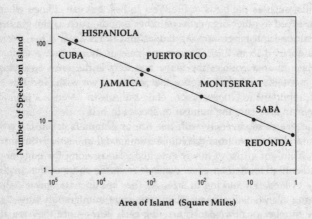

The number of species living on islands increases or decreases with the area of the island. The diversity of reptiles and amphibians in the West Indies, depicted here, is typical: a reduction of 90 percent in area from one island to the next results in a 50 percent loss of species.

extinction rate comes to equal the immigration rate on a large island only after many species have colonized the island, and larger islands have more species at equilibrium than smaller islands do.

The *distance effect* is this: the farther the island from continents and other islands, the fewer the species living on it. Like the area effect, this trend in biogeography can be explained by the basic equilibrium model in a straightforward way. Merely reverse the arrangement of islands, so that now all are the same size but located at varying distances from the continent. As they fill up with species of birds, reptiles, or grasses, the extinction rate on all the islands increases at about the same rate (because they are the same size). But the distant islands fill up more slowly; organisms have farther to travel, and their immigration rate (new species arriving each year) is lower. The extinction rate comes to equal the immigration rate when fewer species are present. Distant islands thus end up at equilibrium with fewer species than nearby islands.

Still theory, even if airtight and plausible, is not enough to put the seal on such complex ecological processes as the growth of species numbers. There must be experiments to confirm the predictions of theory and, in the case of ecology particularly, to expose them to the exhilarating intrusions of natural history. But how can experiments be performed on archipelagoes and entire faunas and floras?

The answer is to miniaturize. In the early 1960s I spent a great deal of time poring over maps of the United States, daydreaming, searching for little islands that might be visited often and somehow manipulated to test the models of island biogeography. I thought a lot about insects, creatures small enough to maintain large populations in tight places. A full-blown bird or mammal fauna might require an island the size of Guernsey or Martha's Vineyard, but aphids and bark beetles can flourish in large numbers on a single tree. I finally settled on the Florida Keys, especially the tiny red mangrove islets that dot the shallow waters of Florida Bay immediately westward. Clumps of the salt-tolerant mangroves vary from a single tree to forests hundreds of hectares in extent. They are present in huge numbers, from the Ten Thousand Islands group of the upper bay to abundant miniature archipelagoes strung along the northern rim of the Lower Keys.

In 1966 Daniel Simberloff, then a graduate student at Harvard University and now a distinguished professor of ecology at Florida State University, joined me in an attempt to turn the mangrove islets into an outdoor laboratory. We needed a series of little Krakataus, islands that could be completely cleared of insects, spiders, and other arthropods and then monitored month by month. To accomplish that much would be to witness recolonization from scratch and to learn unambiguously whether biodiversity is at equilibrium. With the permission of the National Park Service, we selected four tiny islands for our experiment, all clumps of red mangroves about 15 meters across. In order to test the distance effect, we chose one islet only 2 meters from a large island, another 533 meters away, and two more located in between. The 2 meters may seem trivial, being no greater than the height of a guard on the Boston Celtics, but it is the length of 1,000 worker ants which, translated into human terms, is about a mile. We set to work. Crawling over each of the islets from mud bottom to the tops of the trees, examining every millimeter of leaf and bark surface, probing every crack and seam, photographing and collecting, we made as complete a list as possible of the insect and other arthropod species on the four islets. At this point the distance effect was manifest. The nearest islet held the most species, the most distant islet the fewest, and the two intermediate islets numbers in between.

Next we hired a pest-control company in Miami to destroy all the arthropods on the islands, using a method ordinarily employed to fumigate entire buildings. The workmen first covered the islands with a rubberized nylon tent. Then they fumigated the interior with methyl bromide gas at a concentration and at a preselected period of time that was fatal to arthropods but not to mangrove trees. When the tents were removed, we had four empty islands, four little Krakataus.

The recolonization began within days. In less than a year, the faunas had reattained their original levels on the islands. They lined up once again in accordance with the distance effect: from 43 species back to 44 species on the near islet, from 25 back to 22 species on the distant islet, and an equally close return on the middle islets to the numbers prevailing earlier. The numbers held remarkably constant to the end of the second year, when the experiment was discontinued. The equilibrium was also dynamic, with many of the arthropod species colonizing a given islet, vanishing after a month or two, then making a second appearance or else giving way to one or two similar species. The fauna tracked through time was kaleidoscopic, with the total numbers more or less balanced but the composition changing constantly, like travelers in an air terminal.

Depending on the island, only between 7 and 28 percent of the species of the new colonists were the same halfway through the experiment as those present before the fumigation.

The Florida Keys experiment yielded new information on the abilities of different groups of small organisms to immigrate and endure. Spiders rained down on the islets, some of them large, no doubt ballooning over the water on silken threads. But many of the species went quickly extinct. Their distant relatives, the mites, were slower to arrive, blown in air currents like random particles of dust—literally part of the dust—but their individual species persisted longer. Cockroaches, crickets, moths, and ants arrived early and colonized solidly. Centipedes and millipedes, although well established before the fumigation, never made it back during the two years we visited the islets.

The mangrove experiment was inspired by Krakatau and the scientific interest of a land swept clean of all animals. A second method of assessing the balance of diversity would be to reduce the size of islands and observe the decline in numbers of species from a higher equilibrium to a lower one. In the late 1970s Thomas Lovejoy adopted this approach in what was to become the largest biological experiment in history. He took advantage of a Brazilian law that required owners of rain-forest tracts in the Amazon region to leave at least 50 percent of their land covered in forest; the rest they were free to convert into pastureland and farms. With the support of the World Wildlife Fund and the Brazilian government, Lovejoy set out to observe the fate of diversity in patches of forest left behind as the clearing proceeded. He persuaded owners of land along the Boa Vista road north of Manaus to leave square patches of forests varying from 1 hectare to 1,000 hectares in size. A fellow ornithologist, Richard Bierregaard, joined him as field director, and other experts were invited in as guest investigators to move the huge project along. The biologists set out to survey diversity at the experimental sites while the plots were in pristine condition and, later, after they had been turned into islands by the forest clearing. (One of these tracts was Fazenda Dimona, where I sat to watch a storm.)

The enterprise was first called the Minimum Critical Size of Ecosystems Project, MCS for short, because its ultimate goal was to determine the smallest size a rain-forest reserve must be in order to sustain the plant and animal species native to the immediate vicinity. How much land is needed, say, to sustain 99 percent of all the original species for a hundred years? Later the study became part of the Biological Dynamics of Forest Fragments Project, meant eventually to cover habitats all over Brazil. Staff members refer to it as the Forest Fragments Project for short, and many Brazilians call it Projeto Lovejoy. The monitoring near Manaus was begun just before the land was clear-cut in the late 1970s, and it is expected to carry over into the next century.

A mountain of data must be sifted from the Manaus experiment, but even in the first decade, 1979–1989, useful new facts have emerged. The diversity of the smaller "islands" is decreasing the most rapidly, as expected. The extinction of species has been accelerated by the unexpectedly deep penetration of daytime winds, which dry out the forest from the edge in and kill deep-forest trees and shrubs for distances inward of 100 meters or more. Many plant and animal species have disappeared from the smaller plots, but a few others are increasing their numbers. The reasons for the changes are sometimes obvious, but just as often baffling. Army-ant colonies, which require more than 10 hectares to maintain their worker force, quickly disappeared from the 1- and 10-hectare plots. With them went five species of ant birds that make their living by following ant swarms and feeding on the insects driven forward by the 10-meter-wide raiding front. Shade-loving butterflies of the deep forest declined quickly from the wind-drying effect, but other species specialized to live around forest edges and second growth flourished. Large, metallic green and blue euglossine bees, which are premier pollinators of orchids and other plants, were hit hard in the plots of up to 100 hectares. Saki monkeys, which eat fruit, dropped out of the 10-hectare plots. But red howler monkeys, which are leaf eaters and hence able to harvest more food, stayed on. Larger ground-dwelling mammals, including margay cats, jaguars, pumas, pacas, and peccaries, simply walked away from the smaller plots and out of the fauna altogether.

By the late 1980s second-order effects could be seen spreading through the food web. With the peccaries gone, there were no wallows in which temporary forest pools could form. Without the pools, three species of *Phyllomedusa* frogs failed to breed and disappeared. As the mammal and bird populations declined, dung and carrion became scarcer. Scarab beetles that feed on these materials dropped in numbers of species and individuals. The average size of the surviving beetles grew smaller. Bert Klein, who documented these changes, prophesied still more reverberations through a broad segment of the animal community, including carnivorous mites that ride on the beetles and attack fly maggots, with still deeper effects on the disease organisms of mammals and birds:

> A first-order disruption in abundance of some scarabs will undoubtedly result in second-order changes in mite dispersal, which may trigger third-order changes in populations of dung- and carrion-breeding flies. What fourth-order changes could occur because of changes in fly abundance needs further study. By eating and burying dung and carrion, the Scarabaeinae kill nematode larvae and other gastrointestinal parasites of vertebrates. Thus, a change in dung beetle communities may alter the incidence of parasites and disease in some isolated forest fragments or biological reserves.

Through perturbations spreading to third rings of interaction among the species and possibly beyond, the diversity of the smaller forest plots spirals downward to new and still unpredictable levels. We know at least this much: an Amazon forest chopped into many small fragments will become no more than a skeleton of its former self.

Theory confirms common sense with the following theorem: the smaller the average population size of a given species through time, and the more the size fluctuates from generation to generation, the sooner will the population drift all the way down to zero and go extinct. Think of an island with a thousand sparrows on average, varying by chance alone by a hundred individuals either way once or twice a century. Another island holds a hundred sparrows of the same species, and this population also varies by a hundred individuals once or twice in a century. The second population, which is both smaller and experiences a higher degree of fluctuation, faces a shorter life. More precisely, many such populations go extinct sooner than many otherwise comparable larger populations.

The prediction has been proved correct in a meticulous study of one hundred species of landbirds on small islands off the English and Irish coasts conducted by Stuart Pimm, Lee Jones, and Jared Diamond. They found that the lifespan of local island populations of birds does indeed grow shorter as the population size decreases. It also drops when the population fluctuates more widely over time.

In order to see the importance of population size in a broader perspective, imagine that we could protect a local population from catastrophic destruction. The habitat is kept intact, a steady food source is ensured, and no devastating diseases or predators are allowed to sweep the area. The fluctuations in the number of individuals in the population are then based on pure chance in the events of birth and death—how many females are mated that year, how many young survive infancy, and so on. Chance itself is the summed outcome of many other, largely unpredictable events in rainfall, temperature, food supply, and enemy assault. Mathematical models of the history of such steady-state populations reveal that population size and fluctuations can have enormous effects on longevity. A tenfold rise in average population size, say from ten to a hundred individuals, might increase average longevity thousands of times. Expressed in more practical terms, there is threshold below which the population is in eminent danger of extinction from one year to the next. The bright side is that endangered species can often be rescued from this red zone by a relatively modest increase in habitat area and hence average population size.

Because extinction is forever, rare species are the focus of conservation biology. Specialists in this young scientific discipline conduct their studies with the same sense of immediacy as doctors in an emergency ward. They look for quick diagnoses and procedures that can prolong the life of species until more leisurely remedial work is possible. They understand that the populations of a species can be habitually small and frequently vanishing, but if others are born through colonization of new sites at the same rate, and if there are many such populations, then the species as a whole is in no particular danger. Rareness, therefore, requires a many-layered definition in order to be addressed realistically. The essential idea can be expressed by diagnosing three of the most endangered species of birds in North America.

Bachman's warbler. A species is endangered if it occurs over a wide area but is scarce throughout its range. Such is the case of Bachman's warbler (*Vermivora bachmanii*), which is the rarest bird in North America in numbers of individuals per square kilometer of its geographical range. Small, yellow-breasted, olive-green on the back, and black-throated in the male, the warbler once bred in thicket-grown river swamps from Arkansas to South Carolina. Its present breeding range and population size are unknown, and it appears to be close to extinction if not already lost.

Kirtland's warbler. A species is rare if it is densely concentrated but limited to a few small populations restricted to tiny ranges. Kirtland's warbler *(Dendroica kirtlandii)*, with lemon-yellow breast, bluish-gray back streaked with black, and dark mask in the male, is such a case. It is loosely colonial, with a breeding range restricted to jack-pine country in the north-central part of the lower peninsula of Michigan. Between 1961 and 1971 the known population plunged from 1,000 to 400 birds. The decline was apparently due to increased nest parasitism by brown-headed cowbirds *(Molothrus ater)*, which place their eggs in the warbler's nest. Kirtland's warblers are as dense as ever in the localities where they occur, but the progressive restriction of their range has brought them close to extinction.

Red-cockaded woodpecker. A species can be rare even if it has a broad range and is locally numerous, but is specialized to occupy a scarce niche. The red-cockaded woodpecker *(Picoides borealis)*, with zebra back, white breast speckled with black, and each white cheek touched by a carmine speck, is the outstanding example. It ranges across most of the southeastern United States but requires pine forests at least eighty years old. The birds live in small societies composed of a breeding pair and up to several offspring, with the latter helping their parents to protect and rear the younger siblings. Each group requires an average of 86 hectares of woodland to produce an adequate harvest of insect prey. To nest, red-cockaded woodpeckers hollow out cavities in living, mature longleaf pines eighty to one hundred and twenty years old, in which the heartwood has already been destroyed by fungus. These exacting conditions are no longer easy to find in the piney woods of the south. The total size of the woodpecker breeding population was estimated in 1986 to be only 6,000. It was falling steadily, by as much as 10 percent a year in Texas and probably just as fast elsewhere. The species appears doomed unless the cutting of the oldest pine forests is stopped immediately.

Species trapped by specialization and pressed by shrinking habitat form the largest endangered class. The scarcity of Bachman's warbler across the southern United States is no mystery, despite the abundance of riverine swampland in which it can breed. It winters (or wintered) exclusively in the forests of western Cuba and the nearby Isle of Pines, where virtually all the forests have been cleared to grow sugar cane. The bottleneck is the loss of wintering ground and starvation for even the remnant of warblers produced in the lusher summer environment of the United States.

John Terborgh has given a poignant account of his own experience with one of the last Bachman's warblers. In May 1954, as an eighteen-year-old birder (now a foremost ornithologist), he learned of the sighting of a male Bachman's on Pohick Creek in Virginia not far from his home. The song of the Bachman's had been described to him as resembling that of a black-throated green warbler with a downward sweep at the end: *zee-zee-zee-zee-tsew*.

> To my astonishment I walked up to the place that had been described to me and heard it! I had no trouble seeing the bird. A full-plumaged male, it sat on an open branch about 20 feet up and gave me a perfect view while it sang. It hardly stopped singing during the two hours I spent there. Reluctantly, I pulled myself away, wondering whether this was an experience I would ever repeat. It was not.

As other birders were to testify, the male returned to the same spot the next two springs. No female ever joined him. The extraordinary exertions of the Bachman's male were a sign that he was in prime breeding condition, but he was destined to go undiscovered by any female of the same species.

> I imagine that each spring a tiny remnant of birds crossed the Gulf of Mexico and fanned out into a huge area in the Southeast, where they became, so to speak, needles in a haystack. Toward the end, it is likely that most of the males in the population, like the one at Pohick Creek, were never discovered by females. Once this situation developed, there could have been no possible salvation for the species in the wild.

In parallel manner, Kirtland's warbler winters in the pine woodland of two islands in the northern Bahamas, Grand Bahama and Abaco. Terborgh has written that, however zealously the Kirtland's warbler and its habitat may be protected in Michigan, its fate probably lies at the mercy of commercial interests in the Bahamas. Migratory birds as a whole are declining across the United States from the same environmental malady that afflicts the warblers: wintering grounds are being demolished by logging and burning. The prospects are especially grim for species that depend on the rapidly shrinking forests of Mexico, Central America and the West Indies.

Earlier I spoke of specialization, that tender trap of evolutionary opportunism, and how it is affected by natural selection at the species level. A rich resource appears, and a species adapts to use and hold it against all competitors. To keep the edge, the members of the species surrender their ability to compete for other resources. Driven by natural selection, the advantage gained by those members one generation at a time, the species shrinks inside a smaller range. It is then more vulnerable to environmental change. Individual organisms

The rarest songbird: Bachman's warbler of the southeastern United States is on the brink of extinction, if not already gone. This drawing of a singing male is based on one of the last photographs taken.

bearing the specialist genes have triumphed, but in the end the species as a whole will lose the struggle and all its organisms will die. During the Paleozoic era, an entire family of snails, the platycerids, flourished by attaching themselves to the anuses of crinoids, a group of echinoderms called sea lilies. They fed on their hosts' fecal matter, which they were able to appropriate directly and with minimal competition. When the crinoids became extinct, so did all the multitude of ingenious platycerids.

On high, crumbling bluffs above western Florida's Apalachicola River grow the last wild trees of the Florida torreya or stinkingcedar (*Torreya taxifolia*), a small understory conifer. Among them are found relics of a cooler climate, when the advance of the last ice sheet forced boreal elements southward to the southeastern United States. When the glacier retreated 10,000 years ago, most of the plant and animal species spread back, eventually to most of their former wide distributions. The torreya could not expand, in part because of its dependence on rich, moist soils of limestone origin. In the late 1950s a fungus disease struck the small Apalachicola population and brought the species close to extinction.

The streams of the Apalachicola system around the dying torreya stands are occupied by small populations of Barbour's map turtle, a handsome species with large sawteeth running the length of the carapace midline and with curlicues decorating the ventral rim of the shell. The female bears the most unusual feature. She is much larger than the male and has a grotesquely enlarged head. Evolved in only this one river system, the species has not spread beyond, and it is now vulnerable to extinction as the freshwater environments of Florida are increasingly disturbed.

Hidden in the muck-bottomed outflows of springheads of the same region live one-toed amphiumas, dwarf members of a genus of giant salamanders. I visited the habitat of this rare and possibly threatened species on the same day I examined the last stands of the torreyas. Walking with another naturalist in the broiling sun through a turkey-oak flat, one of the most unpromising environments in the eastern United States, we found the springhead I sought, a small, narrow gorge 20 meters deep. It was like an oasis, its walls covered by thick broad-leaved woodland and its interior mercifully cool. A thin stream meandered across the flat muddy bottom. This is where the shy one-toed amphiumas live. They prey on a breed of equally unusual aquatic worms, also limited to this habitat. We found the worms but did not stay to locate the salamanders, because even in daytime the mosquitoes were so ferocious that the oak flat seemed bearable after all.

A small geographical range like those of the Apalachicola endemics carries an extra risk: a single sweep of disease (called *epizootics* to distinguish them from human epidemics), a forest fire, a deep freeze, or a day's work with chain saws can carry the species away. Specialization is perilous even for widespread species, whose local populations, however numerous and far-flung, are individually more likely to dip to extinction, until all happen to vanish.

The fossil record ascribes to this general principle. Recently I studied ants preserved in amber from the Dominican Republic, early Miocene in age, about 20 million years old. Abundant amber, the fossilized gum of trees, is one of the treasures of this Caribbean country. Columbus acquired pieces by trade there during his second

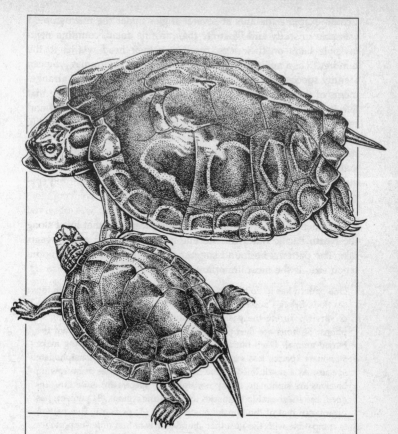

Barbour's map turtle is a threatened species limited to the Apalachicola River system of the Florida panhandle and adjacent portions of Alabama and Georgia. The female of the species is much larger than the male and in addition possesses an outsized head.

voyage in 1493–94, from a mining region still active near present-day Santiago. Ants are among the most abundant insects in the clear golden matrix, many as exquisitely preserved as if they had been set in tinted glass by a master jeweler. From brokers I acquired a total of 1,254 pieces containing specimens. I cut and polished them until

I could examine the ants at several angles under the microscope. I was able to study and illustrate them in fine detail, counting near-invisible hairs on their legs, measuring their head widths to the hundredth of a millimeter, writing out their dental formulas (you can identify species and even individual ants by the shapes and arrangements of their teeth). Comparing the amber species with those that live in tropical America today, I classified some as specialized and relatively rare, because their closest living relatives either take only certain types of prey, such as millipedes or arthropod eggs, or else nest in unusual places. I found that more of the specialized species and their descendants became extinct in the Dominican Republic and elsewhere in the West Indies than was the case for generalized species.

The same trend, of abundance favoring survival, was observed independently by Steven Stanley in mollusk species that lived along the North Pacific rim in Pleistocene times, about two million years ago. The patterns he found suggest that abundance, or total population size, is the most important control on survival.

> One pattern has to do with mode of life for clams, which burrow in the sea floor. Species possessing siphons have enjoyed a much greater rate of survival during the past two million years than species lacking siphons. Siphons are fleshy tubes that channel water to and from the buried animal. Deep burial and a capacity for rapid burrowing make siphonate species less vulnerable to predators than are nonsiphonate species. As a result, most of the highly abundant species of burrowing bivalves are siphonate. Many nonsiphonate species are quite rare. Indeed, the survivorship of species in the Pacific regions (84 percent) has been twice that of the nonsiphonate species (42 percent). This pattern is compatible with the idea that abundance is of first order importance in determining probability of extinction.

The same principle applies widely through the animal kingdom: large body size, like specialization, means smaller populations and earlier extinction. The large mammals of North America and Eurasia were the first to yield to the invasions of human hunters. Wolves, lions, bear, bison, elk, and ibex largely disappeared. Foxes, raccoons, squirrels, rabbits, mice, and voles flourished. In their analysis of the resident landbirds of the British coastal islands, Stuart Pimm and his colleagues found that large-bodied species such as hawks and crows have become locally extinct more frequently than smaller ones such as wrens and sparrows. The greater vulnerability of the larger birds is due in part to their smaller populations, but not entirely. Even when population size is factored out (only populations of the same size are considered, regardless of species), the vulnerability remains. This added weakness of large birds apparently stems from their lower reproductive rates. Hawks and crows raise fewer young than wrens and sparrows. When struck once by high mortality, they are slower to recover; when struck again, they are more likely to fall all the way to extinction. The disadvantage is reversed when populations of both large and small birds are so small as to be on the verge of extinction, when (to be exact) they comprise seven or fewer breeding pairs. Then the greater longevity of the individual larger birds becomes the deciding factor. Hawks live longer than sparrows and are less likely to die off entirely before any one pair can raise offspring to maturity.

As populations decrease to a few individuals, they flirt with extinction through what geneticists term *inbreeding depression*. Imagine the extreme case of a population of birds, say an ill-fated warbler species, whittled down to a single mated pair, brother and sister. They are both heterozygous for a recessive lethal gene. This means that each bird carries a lethal gene on one of its chromosomes and a normal gene at the same site on the corresponding chromosome. The normal gene prevails over the lethal gene, and the birds stay more or less healthy. If they were homozygous for the lethal gene, possessing two instead of one, they would be dead. Brother and sister mate. There is half a chance that any given sperm carries a lethal gene, and half a chance that any given egg carries the same gene: each outcome has the same probability as the flip of a coin coming up tails. The chance that any given offspring gets two lethal genes and dies is the same as throwing two tails with a double flip of a coin: one half (bad sperm) times one half (bad egg) equals one fourth (afflicted offspring). The population gives up a quarter of its reproductive potential for being so small.

Why does inbreeding, as opposed to ordinary haphazard mating with individuals other than close relatives, cause depression of life and reproduction? Siblings, first cousins, and parents and their offspring are so closely related that the lethal recessive genes they carry are likely to be the same. Each human being and each fruit fly, typical organisms in this respect, carries an average of one to several lethal recessive genes. But there are many such genes in the population as a whole, and each one occurs in only one out of hundreds or even thousands of individuals. The chances that two unrelated individuals will carry the same defective gene are very small, even though both carry one kind of defective gene somewhere on their chromosomes. The odds against matching the same gene are so great in human beings that deadly hereditary conditions, such as Tay Sachs disease and cystic fibrosis, are mercifully rare. The chances that one or the other such syndrome will appear is greatly elevated, however, if the child's parents are closely related, and that juxtaposition is more likely to occur if the population is small and closed.

Such is the basic concept of inbreeding depression. But real populations subscribe to it in idiosyncratic and subtle ways. Only a small fraction of deleterious genes are lethal. Most are "sublethal" or "subvital." To varying degrees they interfere with development, reduce strength, and diminish fertility. These are the genes that shorten life and cause sterility in cheetahs and gazelles kept in zoos, and they are the genes that inflict congenital heart defects on cocker spaniels bred too pure.

Conservation biologists have tried to draw danger lines below which a species is at conspicuously higher risk of extinction from genetic depression. They speak loosely of a 50–500 rule of genetic health in populations. When the effective population size falls below 50 and defective genes are present, inbreeding depression becomes common enough to slow population growth. Breeders of domestic animals generally do not worry about the amount of inbreeding depression encountered in populations with an effective size of 50 individual animals or more. But they consider themselves in trouble when the number falls below 50. When the effective population size is below 500, genetic drift (the chance fluctuation of gene percentages) is strong enough to eliminate some genes and reduce the variability of the population as a whole. At the same time, the mutation rate is not high enough to replenish this loss. So the species steadily loses its ability to adapt to changes in the environment. Inbreeding depression, turning the screw generation by generation, shortens the longevity of species. So does the shrinking of genetic reserves over many generations. To express this as concisely as possible: a population of 50 or more is adequate for the short term only, and one of 500 is needed to keep the species alive and healthy into the distant future.

I have used the phrase "effective population size" to make sense of genetic deterioration. It is a measure of considerable importance in the theory of conservation biology. Think of an actual population such as the tree sparrows on Scotland's Isle of May. It could consist of all males, and the effective size would be zero. Or it could be a thousand adults too old to breed plus five healthy females and five healthy males randomly mated; in that case the effective size would be ten. The effective size of a population refers to an idealized population, with random mating of individuals, possessing the same amount of genetic drift as the actual population. There are, in the last imaginary case, 1,010 tree sparrows but the thousand postreproductive individuals don't count. All those tree sparrows are genetically the same as a population of ten birds living by themselves. The effective size declines as sterility rises, by aging or any other cause. It also declines to the extent that the adults forsake random mating and turn to relatives in order to breed. The point is that the age, health, and breeding patterns of individuals have an important effect on the genetic trajectory of a population and eventually its very survival. Even if the woods and fields are swarming with plants and animals of a certain kind, the species might be destined for extinction.

Conservation biologists and geneticists understand such matters in a general way. They have built a loose framework of theory clothed by a smattering of laboratory and zoo studies. They have learned that if the depression of fitness emanates from close inbreeding and the rapid juxtaposition of deleterious genes already present at high levels, the population is in immediate peril. But if inbreeding is gradual, the population has a better chance to pass through the bottleneck. Furthermore, as generations pass, the inbreeding depression will moderate, since natural selection purges the deleterious genes from the population. As the most harmful genes attain the homozygous (double-dose) state in individuals, they are eliminated and their frequency drops throughout the population.

The ruling consideration in the death of a species, however, is not in most cases the bite taken from the population by its defective genes. More important is the size of the population and the manner by which it subdivides and spreads across the terrain. It is risky to say, "Get the species up to an effective population size of 500 and it will be safe." If the species has been reduced to one population in one refuge, a single fire could destroy it, even if it contains 5,000 members. The population could be wiped out by one disease; there could be a local killing freeze; the food species on which it depends might become extinct; the crucial pollinator might vanish. Such events are "demographic accidents"—irregular and drastic reductions in population size caused by environmental change—and they are deadly. For species passing through the narrows of small population size, the Scylla of demographic accident is more dangerous than the Charybdis of inbreeding depression.

Only a few species consist of a single vulnerable population in one locality. They include the giant flightless darkling beetle (*Polposipus herculeanus*), which is restricted to dead trees on tiny Frigate Island in the Seychelles. There is the hau kuahiwi tree (*Hibiscadelphus distans*), which consists of exactly ten 6-meter-high trees growing on a dry rocky cliff on the island of Kauai. And perhaps the most intriguing example of all: the Socorro sowbug (*Thermosphaeroma thermophilum*), an aquatic crustacean that has lost its natural habitat and survives in an abandoned bathhouse in New Mexico. Most species are not like this. In some cases the constituent populations are so isolated from one another that they can never exchange individuals, but more typically the species is arrayed as a metapopulation, a population of populations, among which organisms do occasionally migrate.

Watched across long stretches of time, the species as metapopulation can be thought of as a sea of lights winking on and off across a dark terrain. Each light is a living population. Its location represents a habitat capable of supporting the species. When the species is present in that location the light is on, and when it is absent the light is out. As we scan the terrain over many generations, lights go out as local extinction occurs, then come on again as colonists from lighted spots reinvade the same localities. The life and death of species can then be viewed in a way that invites analysis and measurement. If a species manages to turn on as many lights as go out from generation to generation, it can persist indefinitely. When the lights wink out faster than they are turned on, the species sinks to oblivion.

The metapopulation concept of species existence is cause for both optimism and despair. Even when species are locally extirpated, they often come back quickly, provided the vacated habitats are left intact. But if the available habitats are reduced in sufficient number, the entire system can collapse. All the lights go out even if some intact habitats remain. A few jealously guarded reserves may not be enough. When the number of populations capable of populating empty sites becomes too small, they cannot achieve colonization elsewhere before they themselves go extinct. The system spirals downward out of control, and the entire sea of lights turns dark.

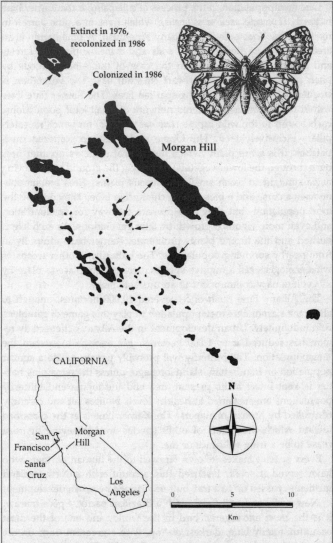

Extinct in 1976, recolonized in 1986

Colonized in 1986

Morgan Hill

CALIFORNIA

San Francisco

Morgan Hill

Santa Cruz

Los Angeles

N

0   5   10

Km

A metapopulation of Bay checkerspot butterflies lives south of San Francisco on serpentine grassland, a habitat represented here in black; the patches occupied in 1987 are indicated by arrows. In metapopulations, the occupancy of suitable environments changes from one year to the next.

One metapopulation in the process of collapsing is the Karner blue butterfly *(Lycaeides melissa samuelis)*, which lives in a pine barren in upstate New York called the Albany Pine Bush. A pine barren is an area of relatively sterile, sandy soils and dunes supporting forests and shrublands—dominated in the case of the Albany woods by pitch pine, scrub oak, and dwarf chestnut oak. The vegetation is frequently burned by lightning-sparked fires. The Karner blue lives within the barren as a scattered network of small local populations, each bound to the wild lupine *(Lupinus perennis)* on which its caterpillars exclusively feed. The lupine itself exists in scattered small patches. It is a fire plant, which means that it grows up after fires burn through the lower vegetation, clearing the ground and allowing more sunlight to reach small herbaceous plants. Fires are simultaneously a curse and a blessing for the Karner blue. They destroy the local population, but they also prepare the way for recolonization and even more vigorous growth by later generations. As each site is burned and the lupine plants proliferate, Karner blue adults fly in from nearby surviving populations. The butterfly, in other words, is what ecologists call a fugitive species, driven from place to place by its evolutionary commitment to an unstable niche.

The Albany Pine Bush once covered 10,000 hectares, enough to allow the Karner blue metapopulation to play this game of gambler's ruin indefinitely. Urban development in the Albany-Schenectady region has reduced it to 1,000 hectares, not enough to sustain the metapopulation. The butterfly will probably perish, as did a similar population on Manhattan Island long ago, unless the remaining habitat is kept intact at its present area and the lupine and butterfly populations are sustained at healthy levels by fires set and carefully controlled by human managers. The Karner blue is at the crossroad toward which thousands of other species are journeying: it must cease to be a truly wild race or die.

Every species makes its own farewell to the human partners who have served it so ill. I started this account with a New Zealand mistletoe, passed on to a frail butterfly ruined by urban development in New York, and will finish with a Brazilian parrot, Spix's macaw. It is the most endangered bird in the world, and one of the most beautiful: totally blue, darkest on top, with a greenish tinge on the belly, black mask around the lemon-yellow eye. The species, *Cyanopsitta spixii*, is so distinctive that it has been placed in its own genus. Never common, it was limited to palm groves and river-edge woodland across southern Pará to Bahia near the center of Brazil. It was

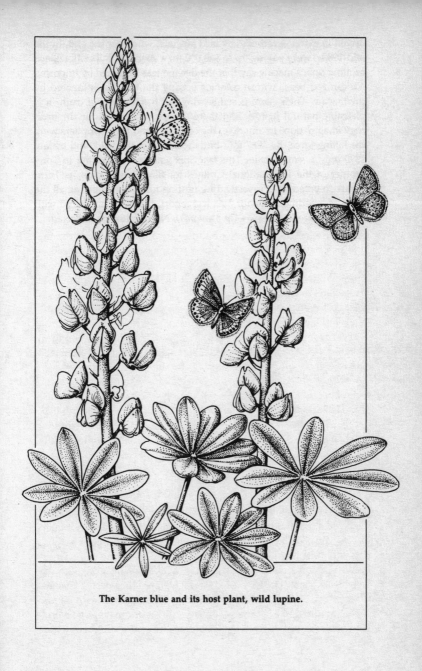

The Karner blue and its host plant, wild lupine.

driven to extreme rareness by bird fanciers, who near the end, in the mid-1980s, were paying up to $40,000 for a single bird. The Brazilians hunting Spix's macaw say that the decline was hastened by imported Africanized bees, whose colonies occupy the tree holes favored by the macaw. Their claim is self-serving but has the ring of truth. It is plausible natural history and therefore too far-fetched for the ordinary imagination. In any case, the collectors and their suppliers were the killing force. By 1987 four birds were left in the wild, and by late 1990 only a single male. This last Spix's macaw, according to Tony Juniper of the International Council for Bird Preservation, is "desperate to breed. It is investigating nesting holes and showing all the signs of breeding behavior." At last report, the male has paired with a female Illiger's macaw (Ara maracana). No hybrids are expected.

Biodiversity Threatened

HIDDEN AMONG the western Andean foothills of Ecuador, a few kilometers from Rio Palenque, there is a small ridge called Centinela. Its name deserves to be synonymous with the silent hemorrhaging of biological diversity. When the forest on the ridge was cut a decade ago, a large number of rare species were extinguished. They went just like that, from full healthy populations to nothing, in a few months. Around the world such anonymous extinctions—call them "centinelan extinctions"—are occurring, not open wounds for all to see and rush to stanch but unfelt internal events, leakages from vital tissue out of sight. Only an accident of timing led to an eyewitness account of the events on Centinela.

The eyewitnesses were Alwyn Gentry and Calaway Dodson, working out of the Missouri Botanical Garden, St. Louis. Gentry and Dodson made their discovery because they are born naturalists. By that I mean they are members of a special cadre of field biologists, those who do not practice science in order to be a success but try to succeed in órder to practice science—at least this kind of science. Even if they have to pay for the trip themselves, they will go into the field to do biology, to blend sun and rain with the findings of evolution and to give memory thereby to places like Centinela.

When Gentry and Dodson visited the ridge in 1978, they were the first to explore it botanically. Centinela is only one of a vast number of little-known spurs and saddles arrayed on either side of the Andes for 7,200 kilometers from Panama to Tierra del Fuego. At middle to high elevations in the tropical latitudes these mountain buttresses are covered by cloud forests. A traverse reveals that they are ecological islands, closed off above by the treeless paramos, surrounded below by the lowland rain forests, and segregated from one another by deep mountain valleys. Like conventional islands in the ocean, they tend to evolve their own species of plants and animals, which are then the endemics of that place, found nowhere else or at most in a few nearby localities. On Centinela Gentry and Dodson discovered about 90 such plant species, mostly herbaceous forms growing under the forest canopy, along with orchids and other epiphytes on the trunks and branches of trees. Several of the species had black leaves, a highly unusual trait and still a mystery of plant physiology.

In 1978 farmers from the valley below were moving in along a newly built private road and were cutting back the ridge forest. This is standard operating procedure in Ecuador. Fully 96 percent of the forests on the Pacific side have been cleared for agriculture, with little notice taken by conservationists outside Ecuador and no constraining policy imposed by local governments. By 1986 Centinela was completely cleared and planted in cacao and other crops. A few of the endemic plants have persisted in the shade of the cacao trees. Several others hold on in the forest of neighboring ridges, which themselves are in danger of clear-cutting. I don't know if any black-leaved plant species survived.

The revelation of Centinela and a growing list of other such places is that the extinction of species has been much worse than even field biologists, myself included, previously understood. Any number of rare local species are disappearing just beyond the edge of our attention. They enter oblivion like the dead of Gray's *Elegy*, leaving at most a name, a fading echo in a far corner of the world, their genius unused.

Extinction has been much greater even among larger, more conspicuous organisms than generally recognized. During the past ten years, scientists working on fossil birds, especially Storrs Olson, Helen James, and David Steadman, have uncovered evidence of massive destruction of Pacific Island landbirds by the first human colonists centuries before the coming of Europeans. The scientists obtain their data by excavating fossil and subfossil bones wherever the dead birds dropped or were thrown, in dunes, limestone sinkholes, lava tubes, crater lake beds, and archaeological middens. On each of the islands the deposits were mostly laid down from 8,000 years ago up to nearly the present, bracketing the arrival of the Polynesians. They leave little room for doubt that in the outer Pacific in particular, from Tonga in the west to Hawaii in the east, the Polynesians extinguished at least half of the endemic species found upon their arrival.

This vast stretch of Pacific islands was colonized by the Lapita people, ancestors of the modern Polynesian race. They emigrated from their homeland somewhere in the fringing islands of Melanesia or Southeast Asia and spread steadily eastward from archipelago to archipelago. With great daring and probably heavy mortality, they traveled in single outrigger or double canoes across hundreds of kilometers of water. Around 3,000 years ago they settled Fiji, Tonga, and Samoa. Stepping from island to island they finally reached Hawaii, with Easter the most remote of the habitable Pacific islands, as recently as 300 A.D.

The colonists subsisted on crops and domestic animals carried in their boats but also, especially in the early days of settlement, whatever edible animals they encountered. They ate fish, turtles, and a profusion of bird species that had never seen a large predator and were easily caught, including doves, pigeons, crakes, rails, starlings, and others whose remains are only now coming to light. Many of the species were endemics, found only on the islands discovered by the Lapita. The voyagers ate their way through the Polynesian fauna. On Eua, in present-day Tonga, twenty-five species lived in the forests when the colonists arrived around 1000 B.C., but only eight survive today. Nearly every island across the Pacific was home to several endemic species of flightless rails before the Polynesian occupation. Today populations survive only on New Zealand and on Henderson, an uninhabited coral island 190 kilometers northeast of Pitcairn. It used to be thought that Henderson was one of the few virgin habitable islands of any size left in the world, never occupied by human beings. But recently discovered artifacts reveal that Polynesians colonized Henderson, then abandoned it, probably because they consumed the birds to less than sustainable levels. On this and other small islands lacking arable soil, birds were the most readily available source of protein. The colonists drove the populations down, erasing some species in the process, then either starved or sailed on.

Hawaii, last of the Edens of Polynesia, sustained the greatest damage measured by lost evolutionary products. When European settlers arrived after Captain Cook's visit in 1778, there were approximately fifty native species of landbirds. In the following two centuries, one third disappeared. Now we know from bone deposits that another thirty-five species identified with certainty, and very likely twenty other species less well documented, had already been extinguished by the native Hawaiians. Among those identified to date are an eagle similar to the American bald eagle, a flightless ibis, and a strange parliament of owls with short wings and extremely long legs. Most remarkable of all were bizarre flightless forms evolved from ducks but possessing tiny wings, massive legs, and bills resembling the beaks of tortoises. Helen James and Storrs Olson record that

> although they were terrestrial and herbivorous, like geese, we now know from the presence of a duck-like syringeal bulla that these strange birds were derived either from shelducks (Tadornini), or more likely from dabbling ducks (Anatini), quite possibly from the genus *Anas.* They may have had an ecological role similar to that of the large tortoises of the Galápagos and islands of the western Indian Ocean. Because we now recognize three genera and four species of these birds, and because they are neither phyletically geese nor functionally ducks, we have coined a new word, *moa-nalo,* as a more convenient general term for all such flightless, goose-like ducks of the Hawaiian Islands.

The surviving native Hawaiian birds are for the most part inconspicuous relicts, small, elusive species restricted to the remnant mountain forests. They are a faint shadow of the eagles, ibises, and moanalos that greeted the Polynesian colonists as the Byzantine empire was born and Mayan civilization reached its zenith.

Centinelan extinctions also occurred on other continents and islands as human populations spread outward from Africa and Eurasia. Mankind soon disposed of the large, the slow, and the tasty. In North America 12,000 years ago, just before Paleo-Indian hunter-gatherers came from Siberia across the Bering Strait, the land teemed with large mammals far more diverse than those in any part of the modern world, including Africa. Twelve millennia back may seem like the Age of Dinosaurs, but it was just yesterday by geological standards. Humanity was stirring then, some eight million people alive and many seeking new land. The manufacture of hooks and harpoons for fishing was widespread, along with the cultivation of wild grains and the domestication of dogs. The construction of the first towns, in the Fertile Crescent, lay only a thousand years in the future.

In western North America, just behind the retreating glacial front, the grasslands and copses were an American Serengeti. The vegetation and insects were similar to those alive in the west today—you could have picked the same wildflowers and netted the same butterflies—but the big mammals and birds were spectacularly different. From one spot, say on the edge of riverine forest looking across open terrain, you could have seen herds of horses (the extinct, pre-Spanish kind), long-horned bison, camels, antelopes of several species, and mammoths. There would be glimpses of sabertooth cats, possibly working together in lionish prides, giant dire wolves, and tapirs. Around a dead horse might be gathered the representatives of a full adaptive radiation of scavenging birds: condors, huge condor-like teratorns, carrion storks, eagles, hawks, and vultures, dodging and threatening one another (we know from the species that survived), the smaller birds snatching pieces of meat and waiting for the body to be whittled down enough to be abandoned by their giant competitors.

Some 73 percent of the large mammal genera that lived in the late Pleistocene are extinct. (In South America the number is 80 percent.) A comparable number of genera of the largest birds are also extinct. The collapse of diversity occurred about the same time that the first Paleo-Indian hunters entered the New World, 12,000 to 11,000 years ago, and then spread southward at an average rate of 16 kilometers a year. It was not a casual, up-and-down event. Mammoths had flourished for two million years to that time and were represented at the end by three species—the Columbian, imperial, and woolly. Within a thousand years all were gone. The ground sloths, another ancient race, vanished almost simultaneously. The last known surviving population, foraging out of caves at the western end of the Grand Canyon, disappeared about 10,000 years ago.

If this were a trial, the Paleo-Indians could be convicted on circumstantial evidence alone, since the coincidence in time is so exact. There is also a strong motive: food. The remains of mammoths, bison, and other large mammals exist in association with human bones, charcoal from fires, and stone weapons of the Clovis culture. These earliest Americans were skilled big-game hunters, and they encountered animals totally unprepared by evolutionary experience for predators of this kind. The birds that became extinct were also those most vulnerable to human hunters. They included eagles and a flightless duck. Still other victims were innocent bystanders: condors, teratorns, and vultures dependent on the newly devastated populations of heavy-bodied mammals.

In defense of the Paleo-Indians, their counsel might argue the existence of another culprit. The end of the Pleistocene was a time not only of human invasion of the New World, but also of climatic warming. As the continental glacier retreated across Canada, forests and grasslands shifted rapidly northward. Changes of this magnitude must have exerted a profound effect on the life and death of local populations. Between 1870 and 1970, by way of comparison, Iceland warmed an average 2°C in the winter and somewhat less in the spring and summer. Two Arctic bird species, the long-tailed duck and the lesser auk, declined to near extinction. At the same time, lapwings, tufted ducks, and several other southern species established themselves on the island and began to breed. There are hints of similar responses during the great Pleistocene decline. Mastodons, for example, were apparently specialized for life in coniferous forests. As this belt of vegetation migrated northward, the proboscideans moved with it. In time they became concentrated along the spruce forest zone in the northeast, then disappeared. Their extinction might have stemmed not only from overkill by hunters but also from fragmentation and reduction of the populations forced by a shrinking habitat.

Let the defense now speak even more forcefully: for tens of millions of years before the coming of man, mammal genera were born and died in large numbers, with the extinction of some accompanied by the origin of others to create a rough long-term balance. The changes were accompanied by climatic shifts much like those in evidence 11,000 years ago, and perhaps they were driven by them. During the last 10 million years, David Webb has pointed out, six major extinction episodes leveled the land mammals of North America. Among them the terminating event of the Pleistocene (the Rancholabrean, named after Rancho La Brea, in California) was not the most catastrophic. The greatest, according to available records,

> was the late Hemphillian (nearly five million years ago) when more than sixty genera of land mammals (of which thirty-five were large, weighing more than 5 kg) disappeared from this continent. The late Rancholabrean extinction pulse (about 10,000 years ago) was the next greatest; over forty genera became extinct, of which nearly all were large mammals . . . Some evidence shows that these extinction episodes were correlated with terminations of glacial cycles, when climatic extremes and instability are thought to have reached their maxima.

In at least two of the great extinction spasms, the large browsing mammals were destroyed as the climate deteriorated and the broad continental savannas gave way to steppes. At the end of the Hemphillian, even grazing mammals such as horses, rhinos, and pronghorns precipitously declined.

It may seem that the debate between experts who favor overkill by humans and those who favor climatic change resembles a replay, in a different theater, of the debate over the end of the Age of Dinosaurs. The Paleo-Indians have replaced the giant meteorite in this new drama. Circumstantial evidence is countered by other circumstantial evidence, while both sides search for a smoking gun. The dispute is the product of neither ideology nor clashing personalities. It is the way science at its best is done.

That said, I will lay aside impartiality. I think the overkill theorists have the more convincing argument for what happened in America 10,000 years ago. It seems likely that the Clovis people spread through the New World and demolished most of the large mammals during a hunters' blitzkrieg spanning several centuries. Some of the doomed species hung on here and there for as long as 2,000 years, but the effect was the same: swift destruction, on the scale of evolution that measures normal lifespans of genera and species in millions of years.

There is an additional reason for accepting this verdict provisionally. Paul Martin, who revived the idea in the mid-1960s (a similar proposal had been made a century earlier for the Pleistocene mammals of Europe), called attention to this important circumstance: when human colonists arrived, not only in America but also in New Zealand, Madagascar, and Australia, and whether climate was changing or not, a large part of the megafauna—large mammals, birds, and reptiles—disappeared soon afterward. This collateral evidence has been pieced together by researchers of various persuasions over many years, and it points away from climate and toward people.

Before the coming of man around 1000 A.D., New Zealand was home to moas, large flightless birds unique to the islands. These creatures had ellipsoidal bodies, massive legs, and long necks topped by tiny heads. The first Maoris, arriving from their Polynesian homeland to the north, found about thirteen species ranging in size from that of large turkeys to giants weighing 230 kilograms or more, the latter among the largest birds ever evolved. There had in fact been a moa radiation, filling many niches. It was of the kind normally occupied by medium-sized and large mammals, of which there were none on New Zealand. The Maoris proceeded to butcher the birds in large numbers, leaving conspicuous moa-hunting sites all over New Zealand. On South Island, where most of the remains occur, the deposits are piled with moa bones dating from 1100 to 1300. During this brief interlude the colonists must have obtained a substantial portion of their diet from cooked moa. The peak kills began on the northern part of the island, the Maori point of entry, and spread slowly to the southern districts. Several Europeans claimed to have seen moas in the early 1800s, but the records cannot be verified. Archaeological and public opinion alike hold the Maori hunters responsible, as declared in the popular New Zealand song:

> No moa, no moa,
> In old Ao-tea-roa.
> Can't get 'em.
> They've et 'em;
> They've gone and there aint no moa!

The moa extinction was only part of the New Zealand carnage. A total of twenty other landbirds, including nine additional flightless species, were also wiped out in short order. The tuatara, only living member of the reptilian order Rhynchocephalia, along with unique frogs and flightless insects, were driven to the edge of extinction. Their demise was partly due to the deforestation and firing of large stretches of land. It was hastened by rats that came ashore with the Maoris and bred in huge numbers, against which the autochthons had few natural defenses. In the 1800s the British settlers came upon a beautiful but already much-damaged archipelago. As elsewhere, they proceeded to reduce its biodiversity still further, with a pernicious ingenuity of their own.

Madagascar, fourth largest island in the world, is a small continent virtually on its own. Fully isolated during a northward drift through the Indian Ocean for 70 million years, it was the theater for a biological tragedy like New Zealand's. Despite the proximity of Africa, the first human colonists came to Madagascar not from that continent but from far-off Indonesia. They arrived around 500 A.D. In the centuries immediately following, the megafauna of the great island vanished. No important climatic change accompanied this event; it appears to have been solely the work of the Malagache pioneers. Six to a dozen elephant birds, large and flightless like the moas, disappeared. They included the heaviest birds of recent geological history,

Aepyornis maximus, a feathered giant almost 3 meters tall with massive legs. Its eggs, the size of soccer balls, can still be pieced together from fragments piled around Malagache archaeological sites. Also erased were seven of the seventeen genera of lemurs, primates most closely related among living mammals to monkeys, apes and men. The lemuroids had undergone a spectacular adaptive radiation on Madagascar. The forms that disappeared were the largest and most interesting of all. One species ran on all fours like a dog, and another had long arms and probably swung through the trees like a gibbon. A third, as big as a gorilla, climbed trees and resembled an oversized koala. Also erased were an aardvark, a pygmy hippopotamus, and two huge land tortoises.

Essentially the same story of destruction was repeated when aboriginal human populations came to Australia about 30,000 years ago, also by way of Indonesia. A number of large mammals soon vanished, including marsupial lions, gigantic kangaroos 2.5 meters (8 feet) tall, and others separately resembling ground sloths, rhinos, tapirs, woodchucks, or, perhaps more accurately expressed, blends of these more familiar types of World Continent fauna. The case for overkill by the aboriginal Australians, however, is complicated by the remote time of their arrival, the longer period during which the extinctions took place, and the scarcity of fossils and kill sites to document the role of hunting. It is also true that Australia experienced a severe arid period from 15,000 to 26,000 years ago, during which the greatest number of animal extinctions occurred. We know that the Australian aboriginals hunted skillfully and burned large stretches of arid land in their search for prey. They still do. Men must have played a role in extinction, but the evidence does not yet allow us to weigh their influence against the drying out of the continent's interior.

In 1989 Jared Diamond summed up for the prosecution in the case of the extinguished megafaunas. Climate, he said, cannot be the principal culprit. He asked: how could changes in climate and vegetation during the retreat of the last glacier lead to mass extinction in North America but not in Europe and Asia? The differences between the land masses were not climatic but the first-time colonization of America, confronting a megafauna with no previous experience of human hunters. And in North America, why did this hecatomb occur at the end of the last glacial cycle, which closed the Quaternary period, but not at the end of the twenty-two glacial cycles preceding it? Again, the difference was the coming of the Paleo-

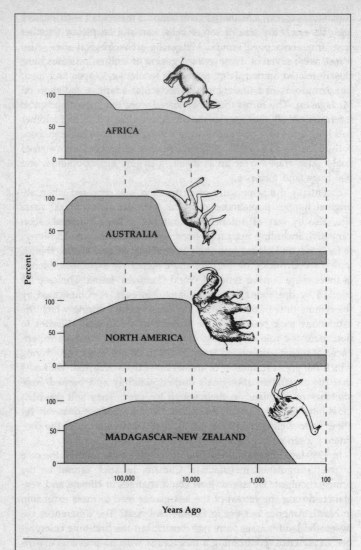

The extinction of large mammals and flightless birds coincided closely with the arrival of humans in North America, Madagascar, and New Zealand, and less decisively earlier in Australia. In Africa, where humans and animals evolved together for millions of years, the damage was less severe.

Indian hunters. How, Diamond pressed, did Australia's reptiles manage to survive the prehistoric human invasions better, as did the smaller mammals and birds? And, finally, why did such large forms as the marsupial wolf and giant kangaroos disappear about the same time from both Australia's arid interior and rain forests, as well as from nearby New Guinea's wet mountain forests?

> Quaternary extinctions were selective in space and time because they appear to have occurred at those places and times where naive animals first encountered humans. It is further argued that they were selective in taxa and in victim size because human hunters concentrate on some species (e.g. large mammals and flightless birds) while ignoring other species (e.g. small rodents). It is argued that Quaternary extinctions befell species in all habitats because humans hunt in all habitats, and human hunters help no species except as an incidental consequence of habitat changes and of removing other species.

"Human hunters help no species." That is a general truth and the key to the whole melancholy situation. As the human wave rolled over the last of the virgin lands like a smothering blanket, Paleo-Indians throughout America, Polynesians across the Pacific, Indonesians into Madagascar, Dutch sailors ashore on Mauritius (to meet and extirpate the dodo), they were constrained by neither knowledge of endemicity nor any ethic of conservation. For them the world must have seemed to stretch forever beyond the horizon. If fruit pigeons and giant tortoises disappear from this island, they will surely be found on the next one. What counts is food today, a healthy family, and tribute for the chief, victory celebrations, rites of passage, feasts. As the Mexican truck driver said who shot one of the last two imperial woodpeckers, largest of all the world's woodpeckers, "It was a great piece of meat."

From prehistory to the present time, the mindless horsemen of the environmental apocalypse have been overkill, habitat destruction, introduction of animals such as rats and goats, and diseases carried by these exotic animals. In prehistory the paramount agents were overkill and exotic animals. In recent centuries, and to an accelerating degree during our generation, habitat destruction is foremost among the lethal forces, followed by the invasion of exotic animals. Each agent strengthens the others in a tightening net of destruction. In the United States, Canada, and Mexico, 1,033 species of fishes are known to have lived entirely in fresh water within recent historical times. Of these, 27 or 3 percent have become extinct within the past hundred years, and another 265 or 26 percent are liable to extinction. They fall into one or the other of the categories utilized by the International Union for Conservation of Nature and Natural Resources (IUCN), which publishes the *Red Data Books:* Extinct, Endangered, Vulnerable, and Rare. The changes that forced them into decline are:

Destruction of physical habitat	73% of species
Displacement by introduced species	68% of species
Alteration of habitat by chemical pollutants	38% of species
Hybridization with other species and subspecies	38% of species
Overharvesting	15% of species

(These figures add up to more than 100 percent because more than one agent impinges on many of the fish populations.) When habitat destruction is defined as both the physical reduction in suitable places to live and the closing of habitats by chemical pollution, then it is found to be an important factor in over 90 percent of the cases. Through a combination of all these factors, the rate of extinction has risen steadily during the past forty years.

In fishes and in all other groups of which we have sufficient knowledge, the depredations were started in prehistory and early historical times and are being pressed with a vengeance by modern generations. Early peoples exterminated most of the big animals on the spot. They also decimated less conspicuous plants and animals on islands and in isolated valleys, lakes, and river systems, where species live in small populations with their backs to the wall. Now it is our turn. Armed with chainsaws and dynamite, we are assaulting the final strongholds of biodiversity—the continents and, to a lesser but growing extent, the seas.

Will it ever be possible to assess the ongoing loss of biological diversity? I cannot imagine a scientific problem of greater immediate importance for humanity. Biologists find it difficult to come up with even an approximate estimate of the hemorrhaging because we know so little about diversity in the first place. Extinction is the most obscure and local of all biological processes. We don't see the last butterfly of its species snatched from the air by a bird or the last orchid of a certain kind killed by the collapse of its supporting tree in some distant mountain forest. We hear that a certain animal or plant is on the edge, perhaps already gone. We return to the last known locality to search, and when no individuals are encountered there year after year we pronounce the species extinct. But hope lingers on. Someone flying a light plane over Louisiana swamps thinks he sees a few ivory-billed woodpeckers start up and glide back down into the foliage. "I'm pretty sure they were ivorybills, not pileated woodpeckers. Saw the white double stripes on the back and the wing bands plain as day." A Bachman's warbler is heard singing somewhere, maybe. A hunter swears he has seen Tasmanian wolves in the scrub forest of Western Australia, but it is probably all fantasy.

In order to know that a given species is truly extinct, you have to know it well, including its exact distribution and favored habitats. You have to look long and hard without result. But we do not know the vast majority of species of organisms well; we have yet to anoint so many as 90 percent of them with scientific names. So biologists agree that it is not possible to give the exact number of species going extinct; we usually turn palms up and say the number is very large. But we can do better than that. Let me start with a generalization: *in the small minority of groups of plants and animals that are well known, extinction is proceeding at a rapid rate, far above prehuman levels. In many cases the level is calamitous: the entire group is threatened.*

To illustrate this principle, I will present a few anecdotes, out of many available: whenever we can focus clearly, we usually see extinction in progress. Then I will take a more theoretical approach, using models of island biogeography, to arrive at an estimate of extinction rates in tropical rain forests, which contain half or more of the world's species of plants and animals. Here are the examples:

• One fifth of the species of birds worldwide have been eliminated in the past two millennia, principally following human occupation of islands. Thus instead of 9,040 species alive today, there probably would have been about 11,000 species if left alone. According to a recent study by the International Council for Bird Preservation, 11 percent or 1,029 of the surviving species are endangered.

• A total of 164 bird species have been recorded from the Solomon Islands in the southwest Pacific. The *Red Data Book* lists only one as recently extinct. But in fact there have been no records for twelve others since 1953. Most of these are ground nesters especially vulnerable to predators. Solomon Islanders who know the birds best have stated that at least some of the species were exterminated by imported cats.

• From the 1940s to the 1980s, population densities of migratory songbirds in the mid-Atlantic United States dropped 50 percent, and many species became locally extinct. One cause appears to be the accelerating destruction of the forests of the West Indies, Mexico, and Central and South America, the principal wintering grounds of many of the migrants. The fate of Bachman's warbler will probably befall other North American summer residents if the deforestation continues.

• About 20 percent of the world's freshwater fish species are either extinct or in a state of dangerous decline. The situation is approaching the critical stage in some tropical countries. A recent search for the 266 species of exclusively freshwater fishes of lowland peninsular Malaysia turned up only 122. Lake Lanao on the Philippine Island of Mindanao is famous among evolutionary biologists for the adaptive radiation of cyprinid fishes that occurred exclusively within the confines of the lake. As many as 18 endemic species in three genera were previously known; a recent search found only three species, representing one of the genera. The loss has been attributed to overfishing and competition from newly introduced fish species.

• The most catastrophic extinction episode of recent history may be the destruction of the cichlid fishes of Lake Victoria, which I described earlier as a paradigm of adaptive radiation. From a single ancestral species 300 or more species emanated, filling almost all the major ecological niches of freshwater fishes. In 1959 British colonists introduced the Nile perch as a sport fish. This huge predator, which grows to nearly 2 meters in length, has drastically reduced the native fish population and extinguished some of the species. It is projected eventually to eliminate more than half of the endemics. The perch affects not only the fishes but the lake ecosystem as a whole. As the alga-feeding cichlids disappear, plant life blooms and decomposes, depleting oxygen in the deeper water and accelerating the decline of cichlids, crustaceans, and other forms of life. A task force of fish biologists observed in 1985, "Never before has man in a single ill advised step placed so many vertebrate species simultaneously at risk of extinction and also, in doing so, threatened a food resource and traditional way of life of riparian dwellers."

• The United States has the largest freshwater mollusk fauna in the world, especially rich in mussels and gill-breathing snails. These species have long been in a steep decline from the damming of rivers, pollution, and the introduction of alien mollusk and other aquatic animals. At least 12 mussel species are now extinct throughout their ranges, and 20 percent of the remainder are endangered. Even where extinction has not yet occurred, the extirpation of local populations is rampant. Lake Erie and the Ohio River system originally held dense populations of 78 different forms; now 19 are extinct and 29 are rare. Muscle [sic] Shoals, a stretch of the Tennessee River in Alabama, once held a fauna of 68 mussel species. Their shells were specialized for life in riffles or shoals, shallow streams with sandy gravel bottoms and rapid currents. When Wilson Dam was constructed in the early 1920s, impounding and deepening the water, 44 of the species were extinguished. In a parallel development, impoundment and pollution have combined to extinguish two genera and 30 species of gill-breathing snails in the Tennessee and nearby Coosa rivers.

• Freshwater and land mollusks are generally vulnerable to extinction because so many are specialized for life in narrow habitats and unable to move quickly from one place to another. The fate of the tree snails of Tahiti and Moorea illustrates the principle in chilling fashion. Comprising 11 species in the genera *Partula* and *Samoana*, a miniature adaptive radiation in one small place, the snails were recently exterminated by a single species of exotic carnivorous snail. It was folly in the grand manner, a pair of desperate mistakes by people in authority, which unfolded as follows. First, the giant African snail *Achatina fulica* was introduced to the islands as a food animal. Then, when it multiplied enough to become a pest, the carnivorous snail *Euglandina rosea* was introduced to control the *Achatina*. *Euglandina* itself multiplied prodigiously, advancing along a front at 1.2 kilometers a year. It consumed not only the giant African snail but every native tree snail along the way. The last of the wild tree snails became extinct on Moorea in 1987. On nearby Tahiti the same sequence is now unfolding. And in Hawaii the entire endemic tree-snail genus *Achatinella* is endangered by *Euglandina* and habitat destruction. Twenty-two species are extinct and the remaining 19 are endangered.

• A recent survey by the Center for Plant Conservation revealed that between 213 and 228 plant species, out of a total of about 20,000, are known to have become extinct in the United States. Another 680 species and subspecies are in danger of extinction by the year 2000. About three fourths of these forms occur in only five places: California, Florida, Hawaii, Puerto Rico, and Texas. The predicament of the most endangered species is epitomized by *Banara vanderbiltii*. By 1986 this small tree of the moist limestone forests of Puerto Rico was down to two plants growing on a farm near Bayamon. At the eleventh hour, cuttings were obtained and are now successfully growing in the Fairchild Tropical Garden in Miami.

• In western Germany, the former Federal Republic, 34 percent of 10,290 insect and other invertebrate species were classified as threatened or endangered in 1987. In Austria the figure was 22 percent of 9,694 invertebrate species, and in England 17 percent of 13,741 insect species.

• The fungi of western Europe appear to be in the midst of a mass extinction on at least a local scale. Intensive collecting in selected sites in Germany, Austria, and the Netherlands have revealed a 40 to 50 percent loss in species during the past sixty years. The main cause of the decline appears to be air pollution. Many of the vanished species are mycorrhizal fungi, symbiotic forms that enhance the absorption of nutrients by the root systems of plants. Ecologists have long wondered what would happen to land ecosystems if these fungi were removed, and we will soon find out.

For species on the brink, from birds to fungi, the end can come in two ways. Many, like the Moorean tree snails, are taken out by the metaphorical equivalent of a rifle shot—they are erased but the ecosystem from which they are removed is left intact. Others are destroyed by a holocaust, in which the entire ecosystem perishes.

The distinction between rifle shots and holocausts has special merit in considering the case of the spotted owl (*Strix occidentalis*) of the United States, an endangered form that has been the object of intense national controversy since 1988. Each pair of owls requires about 3 to 8 square kilometers of coniferous forest more than 250 years old. Only this habitat can provide the birds with both enough large hollow trees for nesting and an expanse of open understory for the effective hunting of mice and other small mammals. Within the range of the spotted owl in western Oregon and Washington, the suitable habitat is largely confined to twelve national forests. The controversy was engaged first within the U.S. Forest Service and then the public at large. It was ultimately between loggers, who wanted to continue cutting the primeval forest, and environmentalists determined to protect an endangered species. The major local industry around the owl's range was affected, the financial stakes were high, and the confrontation was emotional. Said the loggers: "Are we really expected to sacrifice thousands of jobs for a handful of birds?" Said the environmentalists: "Must we deprive future generations of a race of birds for a few more years of timber yield?"

Overlooked in the clamor was the fate of an entire habitat, the old-growth coniferous forest, with thousands of other species of plants, animals, and microorganisms, the great majority unstudied and unclassified. Among them are three rare amphibian species, the tailed frog and the Del Norte and Olympic salamanders. Also present is the western yew, *Taxus brevifolia*, source of taxol, one of the most potent anticancer substances ever found. The debate should be framed another way: what else awaits discovery in the old-growth forests of the Pacific Northwest?

The cutting of primeval forest and other disasters, fueled by the demands of growing human populations, are the overriding threat to biological diversity everywhere. But even the data that led to this conclusion, coming as they do mainly from vertebrates and plants, understate the case. The large, conspicuous organisms are the ones most susceptible to rifle shots, to overkill and the introduction of competing organisms. They are of the greatest immediate importance to man and receive the greater part of his malign attention. People hunt deer and pigeons rather than sowbugs and spiders. They cut roads into a forest to harvest Douglas fir, not mosses and fungi.

Not many habitats in the world covering a kilometer contain fewer than a thousand species of plants and animals. Patches of rain forest and coral reef harbor tens of thousands of species, even after they have declined to a remnant of the original wilderness. But when the *entire* habitat is destroyed, almost all of the species are destroyed. Not just eagles and pandas disappear but also the smallest, still uncensused invertebrates, algae, and fungi, the invisible players that make up the foundation of the ecosystem. Conservationists now generally recognize the difference between rifle shots and holocausts. They place emphasis on the preservation of entire habitats and not only the charismatic species within them. They are uncomfortably aware that the last surviving herd of Javan rhinoceros cannot be saved if the remnant woodland in which they live is cleared, that harpy eagles require every scrap of rain forest around them that can be spared from the chainsaw. The relationship is reciprocal: when star species like rhinoceros and eagles are protected, they serve as umbrellas for all the life around them.

And so to threatened and endangered species must be added a

growing list of entire ecosystems, comprising masses of species. Here are several deserving immediate attention:

Usambara Mountain forests, Tanzania. Varying widely in elevation and rainfall, the Usambaras contain one of the richest biological communities in East Africa. They protect large numbers of plant and animal species found nowhere else, but their forest cover is declining drastically, having already been cut to half, some 450 square kilometers, between 1954 and 1978. Rapid growth of human populations, more extensive logging, and the takeover of land for agriculture are pressing the last remaining reserves and thousands of species toward extinction.

San Bruno Mountain, California. In this small refuge surrounded by the San Francisco metropolis live a number of federally protected vertebrates, plants, and insects. Some of the species are endemics of the San Francisco peninsula, including the San Bruno elfin butterfly and the San Francisco garter snake. The native fauna and flora are threatened by offroad vehicular traffic, expansion of a quarry, and invasion by eucalyptus, gorse, and other alien plant species.

Oases of the Dead Sea Depression, Israel and Jordan. These humid refuges in a quintessentially desert area, called *ghors*, are isolated tropical ecosystems sustained by freshwater springs. They contain true pockets of an ancient African fauna and flora cut off by the dry terrain of the Jordan Rift Valley. Species that flourish thousands of kilometers to the south are joined here by others restricted to the vicinity of ghors or even to single springs. In 1980 I walked most of the length of Ein Gedi, one of these sites, through the lush bankside vegetation, marveling at the crystalline water of the spring-fed brook, with its endemic cichlid fish and emerald algae. I studied large weaver ants that nest in the banks—a little slice of Africa an hour's drive from Jerusalem. Climbing away from the bank trail for a hundred meters, I was back in the desert terrain of the Middle East. The ghors are of exceptional scientific interest because they bring an African fauna and flora into direct contact with a different set of species that together range from Europe across the Middle East to temperate Asia. The oases are threatened by overgrazing, mining, and commercial development. In an exquisitely symbolic reflection of the region's politics, several are used as minefields.

If species vanish en masse when their isolated habitats collapse, they die even more catastrophically when entire systems are obliterated. The logging of a mountain ridge in the Andes may extinguish scores of species, but logging all such ridges will erase hundreds of thousands. Such broad areas were labeled "hot spots" by Norman Myers in 1988. The emergency-care cases of global conservation, they are defined as areas that both contain large numbers of endemic species and are under extreme threat; their major habitats have been reduced to less than 10 percent of the original cover or are destined to fall that low within one to several decades. Myers has listed eighteen hot spots. Although they collectively occupy a tiny amount of space, only half a percent of the earth's land surface, they are the exclusive home of one fifth of the world's plant species. The hot spots comprise a far-flung array of forests and Mediterranean-type scrubland and are represented on every continent except Antarctica. Each deserves special and immediate mention.

California floristic province. This familiar Mediterranean-climate domain, stretching from southern Oregon to Baja California and recognized by botanists as a separate evolutionary center, contains one fourth of all the plant species found in the United States and Canada combined. Half, or 2,140 species, are found nowhere else in the world. Their environment is being rapidly constricted by urban and agricultural development, especially along the central and southern coasts of California.

Central Chile. South America's preeminent Mediterranean vegetation contains 3,000 plant species, slightly over half of the entire Chilean flora, crowded into only 6 percent of the national territory. The surviving cover is only one third that of the original and unfortunately is located in the most densely populated part of the country. It is being pressed especially hard by rural families, who rely on natural vegetation for fuel and livestock fodder.

The Colombian Chocó. The forest of Colombia's coastal plain and low mountains extends the entire length of the country. The Chocó, as the region is called after the state it includes, is drenched with extreme rainfall and blessed with one of the richest but least explored floras in the world. At present, 3,500 plant species are known but as many as 10,000 may grow there, of which one fourth are estimated to be endemic and a smaller but still substantial fraction are new to science. Since the early 1970s, the Chocó has been relentlessly

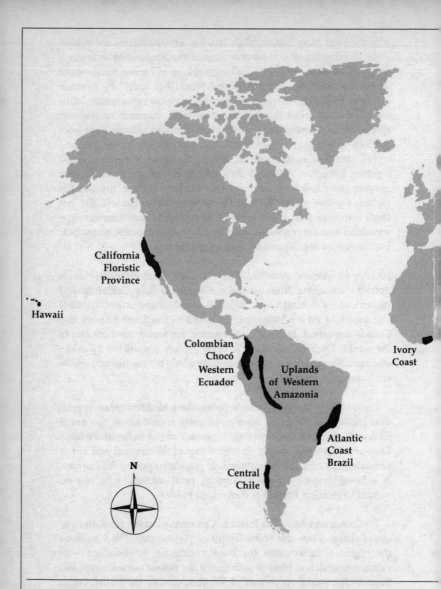

Hot spots are habitats with many species found nowhere else and in greatest danger of extinction from human activity. The 18 hot spots identified here are forests and Mediterranean scrubland well enough known to be included with certainty. But the map, based on preliminary study, is far from complete. Other

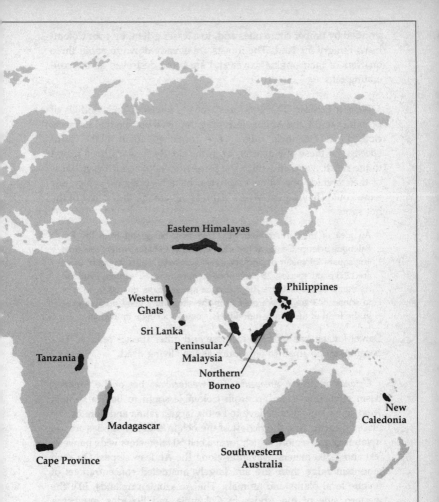

Forests and Heathland: Hot-Spot Areas

forest types not shown are endangered, as well as a large number of lakes, river systems, and coral reefs. The broader areas depicted, such as the coastal forests of Brazil and the Philippines, actually consist of many smaller hot spots scattered across local mountain ridges, valleys, and islands.

invaded by timber companies and, to a lesser extent, by poor Colombians hungry for land. The forests are already down to about three quarters of their original cover and are being destroyed at an accelerating rate.

Western Ecuador. The wet forests of the lowlands and foothills of Ecuador west of the Andes, including the small portion that formerly clothed the Centinela ridge, once contained about 10,000 plant species. Of these one quarter, as in the closely similar Chocó region to the north, were endemic. The forests, so notable for the richness of their orchids and other epiphytes, have been almost completely wiped out. They are, in Myers' expression, among the hottest of the hot spots.

> An idea of the former biotic diversity may be gained from the Rio Palenque Science Center at the southern tip of the area, where less than one square kilometer of primary forest survives. In this fragment there are 1200 plant species, 25 percent of them endemic to western Ecuador. As many as 100 of these Rio Palenque species have proved to be new to science; 43 are known only from the site, and a good number exist in the form of just a few individuals, some as a single individual.

Daniel Janzen has referred to these and other species reduced to a population too small to reproduce as the "living dead."

Uplands of western Amazonia. The western reaches of the Amazon Basin stretching in an arc from Colombia south to Bolivia contain what some biologists believe to be the largest fauna and flora of any place on earth. And the richest of the rich in endemic species are the uplands of the region, which form a belt 50 kilometers wide between 500 and 1,500 meters' elevation along the Andean slopes. On each mountain ridge there are still largely unstudied concentrations of unique local plants and animals. The Amazonian uplands, like the western side of the Andes in Colombia and Ecuador, are being rapidly settled. In Ecuador's sector alone the population has grown from 45,000 to about 300,000 during the past forty years. About 65 percent of the upland forests have already been cleared or converted into palm-oil plantations. The loss is projected to approach 90 percent by the year 2000.

Atlantic coast of Brazil. A unique rain forest once reached from Recife southward through Rio de Janeiro to Florianópolis, of which the

Forests of Western Ecuador 1938–1988

EQUATOR

ECUADOR

SOUTH AMERICA

N

Forest Cover 1938

Quito

Guayaquil

Forest Cover 1958

Quito

Guayaquil

Forest Cover 1988

Quito

Guayaquil

More than 90 percent of the forests of western Ecuador have been destroyed during the past four decades. The loss is estimated to have extinguished or doomed over half of the species of the area's plants and animals. Many other biologically diverse areas of the world are under similar assault.

young Charles Darwin once wrote, "twiners entwining twiners—tresses like hair—beautiful lepidoptera—silence—hosanna—silence well exemplified—lofty tree . . . Wonder, astonishment, & sublime devotion, fill & elevate the mind." That was 1832 when, as naturalist on the *Beagle*, Darwin first put ashore in South America and jotted his impressions in a notebook. The Atlantic forests originally covered about a million square kilometers. Geographically isolated from the Amazonian forests to the north and west, they contain one of the most diverse and distinctive biotas in the world. But Brazil's south Atlantic coast is also the agriculturally most productive and densely populated part of the country. The forests have been reduced to less than 5 percent of the original cover, and that part survives mostly in steep mountainous regions. A good part of this remnant is protected as parks and reserves, a last glimpse of Eden for future generations teeming around it.

Southwestern Ivory Coast. The towering rain forest of the Ivory Coast and adjacent areas of Liberia, a distinct botanical province of West Africa, once covered 160,000 square kilometers. Unrestricted logging and slash-and-burn farming have reduced it to 16,000 square kilometers. What remains is being cleared at the rate of up to 2,000 square kilometers a year. Only the Taï National Park, with 3,300 square kilometers, is officially protected, and even this solitary reserve is under pressure from illegal logging and gold prospecting.

Eastern arc forests of Tanzania. The Usambara forest, described earlier, is one of nine sections of montane forests strung across eastern Tanzania. Isolated to some extent since prehuman times, these habitats are the sites of profuse local evolution. They are the native home, for example, of 18 of the known 20 species of African violets and 16 of the species of wild coffee. The forests are down to half their original cover and shrinking fast from the incursions of Tanzania's exploding population.

Cape floristic province of South Africa. At the southern tip of Africa is a specialized heathland called *fynbos*, which is graced with one of the world's most unusual and diverse floras. In the 89,000 square kilometers of the environment still surviving, 8,600 plant species can be found. Of these, 73 percent exist nowhere else in the world. A third of the fynbos has been lost to agriculture, development, and the incursion of exotic plant species. The remainder is being rapidly fragmented and degraded. Most of the native species occur in local areas of a square kilometer or less in extent. At least 26 are known to be extinct and another 1,500 are rare and threatened, a total exceeding the entire flora of the British Isles. Unless swift action is taken, South Africa will lose a large part of its greatest natural heritage.

Madagascar. Madagascar, the most isolated of the great islands of the world, has a fauna and flora independently evolved to a corresponding degree: 30 primates, all lemurs; reptiles and amphibians that are 90 percent endemic, including two thirds of all the chameleons of the world; and 10,000 plant species of which 80 percent are endemic, including a thousand kinds of orchids. The impoverished Malagasy people have relied heavily on slash-and-burn agriculture on poor rain forest soils to sustain their growing populations, causing them to grind their way through and destroy most of the world-class biological environment they inherited. In 1985 the forest remaining intact was down to a third of the cover encountered by the first colonists fifteen centuries ago. The destruction is accelerating along with population growth, with most of the loss having occurred since 1950.

Lower slopes of the Himalayas. A girdle of lush mountain forest encircles the southern and eastern edges of the Himalayas, from Sikkim in northern India across Nepal and Bhutan to the western provinces of China. It comprises a complex mixture of tropical species of southern origin and temperate species from the north. A seemingly endless succession of deep valleys and knife-edge ridges breaks the fauna and flora into large numbers of local assemblages, containing for example about 9,000 plant species of which 39 percent are limited to the region as a whole. The original extent of the forests was roughly 340,000 square kilometers. Occurring in or near some of the most densely populated regions of the world, the forests are down by two thirds and disappearing quickly through unregulated logging and conversion to farmland.

Western Ghats of India. Along the seaward slopes of the Western Ghat mountains, extending the length of peninsular India, is a zone of tropical forest covering about 17,000 square kilometers. It is home to 4,000 known plant species, of which 40 percent are endemic. The pressure from the expanding local populations is intense, and clearing for timber and agriculture has been rapid. About a third of the cover is gone already, and the remainder is disappearing at a rate of 2–3 percent a year.

Sri Lanka. The wet forests of this island off the southern tip of India are relics of an ancient, largely vanished floristic province that once covered all of the Indian peninsula. The Sri Lankan remnant itself contains over a thousand plant species, of which half are endemic. With a population density of 260 persons per square kilometer and timber and agricultural land in heavy demand, the forest cover has been reduced to slightly less than 10 percent of its original area. Much of the primary forest growth is limited to a 56-square-kilometer tract within the Sinharaja forest near the southwestern corner of the island. This sector is also the most densely settled part of the island. To make matters worse, most of the local people depend on shifting cultivation and forest products for their livelihood.

Peninsular Malaysia. Most of the Malay peninsula was once covered by tropical forest. It contained at least 8,500 plant species, of which as many as a third were endemic. By the mid-1980s half the forest was gone. Almost all of the remaining lowland sector, the richest repository of diversity, had been degraded to some degree. About a half of the endemic tree species are now classified as endangered or extinct.

Northwestern Borneo. In earlier times Borneo was equated in lore with the perfect image of vast pristine jungle. That image has mostly faded. The forest is being stripped back swiftly, and many of the resident 11,000 plant and uncounted animal species are under siege. The northern third of the island, where biodiversity is deep and plant endemicity approaches 40 percent, has been extensively cleared by logging. In the state of Sarawak, part of Malaysia, the forest cover has been reduced by nearly a half, and most of the remainder has been consigned to timber companies.

The Philippines. This island nation is at the edge of a full-scale biodiversity collapse. Isolated from the Asian mainland but close enough to Indonesia to receive many plant and animal colonists, fragmented into 7,100 islands in a pattern that promotes species formation, the Philippines had evolved a very large fauna and flora with high levels of endemicity. In the past fifty years, two thirds of the forest has been cleared, including all but 8,000 square kilometers of the original lowland cover. From island to island intensive logging was pursued until it became uneconomical, to be followed in lockstep by full agriculture settlement. The demand for new land by a growing population endangers the remaining upland forest. Preserves are planned for 6,450 square kilometers, 2 percent of the nation's land surface. At best the ultimate losses will be heavy. As I write, the Philippine or monkey-eating eagle, majestic symbol of the nation's fauna, is down to 200 or fewer individuals.

New Caledonia. My favorite island: far enough off the east coast of Australia to spawn a unique fauna and flora; large enough to accommodate large numbers of animals and plants; and close enough to the Melanesian archipelagoes to the north to have received elements from that different biogeographical realm. For the naturalist, New Caledonia is a melting pot and a place of mystery. One of the finest days of my life was spent climbing Mt. Mou and then hiking along the summit ridge in mist-shrouded araucaria forest, where I found a pure native biota, not a single species of which I had ever seen in the wild before. The forests of New Caledonia hold 1,575 species of plants, of which an astonishing 89 percent are endemic. The New Caledonians, including the colonial French, have exploited the environment with abandon, logging, mining, and setting brushfires that push back the edges of the drier woodlands. Less than 1,500 square kilometers of undisturbed forests survive, covering 9 percent of the island. To see New Caledonia as it was, you must climb to mountain slopes too remote or steep for the loggers to clear.

Southwestern Australia. The extensive heathland west of the Nullarbor Plain evolved in a Mediterranean climate and state of isolation similar to those of the South African fynbos. It also resembles the fynbos in physical appearance and rivals it in diversity, harboring 3,630 plant species of which 78 percent are found nowhere else in the world. When I visited it in 1955, the environment was in nearly pristine condition. You could stand in the midst of the waist-high scrub in many places and see an unbroken horizon in all directions. In spring the flowers bloomed in splendid profusion. The cover has been reduced by half since then, mostly through agricultural conversion. It is being degraded further by mining operations, invasion of exotic weeds, and frequent wildfires. One quarter of its species are now classified as rare or threatened.

These are the eighteen hot spots, but the list is not closed. There are other candidates among forested regions, including the remnant rain forests of Mexico, Central America, West Indies, Liberia, Queensland, and Hawaii. To them can be added a large assemblage of entirely different habitats: the Great Lakes of East Africa and their counterpart in Siberia, Lake Baikal; virtually every river drainage system in the world near heavily populated regions, from the Tennessee to the Ganges and even some of the tributaries of the Amazon; the Baltic and Aral seas, the latter dying not just as an ecosystem but as a body of water; and a myriad of isolated tracts of species-rich tropical deciduous forests, grasslands, and deserts.

Then there are the coral reefs. These fortresses of biological diversity in the shallow tropical seas are giving way to a combination of natural and human assaults. The reefs have a permanent look but are highly dynamic in composition. Subject to the vagaries of weather and climate, they have always experienced local advances and retreats. Hurricanes periodically turn portions of the Caribbean reefs into rubble, but they grow back. El Niño events, the warming of water currents in the equatorial eastern Pacific, cause widespread mortality. The 1982–83 phenomenon, strongest recorded during the past two centuries, killed huge quantities of coral along the coasts of Costa Rica, Panama, Colombia, and Ecuador.

In normal circumstances, the reefs recover from natural destruction within a few decades. But now these natural stresses are being augmented by human activity, and the coral banks are being steadily degraded with less chance for regeneration. Reefs of twenty countries are affected around the world, from the Florida Keys and the West Indies to the Gulf of Panama and the Galápagos Islands, from Kenya and the Maldives east across a large swath of tropical Asia and south to Australia's Great Barrier Reef. In some places the reduction of reef area approaches 10 percent. Off Florida's Key Largo it is 30 percent, with most of the damage having occurred since 1970. In no particular order the principal causes are pollution (the oil spill during the Persian Gulf war being a disastrous example), accidental grounding of freighters, dredging, mining for coral rock, and harvesting of the more attractive species for decoration and amateur collections.

The decline of the reefs has been accompanied by coral bleaching. The loss of pigment is due to the failure of the zooxanthellae, the single-celled algae that live in the tissues of the coral animals and share a large fraction of the energy fixed by photosynthesis. The algae either die or lose much of the photosynthetic pigment they hold in their own cells. Like etiolated and dwarfed green plants germinated in the dark, the corals are as sickly as they look, and unless the process is reversed they die. Bleaching is a generalized stress reaction. It results variously from excessive heat or cold, chemical pollution, or dilution by fresh water, all of which are promoted by human activity.

During the 1980s, coral-reef bleaching occurred over a large part of the tropics. Rapid change most often proceeded in places where water temperatures rose conspicuously. It has been estimated that if the shallow tropical seas warm by as little as one or two degrees Celsius over the next century, many coral species would become extinct (three were lost from the eastern Pacific during the El Niño event of 1982–83 alone) and some reefs might disappear altogether. It is therefore possible that the bleaching of the last decade was the first step toward a catastrophe foretold by the rising levels of carbon dioxide in the atmosphere—possible, but still unproved. Coral bleaching in the 1980s occurred in some localities around the world but not in others. It was probably due to a variety of causes of which warming was only one. As they await further developments, marine biologists are inclined to agree that the greatest immediate peril for coral reefs comes from physical damage and pollution, not a worldwide warming trend.

But the long-term danger from climatic change looms in the decades ahead, for most ecosystems. If even the more modest projections of global warming prove correct, the world's fauna and flora will be trapped in a vise. On one side they are being swiftly reduced by deforestation and other forms of direct habitat destruction. On the other side they are threatened by the greenhouse effect. Whereas habitat loss on the land is most destructive to tropical biotas, climatic warming is expected to have a greater impact on the biotas of the cold-temperate and polar regions. A poleward shift of climate at the rate of 100 kilometers or more each century, equal to one meter or more a day, is considered at least a possibility. That rate of progression would soon leave wildlife preserves behind in a warmer regime, and many animal and plant species simply could not depart from the preserves and survive. The fossil record supports this forecast of limited dispersal. As the last continental ice sheet retreated from North America 9,000 years ago, spruce managed to spread at a rate of 200 kilometers a century, but the ranges of most other tree species spread at rates of only 10 to 40 kilometers. This history suggests that unless transplantings of entire ecosystems are undertaken, many thousands of native species are likely to be dislocated. How many will adapt to the changing climate, not having emigrated northward, and how many will become extinct? No one knows the answer.

It seems to follow that the organisms of the tundra and polar seas have no place to go even with a modest amount of global warming; the north and south poles are the end of the line. All the species of the high latitudes, reindeer moss to polar bears, risk extinction.

In another arena, large numbers of species around the world, at all latitudes, are restricted to low-lying coastal areas that will be flooded as the sea rises from the melting of polar ice. Various estimates have bracketed the rise somewhere between half a meter and two meters. In the United States, Florida will be the hardest hit region biologically. More than half of the rare animals and plants specialized for existence on the extreme coastal fringe live there. In the western Pacific many atolls, and even two small island nations, Kiribati and Tuvalu, would be largely covered by the sea.

Human demographic success has brought the world to this crisis of biodiversity. Human beings—mammals of the 50-kilogram weight class and members of a group, the primates, otherwise noted for scarcity—have become a hundred times more numerous than any other land animal of comparable size in the history of life. By every conceivable measure, humanity is ecologically abnormal. Our species appropriates between 20 and 40 percent of the solar energy captured in organic material by land plants. There is no way that we can draw upon the resources of the planet to such a degree without drastically reducing the state of most other species.

An awful symmetry of another kind binds the rise of humanity to the fall of biodiversity: the richest nations preside over the smallest and least interesting biotas, while the poorest nations, burdened by exploding populations and little scientific knowledge, are stewards of the largest. In 1950 the industrialized nations held a third of the world's population. The proportion fell to a quarter by 1985 and is expected to decline further to a sixth by 2025, when the total world population will have risen by 60 percent to 8 billion. One cannot help being struck with an irony, that if nineteenth-century technology had been born midst tropical rain forests instead of temperate-zone oaks and pines, there would be very little biodiversity left for us to save.

But what precisely is the magnitude of the crisis—how many species are disappearing? Biologists cannot tell in absolute terms because we do not know to the nearest order of magnitude how many species exist on earth in the first place. Probably fewer than 10 percent have even been given a scientific name. We cannot estimate the percentage of species going extinct each year around the world in most habitats, including coral reefs, deserts, and alpine meadows, because the requisite studies have not been made.

It is possible, though, to get a handle on the richest environment of all, the tropical rain forests, and to make a rough estimate of the extinction rates of species there. That much is possible because, thanks to the efforts of the Food and Agriculture Organization of the United Nations and a few pioneer researchers, such as Norman Myers, the rate of destruction of the rain forests has been ascertained. From the loss in forest area we can infer the rates at which species are being extinguished or doomed. And since the tropical forests contain more than half the species of plants and animals on earth, estimates pertaining to them allow us to make a rough qualitative assessment of the general severity of the biodiversity crisis.

Before attempting this projection, I am obliged to say something about the regenerative powers of rain forests. Despite their extraordinary richness, despite their reputation for exuberant growth ("the jungle quickly reclaimed the settlement as though nothing had existed there before"), these forests are among the most fragile of all habitats. Many of them grow on "wet deserts"—an unpromising soil base washed by heavy rains. Two thirds of the area of the forest surface worldwide consists of tropical red and yellow earths, which are typically acidic and poor in nutrients. High concentrations of iron and aluminum form insoluble compounds with phosphorus, decreasing the availability of that element to plants. Calcium and potassium are leached from the soil soon after their compounds are dissolved in the rain water. Only a tiny fraction of the nutrients filters deeper than 5 centimeters (2 inches) beneath the soil surface.

During the 150 million years of their existence, rain-forest trees have nevertheless evolved to grow thick and tall. At any given time, most of the carbon and a substantial fraction of the nutrients of the ecosystem are locked up in the tissue and dead wood of the vegetation. So the litter and humus on the ground are, in most cases, as thin as in any forests in the world. Here and there, patches of bare earth show through. At every turn there are signs of rapid decomposition by termites and fungi. When the forest is cut and burned, the ash and decomposing vegetation flush enough nutrients into the soil to support vigorous new herbaceous and shrubby growth for two or three years. Then the nutrients decline to levels too low to support healthy crops and forage. Farmers must add artificial fertilizer or move on to the next patch of rain forest, perpetuating the cycle of slash-and-burn.

The regeneration of rain forests is also limited by the fragility of the seeds of its trees. Those of most species germinate within a few days or weeks. They have little time to be carried by animals or water currents across the stripped land into sites favorable for growth. Most sprout and die in the hot, sterile soil of the clearings. The monitoring of logged sites indicates that regeneration of a mature forest may take centuries. Even though the forest at Angkor, for example, dates back to the abandonment of the Khmer capital in 1431, it is still structurally different from even older forests in the same region. The process of rain-forest regeneration is generally so slow, particularly after agricultural development, that few projections of its progress have been possible. In some areas, where the greatest damage is combined with low soil fertility and no native forest exists nearby to provide seeds, restoration might never occur without human intervention.

The ecology of rain forests stands in sharp contrast to that of northern temperate forests and grasslands. In North America and Eurasia, organic matter is not locked up so completely in the living vegetation. A large portion lies relatively fallow in the deep litter and humus of the soil. Seeds are more resistant to stress and able to lie dormant for long periods of time until the right conditions of temperature and humidity return. That is why it is possible to cut and burn large portions of the forest and grassland, graze cattle or grow crops for years on the land, and then see the vegetation grow back to nearly the original state a century after abandonment. Ohio, in a word, is not the Amazon. On a global scale, the north has been luckier than the south.

In 1979 tropical rain forests were down to about 56 percent of the prehistoric cover. Surveys made by satellite, by low-altitude overflights, and on the ground disclosed that the remainder, along with the much less extensive monsoon forests, was being removed at the rate of approximately 75,000 square kilometers, or 1 percent of the cover a year. Removal means that the forest is completely destroyed, with hardly a tree standing, or else degraded so severely that most of the trees die a short time later. The main causes of deforestation continue to be small-scale farming, especially slash-and-burn cultivation that leads to permanent agricultural settlement; only somewhat less important are commercial logging and cattle ranching.

During the 1980s the rate of deforestation was accelerated everywhere. It soared to tragic proportions in the Brazilian Amazon. There the people recognize three seasons, the dry, the wet, and the *queimadas*, or burnings. During the last brief period, armies of small farmers and peons employed by land barons set fires to clear the land of fallen trees and brush. About 50,000 square kilometers in four states of the Amazon (Acre, Mato Grosso, Pará, Rondonia) were cleared and burned during four months, July through October, in 1987. A similar amount was destroyed the following year. Deforestation was driven by government-sponsored road building and settlement, sanctioned as government policy. It approached holocaust proportions, with effects spreading outward across larger parts of Brazil. "At night, roaring and red," observed the journalist Marlise Simons, "the forest looks to be at war." According to a report of the Institute for Space Research, "The dense smoke produced by the Amazonian burnings, at the height of the season, spread over millions of square kilometers, bringing health problems to the population, shutting down airports, hampering air traffic, causing various accidents on riverways and on roads, and polluting the earth's atmosphere in general." Global pollution did indeed occur. The Brazilian fires manufactured carbon dioxide containing more than 500 million tons of carbon, 44 million tons of carbon monoxide, over 6 million tons of particles, and a million tons of nitrogen oxides and other pollutants. Much of this material reached the upper atmosphere and traveled in a plume eastward across the Atlantic.

By 1989 the tropical rain forests of the world had been reduced to about 8 million square kilometers, or slightly less than half of the prehistoric cover. They were being destroyed at the rate of 142,000 square kilometers a year, or 1.8 percent of the standing cover, nearly double the 1979 amount. The loss is equal to the area of a football field every second. Put another way, in 1989 the surviving rain forests occupied an area about that of the contiguous forty-eight states of the United States, and they were being reduced by an amount equivalent to the size of Florida every year.

What impact does this destruction have on biodiversity in the tropical forests? In order to set a lower limit above which the species extinction rate can be reasonably placed, I will employ what we know about the relation between the area of habitats and the numbers of species living within them. Models of this kind are used routinely in science when direct measurements cannot be made. They yield first approximations that can be improved stepwise as better models are devised and more data added.

The first model is based on the widely observed area-species curve earlier given, $S = CA^z$, where S is the number of species, A is the area of the place where the species live, and C and z are constants that vary from one group of organisms to another and from one place to another. For purposes of calculating the rate of species extinction, C can be ignored; z is what counts. In the great majority of cases the value of z falls between 0.15 and 0.35. The exact value depends on the kind of organism being considered and on the habitats in which the organisms are found. When species are able to disperse easily from one place to another, z is low. Birds have a low z value, land snails and orchids a high z value.

The higher the z value, the more the species numbers will eventually fall after the area is reduced. I say "eventually fall": whereas some doomed species may vanish quickly when a forest is trimmed back or a lake partly drained, other species decline slowly and linger a while before disappearing. In more precise language, when an area is reduced, the extinction rate rises and stays above the original background level until the species number has descended from a higher equilibrium to a lower equilibrium. The rule of thumb, to make the result immediately clear, is that when an area is reduced to one tenth of its original size, the number of species eventually drops to one half. This corresponds to a z value of 0.30 and is actually close to the number often encountered in nature.

In 1989 the area of the combined rain forests was declining by 1.8 percent each year, a rate that can be reasonably assumed to have continued into the early 1990s. At the typical z value, 0.30, each year's area reduction can be expected to reduce the number of species by 0.54 percent. Let us try to bracket the extinction rate for most kinds of organisms by estimating the minimal and maximal numbers possible. At the lowest likely z value, 0.15, this extinction rate would be 0.27 percent a year; at the highest likely z value, 0.35, the extinction rate would be 0.63 percent. *Very roughly, then, reduction in the area of tropical rain forest at the current rate can be expected to extinguish or doom to extinction about half a percent of the species in the forests each year.* More precisely, groups with a low z value will be affected the least, those with high z values the most. If most groups of organisms have low z values, the overall extinction rate will be closer to 0.27 percent; if most have high z values, the overall extinction rate will approach 0.63 percent. Not enough data exist to guess where the true overall value falls between these extremes.

If destruction of the rain forest continues at the present rate to the year 2022, half of the remaining rain forest will be gone. The total extinction of species this will cause will lie somewhere between 10 percent (based on a z value of 0.15) and 22 percent (based on a z value of 0.35). The "typical" intermediate z value of 0.30 would lead to a cumulative extinction of 19 percent over that span of time. Roughly, then, if deforestation continues for thirty more years at the present rate, one tenth to one quarter of the rain-forest species will disappear. If the rain forests are as rich in diversity as most biologists think, their reduction alone will eliminate 5 to 10 percent or more—probably considerably more—of all the species on earth in thirty years. When other species-rich but declining habitats are added, including heathland, dry tropical forests, lakes, rivers, and coral reefs, the toll mounts steeply.

The area-species relation accounts for a great deal of extinction, but not for all of it. We need a second model. As the last trees are cut, the last patch turned into a pasture or cornfield, the area-species curve plunges off the extrapolated line down to zero. So long as a small remnant of forest exists somewhere, say on a ridge in western Ecuador, a substantial number of species will hang on, most in tiny populations. Some may be doomed unless heroic efforts are made to culture and transplant them to new sites. But for the moment they hang on. When the last bit of forest or other natural habitat is removed and the area falls from 1 percent to zero, a great many species immediately perish. Such is the condition of legions of Centinelas around the world, the silent extinctions occurring as the last trees are felled. When Cebu in the Philippines was completely logged, nine of the ten bird species unique to the island were extinguished, and the tenth is in danger of joining them. We don't know how to assess global species loss from all these small-scale total extinctions. One thing is certain: because they do occur, the estimation of global rates based purely on the area-species curve must be on the low side. Consider the impact of removing the final few hundred square kilometers of natural reserves: in most cases, more than half of the original species would vanish immediately. If these were the refuges of species found nowhere else, the circumstance for so many rain-forest animals and plants, the loss in diversity would be immense.

The concept of a world peppered by miniature holocausts can be extended. Take the extreme imaginary case in which all species dwelling in the rain forests were local in distribution, limited to a few square kilometers, in the manner of the endemic plant species of

Centinela. As the forest is cut back, the percentage loss in species approaches but never quite equals the percentage loss in forest area. In the next thirty years, the world would lose not only half of its forest cover but nearly half of the forest species. Fortunately, this assumption is excessive. Some species of animals and plants dwelling in rain forests have wide geographical distributions. So the rate of species extinction is less than the reduction in area.

It follows that the amount of species loss from halving of the rain-forest area will be greater than 10 percent and less than 50 percent. But note that this range of percentages is the loss expected from the area effect only, and it is still on the low side. A few species in the remnant patches will also be lost by rifle-shot extinction, the hunting out of rare animals and plants in the manner of Spix's macaw and the New Zealand mistletoe. Others will be erased by new diseases, alien weeds, and animals such as rats and feral pigs. That secondary loss will intensify as the patches grow smaller and more open to human intrusion.

No one has any idea of the combined magnitude of these additional destructive forces in all habitats. Only the minimal in the case of tropical rain forests—10 percent extinction with a halving of area—can be drawn with confidence. But because of the generally higher z values prevailing and the additional and still unmeasured extinction factors at work, the real figure might easily reach 20 percent by 2022 and rise as high as 50 percent or more thereafter. A 20 percent extinction in total global diversity, with all habitats incorporated, is a strong possibility if the present rate of environmental destruction continues.

How fast is diversity declining? The firmer numbers I have given are the estimates of species extinctions that will *eventually* occur as rain forests are cut back. How long is "eventually"? When a forest is reduced from, say, 100 square kilometers to 10, some immediate extinction is likely. Yet the new equilibrium described by the equation $S = CA^z$ will not be reached all at once. Some species will linger on in dangerously reduced populations. Elementary mathematical models predict that the number of species in the 10-kilometer-square plot will decline at a steadily decelerating rate, swiftly at first, then slowing as the new and lower equilibrium is approached. The reasoning is simple: at first there are many species destined for extinction, which therefore vanish at a high overall rate; later only a few are endangered and the rate slows. In ideal form, with species going extinct independently of one another, this course of events is called *exponential decay.*

Employing the exponential-decay model, Jared Diamond and John Terborgh approached the problem in the following way. They took advantage of the fact that rising sea levels at the end of the Ice Age 10,000 years ago cut off small land masses that had once been connected to South America, New Guinea, and the main islands of Indonesia. When the sea flowed around them, these land masses became "land-bridge islands." The islands of Tobago, Margarita, Coiba, and Trinidad were originally part of the South and Central American mainland and shared the rich bird fauna of that continent. In a similar manner, Yapen, Aru, and Misool were connected to New Guinea and shared its fauna before becoming islands fringing its coast. Diamond and Terborgh studied birds, which are good for measuring extinction because they are conspicuous and easily identified. Both investigators arrived at the same conclusion: after submergence of the land bridges, the smaller the land-bridge island, the more rapid the loss. The extinctions were regular enough to justify use of the exponential-decay model. Extending the analysis in the American tropics, Terborgh turned to Barro Colorado Island, which was created by the formation of Gatun Lake during the construction of the Panama Canal. In this case the clock started ticking not 10,000 years ago but fifty years before the study. Applying the land-bridge decay equation to an island of this size, 17 square kilometers, Terborgh predicted an extinction of 17 bird species during the first 50 years. The actual number known to have vanished during that time is 13, or 12 percent of 108 breeding species originally present.

For a process as complex as the decline of biodiversity, the conformity of the Barro Colorado bird data to the same equation based on much larger islands and longer times, even if just within a factor of two, seemed too good to be true. But several other studies of new islands have produced similar results, which are at least consistent with the decay models, and depressing. The islands are patches of forest isolated in cleared agricultural land. When the islands are in the range of 1 to 25 square kilometers, the extinction rate of bird species during the first hundred years is 10 to 50 percent. Also, as predicted by theory, the extinction rate is highest in the smaller patches and rises steeply when the area drops below a square kilometer. Three patches of subtropical forest in Brazil surrounded by agricultural land for about a hundred years varied in area from 0.2 to 14 square kilometers; the resident bird species suffered a 14 to 62 percent extinction, in reverse order. On the other side of the world, a 0.9-square-kilometer forest patch, the Bogor Botanical Garden, was also isolated by clearing. In the first fifty years it lost 20 of its 62 breeding bird species. Still another example in a different environment: comparable rates of local extinction of bird species occurred in the wheat belt of southwestern Australia when 90 percent of the original eucalyptus woodland was removed and the remainder was broken into fragments.

There is no way to measure the absolute amount of biological diversity vanishing year by year in rain forests around the world, as opposed to percentage losses, even in groups as well known as the birds. Nevertheless, to give an idea of the dimension of the hemorrhaging, let me provide the most conservative estimate that can be reasonably based on our current knowledge of the extinction process. I will consider only species being lost by reduction in forest area, taking the lowest z value permissible (0.15). I will not include overharvesting or invasion by alien organisms. I will assume a number of species living in the rain forests, 10 million (on the low side), and I will further suppose that many of the species enjoy wide geographical ranges. Even with these cautious parameters, selected in a biased manner to draw a maximally optimistic conclusion, the number of species doomed each year is 27,000. Each day it is 74, and each hour 3.

If past species have lived on the order of a million years in the absence of human interference, a common figure for some groups documented in the fossil record, it follows that the normal "background" extinction rate is about one species per one million species a year. Human activity has increased extinction between 1,000 and 10,000 times over this level in the rain forest by reduction in area alone. Clearly we are in the midst of one of the great extinction spasms of geological history.

CHAPTER THIRTEEN

Unmined Riches

BIODIVERSITY is our most valuable but least appreciated resource. Its potential is brilliantly illustrated by the maize species *Zea diploperennis*, a wild relative of corn discovered in the 1970s by a Mexican college student in the west central state of Jalisco, south of Guadalajara. The new species is resistant to diseases and unique among living forms of maize in possessing perennial growth. Its genes, if transferred into domestic corn (*Zea mays*), could boost domestic production around the world by billions of dollars. The Jalisco maize was found just in time, however. Occupying no more than 10 hectares (25 acres) of mountain land, it was only a week away from extinction by machete and fire.

It can be safely assumed that a vast array of other beneficent but still unknown species exist. A rare beetle sitting on an orchid in a remote valley of the Andes might secrete a substance that cures pancreatic cancer. A grass down to twenty plants in Somalia could provide green cover and forage for the saline deserts of the world. No way exists to assess this treasure house of the wild except to grant that it is immense and that it faces an uncertain future.

For a start we need to reclassify environmental problems in a way that more accurately reflects reality. There are two major categories, and two only. One is alteration of the physical environment to a state uncongenial to life, the now familiar syndrome of toxic pollution, loss of the ozone layer, climatic warming by the greenhouse effect, and depletion of arable land and aquifers—all accelerated by the continued growth of human populations. These trends can be reversed if we have the will. The physical environment can be guided back and held rock-steady in a state close to optimum for human welfare.

The second category is the loss of biological diversity. Its root cause is the despoliation of the physical environment, but is otherwise radically different in quality. Although the loss cannot be redeemed, its rate can be slowed to the barely perceptible levels of prehistory. If what is left is a lesser biotic world than the one humanity inherited, at least an equilibrium will have been reattained in the birth and death of species. There is in addition a positive side not shared by the reversal of physical deterioration: merely the attempt to solve the biodiversity crisis offers great benefits never before enjoyed, for to save species is to study them closely, and to learn them well is to exploit their characteristics in novel ways.

A revolution in conservation thinking during the past twenty years, a New Environmentalism, has led to this perception of the practical value of wild species. Except in pockets of ignorance and malice, there is no longer an ideological war between conservationists and developers. Both share the perception that health and prosperity decline in a deteriorating environment. They also understand that useful products cannot be harvested from extinct species. If dwindling wildlands are mined for genetic material rather than destroyed for a few more boardfeet of lumber and acreage of farmland, their economic yield will be vastly greater over time. Salvaged species can help to revitalize timbering, agriculture, medicine, and other industries located elsewhere. The wildlands are like a magic well: the more that is drawn from them in knowledge and benefits, the more there will be to draw.

The old approach to the conservation of biodiversity was that of the bunker. Close off the richest wildlands as parks and reserves, post guards. Let the people work out their problems in the unreserved land, and they will come to appreciate the great heritage preserved inside, much as they value their cathedrals and national shrines. Parks and guards are necessary, without doubt. The approach has worked to some extent in the United States and Europe, but it cannot succeed to the desired degree in the developing countries. The reason is that the poorest people with the fastest-growing populations live next to the richest deposits of biological diversity. One Peruvian farmer clearing rain forest to feed his family, progressing from patch to patch as the soil is drained of nutrients, will cut more kinds of trees than are native to all of Europe. If there is no other way for him to make a living, the trees will fall.

Proponents of the New Environmentalism act on this reality. They recognize that only new ways of drawing income from land already cleared, or from intact wildlands themselves, will save biodiversity from the mill of human poverty. The race is on to develop methods, to draw more income from the wildlands without killing them, and so to give the invisible hand of free-market economics a green thumb.

This revolution has been accompanied by another, closely related change in thinking about biodiversity: the primary focus has moved from species to the ecosystems in which they live. Star species such as pandas and redwoods are no less esteemed than before, but they are also viewed as protective umbrellas over their ecosystems. The ecosystems for their part, containing thousands of less-conspicuous species, are assigned equivalent value, enough to justify a powerful effort to conserve them, with or without the star species. When the last tiger on Bali was shot in 1937, the rest of the island's diversity lost none of its importance.

The humble and ignored are in fact often the real star species. An example of a species lifted from obscurity to fame by its biochemistry is the rosy periwinkle (Catharanthus roseus) of Madagascar. An inconspicuous plant with a pink five-petaled flower, it produces two alkaloids, vinblastine and vincristine, that cure most victims of two of the deadliest of cancers, Hodgkin's disease, mostly afflicting young adults, and acute lymphocytic leukemia, which used to be a virtual death sentence for children. The income from the manufacture and sale of these two substances exceeds $180 million a year. And that brings us back to the dilemma of the stewardship of the world's biological riches by the economically poor. Five other species of periwinkles occur on Madagascar. One, Catharanthus coriaceus, is approaching extinction as the last of its natural habitat, in the Betsileo region of the central highlands, is cleared for agriculture.

Few are aware of how much we already depend on wild organisms for medicine. Aspirin, the most widely used pharmaceutical in the world, was derived from salicylic acid discovered in meadowsweet (Filipendula ulmaria) and later combined with acetic acid to create acetylsalicylic acid, the more effective painkiller. In the United States a quarter of all prescriptions dispensed by pharmacies are substances

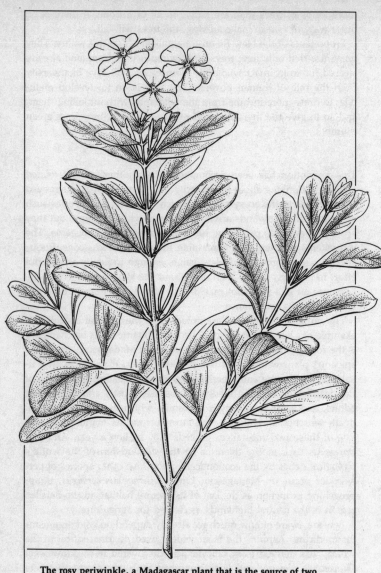

The rosy periwinkle, a Madagascar plant that is the source of two
alkaloid chemicals with powerful anticancer activity.

extracted from plants. Another 13 percent come from microorganisms and 3 percent more from animals, for a total of over 40 percent that are organism-derived. Yet these materials are only a tiny fraction of the multitude available. Fewer than 3 percent of the flowering plants of the world, about 5,000 of the 220,000 species, have been examined for alkaloids, and then in limited and haphazard fashion. The anti-cancer potency of the rosy periwinkle was discovered by the merest chance, because the species happened to be widely planted and under investigation for its reputed effectiveness as an antidiuretic.

The scientific and folkloric record is strewn with additional examples of plants and animals valued in folk medicine but still unaddressed in biomedical research. The neem tree (*Azadirachta indica*), a relative of mahogany, is a native of tropical Asia virtually unknown in the developed world. The people of India, according to a recent report of the U.S. National Research Council, treasure the species. "For centuries, millions have cleaned their teeth with neem twigs, smeared skin disorders with neem-leaf juice, taken neem tea as a tonic, and placed neem leaves in their beds, books, grain bins, cupboards, and closets to keep away troublesome bugs. The tree has relieved so many different pains, fevers, infections, and other complaints that it has been called the 'village pharmacy.' To those millions in India neem has miraculous powers, and now scientists around the world are beginning to think they may be right."

One should never dismiss the reports of such powers as superstition or legend. Organisms are superb chemists. In a sense they are collectively better than all the world's chemists at synthesizing organic molecules of practical use. Through millions of generations each kind of plant, animal, and microorganism has experimented with chemical substances to meet its special needs. Each species has experienced astronomical numbers of mutations and genetic recombinations affecting its biochemical machinery. The experimental products thus produced have been tested by the unyielding forces of natural selection, one generation at a time. The special class of chemicals in which the species became a wizard is precisely determined by the niche it occupies. The leech, which is a vampire annelid worm, must keep the blood of its victims flowing once it has bitten through the skin. From its saliva comes the anticoagulant called hirudin, which medical researchers have isolated and used to treat hemorrhoids, rheumatism, thrombosis, and contusions, conditions where clotting blood is sometimes painful or dangerous. Hirudin readily dissolves blood clots that threaten skin transplants. A second substance obtained from the saliva of the vampire bat of Central and South America is being developed to prevent heart attacks. It opens clogged arteries twice as fast as standard pharmaceutical remedies, while restricting its activity to the area of the clot. A third substance called kistrin has been isolated from the venom of the Malayan pit viper.

The discovery of such materials in wild species is but a fraction of the opportunities waiting. Once the active component is identified chemically, it can be synthesized in the laboratory, often at lower cost than by extraction from raw harvested tissue. In the next step, the natural chemical compound provides the prototype from which an entire class of new chemicals can be synthesized and tested. Some of these less-than-natural substances may prove even more efficient on human subjects than the prototype, or cure diseases never confronted with chemicals of their structural class in nature. Cocaine, for example, is used as a local anesthetic, but it has also served as a blueprint for the laboratory synthesis of a large number of specialized anesthetics that are more stable and less toxic and addictive than the natural product. Here is a brief list of pharmaceuticals derived from plants and fungi:

Drug	Plant source	Use
Atropine	Belladonna (Atropa belladonna)	Anticholinergic
Bromelain	Pineapple (Ananas comosus)	Controls tissue inflammation
Caffeine	Tea (Camellia sinensis)	Stimulant, central nervous system
Camphor	Camphor tree (Cinnamomium camphora)	Rubefacient
Cocaine	Coca (Erythroxylon coca)	Local anesthetic
Codeine	Opium poppy (Papaver somniferum)	Analgesic
Colchicine	Autumn crocus (Colchicum autumnale)	Anticancer agent
Digitoxin	Common foxglove (Digitalis purpurea)	Cardiac stimulant
Diosgenin	Wild yams (Dioscorea species)	Source of female contraceptive
L-Dopa	Velvet bean (Mucuna deeringiana)	Parkinson's disease suppressant

Ergonovine	Smut-of-rye or ergot (*Claviceps purpurea*)	Control of hemorrhaging and migraine headaches
Glaziovine	*Ocotea glaziovii*	Antidepressant
Gossypol	Cotton (*Gossypium* species)	Male contraceptive
Indicine N-oxide	*Heliotropium indicum*	Anticancer (leukemias)
Menthol	Mint (*Menta* species)	Rubefacient
Monocrotaline	*Crotalaria sessiliflora*	Anticancer (topical)
Morphine	Opium poppy (*Papaver somniferum*)	Analgesic
Papain	Papaya (*Carica papaya*)	Dissolves excess protein and mucus
Penicillin	Penicillium fungi (esp. *Penicillium chrysogenum*)	General antibiotic
Pilocarpine	*Pilocarpus* species	Treats glaucoma and dry mouth
Quinine	Yellow cinchona (*Cinchona ledgeriana*)	Antimalarial
Reserpine	Indian snakeroot (*Rauvolfia serpentina*)	Reduces high blood pressure
Scopolamine	Thornapple (*Datura metel*)	Sedative
Strychnine	Nux vomica (*Strychnos nuxvomica*)	Stimulant, central nervous system
Taxol	Pacific yew (*Taxus brevifolia*)	Anticancer (esp. ovarian cancer)
Thymol	Common thyme (*Thymus vulgaris*)	Cures fungal infection
D-tubocurarine	*Chondrodendron* and *Strychnos* species	Active component of curare; surgical muscle relaxant
Vinblastine, vincristine	Rosy periwinkle (*Catharanthus roseus*)	Anticancer

The same bright prospect exists with wild plants that can serve as food. Very few of the species with potential economic importance actually reach world markets. Perhaps 30,000 species of plants have edible parts, and throughout history a total of 7,000 kinds have been grown or collected as food but, of the latter, 20 species provide 90 percent of the world's food and just three—wheat, maize, and rice—

supply more than half. This thin cushion of diversity is biased toward cooler climates, and in most parts of the world it is sown in mono-cultures sensitive to disease and attacks from insects and nematode worms.

Fruits illustrate the pattern of underutilization. A dozen temperate-zone species—apples, peaches, pears, strawberries, and so on down the familiar roster—dominate the northern markets and are also used heavily in the tropics. In contrast, at least 3,000 other species are available in the tropics, and of these 200 are in actual use. Some, like cherimoyas, papayas, and mangos, have recently joined bananas as important export products, while carambolas, tamarindos, and co-quitos are making a promising entry. But most consumers in the north have yet to savor lulos (the "golden fruit of the Andes"), mamones, rambutans, and the near-legendary durians and mango-steens, esteemed by aficionados as the premier fruits of the world. Here are other plant foods that could be developed:

Species	Location	Use
Arracacha (Arracacia xanthorrhiza)	Andes	Carrot-like tubers with delicate flavor
Amaranths (3 species of Amaranthus)	Tropical and Andean America	Grain and leafy vegetable; livestock feed; rapid growth, drought-resistant
Buffalo gourd (Curcurbita foetidissima)	Deserts of Mexico and southwestern United States	Edible tubers, source of edible oil; rapid growth in arid land unusable for conventional crops
Buriti palm (Mauritia flexuosa)	Amazon lowlands	"Tree of life" to Amerindians; vitamin-rich fruit; pith as source for bread; palm heart from shoots
Guanabana (Annona muricata)	Tropical America	Fruit with delicious flavor; eaten raw or in soft drinks, yogurt, and ice cream
Lulo (Solanum quitoense)	Colombia, Ecuador	Fruit prized for soft drinks
Maca (Lepidium meyenii)	High Andes	Cold-resistant root vegetable resembling radish, with distinctive flavor; near extinction

Spirulina (*Spirulina platensis*)	Lake Tchad, Africa	Cyanobacterium producing vegetable supplement; very nutritious; rapid growth in saline waters
Tree tomato (*Cyphomandra betacea*)	South America	Elongated fruit with sweet taste
Ullucu (*Ullucus tuberosus*)	High Andes	Potato-like tubers, leafy part a nutritious vegetable; adapted to cold climates
Uvilla (*Pouroma cecropiaefolia*)	Western Amazon	Fruit eaten raw or made into wine; fast-growing and robust
Wax gourd (*Benincasa hispida*)	Tropical Asia	Melon-like flesh used as vegetable, soup base, and dessert; rapid growth, several crops each year

Our narrow diets are not so much the result of choice as of accident. We still depend on the plant species discovered and cultivated by our neolithic ancestors in the several regions where agriculture began. These cradles of agriculture include the Mediterranean and Near East, Central Asia, the horn of Africa, the rice belt of tropical Asia, the uplands of Mexico and Central America, and middle to high elevations in the Andes. A few favored crops were spread around the world, woven into almost all existing cultures. Had the European settlers of North America not followed the practice, had they stayed resolutely with the cultivated crops native to the new land, citizens of the United States and Canada today would be living on sunflower seeds, Jerusalem artichokes, pecans, blueberries, cranberries, and muscadine grapes. Only these relatively minor foods originated on the continent north of Mexico.

Yet even when stretched to the limit of the neolithic crops, modern agriculture is only a sliver of what it could be. Waiting in the wings are tens of thousands of unused plant species, many demonstrably superior to those in favor. One potential star species that has emerged from among the thousands is the winged bean (*Psophocarpus tetragonolobus*) of New Guinea. It can be called the one-species supermarket. The entire plant is palatable, from spinach-like leaves to young pods usable as green beans, plus young seeds like peas and tubers that, boiled, fried, baked or roasted, are richer in protein than

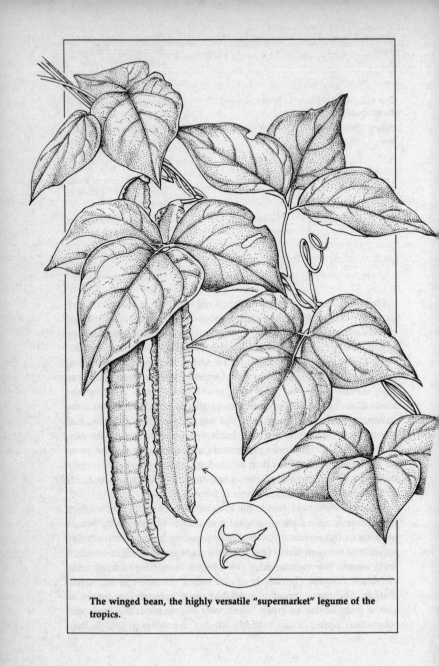

The winged bean, the highly versatile "supermarket" legume of the tropics.

potatoes. The mature seeds resemble soybeans. They can be cooked as they are or ground into flour or liquified into a caffeine-free beverage that tastes like coffee. Moreover, the plant grows at a phenomenal pace, reaching a length of 4 meters in a few weeks. Finally, the winged bean is a legume; it harbors nitrogen-fixing nodules in its roots and has little need for fertilizer. Apart from its potential as a crop, it can be used to raise soil fertility for other crops. With a small amount of genetic improvement through selective breeding, the winged bean could raise the standard of living of millions of people in the poorest tropical countries.

From the mostly unwritten archives of native peoples has come a wealth of information about wild and semicultivated crops. It is a remarkable fact that with a single exception, the macadamia nut of Australia, every one of the fruits and nuts used in western countries was grown first by indigenous peoples. The Incas were arguably the all-time champions in creating a reservoir of diverse crops. Without the benefit of wheels, money, iron, or written script, these Andean people evolved a sophisticated agriculture based on almost as many plant species as used by all the farmers of Europe and Asia combined. Their abounding crops, tilled on the cool upland slopes and plateaus, proved especially suited for temperate climates. From the Incas have come lima beans, peppers, potatoes, and tomatoes. But many other species and strains, including a hundred varieties of potatoes, are still confined to the Andes. The Spanish conquerors learned to use a few of the potatoes, but they missed many other representatives of a vast array of cultivated tuberous vegetables, including some that are more productive and savory than the favored crops. The names are likely to be unfamiliar: achira, ahipa, arracacha, maca, mashua, mauka, oca, ulloco, and yacon. One, maca, is on the verge of extinction, limited to 10 hectares in the highest plateau region of Peru and Bolivia. Its swollen roots, resembling brown radishes and rich in sugar and starch, have a sweet, tangy flavor and are considered a delicacy by the handful of people still privileged to consume them.

Another premier native crop of the Americas is amaranth. It is only now coming into the markets of the United States, mostly as a cereal supplement. Out of 60 wild species available to them, Indians from Mexico to South America cultivated three species widely during pre-Columbian times. Amaranth seeds yield a nutritious grain, and the young leaves when cooked become a palatable spinach-like green. The plants grow so well in cool, dry climates that they were favored as much as corn in Mexico at the time of the Conquest.

A cultivated amaranth, one of the principal food plants of the Amer-indians and a crop of outstanding worldwide potential.

Amaranth might have become one of the world's several leading crops after the Spanish conquest except for a bizarre historical circumstance, described by Jean Marx:

> Five hundred years ago, amaranth grain was a staple of the Aztec diet and an integral part of their religious rites. The Aztecs made idols out of a paste composed of ground, toasted amaranth seeds mixed with the blood of the human sacrifice victims. During the religious festivals, the idols were broken into pieces that were consumed by the faithful, a practice that the Spanish conquistadors considered a perverse parody of the Catholic Eucharist. When the Spanish subjugated the Aztecs in 1519, they banned the Aztec religion and with it the cultivation of amaranth.

Prejudice and inertia have always slowed the advance of agriculture. The mystery of untapped wild species is illustrated as in a parable by the case of natural sweeteners. A plant has been found in West Africa, the katemfe (*Thaumatococcus daniellii*) that produces proteins 1,600 times sweeter than sucrose. A second West African plant, the serendipity berry (*Dioscoreophyllum cumminsii*), yields a substance 3,000 times sweeter. The parable is the following: where among wild species do such progressions end? Human ingenuity has never been stretched to find the answer in this or any other domain of practical application. Consider a second, equally instructive case. The Amazonian babassu palm (*Orbignya phalerata*), even though still harvested in the wild and semiwild states, gives the world's highest known yield of vegetable oil. A stand of 500 trees produces about 125 barrels a year from huge, 100-kilogram masses of fruit. Different parts of the tree are used by local peoples to make feedcakes for livestock, pulpwood, thatching materials for roots and baskets, and finally charcoal. The babassu has not been bred to bring it to fuller commercial use, nor has it been planted extensively away from the fertile upland soils and alluvial bottomlands on which it originally grew as a wild plant.

Another frontier awaiting capital investment is saline agriculture, using salt-tolerant plants to cultivate land not previously arable. In an experimental farm in Mexico, farmers have begun to use seawater to irrigate salicornia, a native of saltflats. The small, succulent plants produce an oil resembling that of safflower. They yield two tons of oil seeds per hectare annually, leaving a residual straw that can be used to feed livestock. In Pakistan, kallar grass is grown in soil saturated with saltwater, then harvested as animal fodder. In the forbidding Atacama Desert of northern Chile, where seven years may pass without rain, the tamarugo tree sends roots through a meter of salt to tap brackish water deep within the desert soil. This extraordinary plant can create open woodland and ground vegetation in otherwise sterile wastelands. Sheep reared in tamarugo forests grow about as rapidly as those reared in high-quality pastures elsewhere in the world.

The history of animal husbandry has been just as haphazard as that of agriculture. Like crop plants, the animals of the barnyard and range are mostly limited to those first domesticated by our neolithic ancestors 10,000 years ago in the temperate zones of Europe and Asia. We have been stuck with a narrow range of ungulate mammals, horses, cattle, donkeys, camels, pigs, and goats, ill suited for most habitats of the world and often spectacularly destructive of the natural environment. In many cases these species are locally inferior in yield to wild species that humanity has left unattended.

A good example of wildlife superiority is provided by Amazon river turtles of the genus *Podocnemis*. The seven known species are highly regarded as a protein source by local people. The meat is of excellent quality and the base of a pleasing native cuisine. As the river banks have been more thickly settled, the turtles have been overhunted and several species are now endangered. But they are also easy to cultivate. Each female lays a clutch of up to 150 eggs, and the young grow rapidly. One species, the giant *Podocnemis expansa*, reaches a length of nearly a meter and a weight of 50 kilograms. It can be confined in cement tanks and natural ponds along the broad flood plains while being fed on aquatic vegetation and fruit, all at minimal cost. Under these conditions the turtle produces each year about 25,000 kilograms of meat per hectare (22,000 pounds per acre), more than 400 times the yield of cattle raised in nearby pasture cut from surrounding forests. Since floodplains compose 2 percent of the land surface of the Amazon region, the commercial potential of the species is enormous. It carries far less cost to the environment than the cattle and other exotic animals now being thrust upon the land with disastrous result.

Similar advantages are offered by the green iguana, the "chicken of the trees." A large lizard with light and tasty flesh, it has been favored as a delicacy for centuries by farmers in the humid regions of Central and South America. To be sure the iguana is a lizard, and some may flinch at the idea of eating a reptile. But it is all a matter of cultural perspective. In a phylogenetic sense, chickens and other

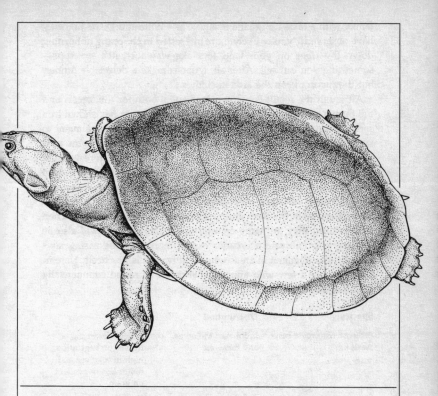

The giant Amazon river turtle, an easily cultured species that far outstrips cattle as a potential source of meat on the river floodplains.

birds are just hot reptiles with wings, and in any case our cuisines are already filled with creatures far more visually horrifying, from lobsters to thresher sharks.

But I digress. Iguanas have been rendered scarce over most of their range by overhunting and each animal now fetches $25 on the Panamanian black market. Even though they are protected by law in several Latin American countries, the big reptiles are declining because of the accelerating destruction of their forest habitat. If farmers were to leave more forest standing, there would be more iguanas for the stew pot. "But if you're a farmer with a family to feed, even a family with a taste for iguana meat," Chris Wille and Diane Jukofsky have observed, "you're likely more interested in chopping or burning down the trees on your land to make way for cattle or crops—something you can sell. After all, iguanas make a delectable dinner, but they won't keep the kids in clothes."

Although the result is a downward spiral both for the forests and the farmers, it can be reversed. As Dagmar Werner has shown in a series of impressive field experiments, the iguanas can be made to yield up to ten times the amount of meat as cattle on the same land if managed carefully, while leaving a large part of the forest intact. The trick is to cultivate a breeding stock, incubate the eggs, then protect the hatchlings during their earliest and most vulnerable growth before releasing them into the forest. The iguanas are left to feed on leaves in the tree canopies, perhaps helped along by kitchen scraps, until they are large enough to be harvested. It will also be necessary to cultivate a broader export market, while easing laws protecting the iguana in areas where cultivation is practiced. Here in summary are a few wild animals that could be raised commercially for food products:

Species	Distribution	Uses
Babirusa (*Babyrousa babyrussa*)	Indonesia: Moluccas and Sulawesi	A deep-forest pig, thrives on vegetation high in cellulose and hence less dependent on grain
Capybara (*Hydrochoeris hydrochoeris*)	South America	World's largest rodent; meat esteemed; easily ranched in open habitats near water
Chachalacas (*Ortalis*, many species)	South and Central America	Birds, potentially tropical chickens, thrive in dense populations, adaptable to human habitations, fast-growing
Gaur (*Bos gaurus*)	India to Malay peninsula	Threatened relative of domestic cattle; alternative cattle species
Green iguana (*Iguana iguana*)	American tropics	Chicken of the trees: traditional native food for 7,000 years; rapid growth; low rearing costs

Guanaco (*Lama guanicoe*)	Andes to Patagonia	Threatened species related to llama; excellent source of meat, fur, and hides; can be profitably ranched
Olive ridley sea turtles (*Lepidochelys olivacea*)	Beaches of India and Pacific coast of Mexico and Central America	Turtles emerge from sea to lay eggs; egg harvesting productive when beaches protected
Paca (*Cuniculus paca*)	American tropics	Large rodents, flesh esteemed; usually caught in wild but can be maintained in small packs in forested areas
Pigmy hog (*Sus salvanus*)	Northeastern India	One of most endangered mammal species on earth; potential source of new genes for domestic pig
Sand grouse (*Pterocles*, many species)	Deserts of Africa and Asia	Pigeon-like birds adapted to harshest deserts; domestication a possibility
Vicuna (*Lama vicugna*)	Central Andes	Threatened species related to llama; valuable source of meat, fur, and hides; can be profitably ranched

The goal of all such innovations is to increase productivity and wealth with a minimal disturbance of natural ecosystems and loss of biological diversity. Chosen and managed wisely, the exotic becomes the familiar and favored—and remains environmentally benign.

To river turtles and iguanas in the category of potential elites add the babirusa, a pig-like animal inhabiting the rain forests of Sulawesi, the Sula and Togian Islands, and Buru in eastern Indonesia. The babirusa is a bizarre creature of the kind normally seen only in zoos—slender, gray-skinned and mostly hairless, with males whose upper canines grow upward as tusks, piercing the flesh of the snout, and curving back toward the forehead without ever entering the mouth. The closest known relatives of the babirusa, all extinct, once roamed the forests of Europe. An adult is larger than most men, weighing up to 100 kilograms. Despite its resemblance to a Hindu demon, the species has been tamed by Indonesian forest peoples and serves as an important source of meat. Its most promising commercial feature, however, is its status as a possible ruminant pig. Its stomach is enlarged and chambered like that of sheep, a unique trait that apparently enables it to feed extensively on leaves and other vegetation heavy in cellulose. With luck the babirusa might enter the ranks of domestic pigs elsewhere in the world, sustained on an inexpensive and universally available fodder.

The goals of economic growth and conservation might both be served by cultivating species within their natural ecosystems, in the manner of river turtles, iguanas, and babirusas, or by the transfer of hardy species to marginal lands possessing few endemic species. The greatest potential expansion in production is by aquaculture, the rearing of fish, oysters and other mollusks, and other marine and freshwater organisms in artificial ponds or, in the case of mollusks, on the surfaces of support racks set up in estuaries. More than 90 percent of the fish consumed by human beings worldwide is obtained by the hunting of wild species in fully natural environments. This primitive industry prevails despite the fact that sophisticated aquaculture techniques are available, and fish in particular have been reared in ponds and other enclosed structures for 4,000 years. If pressed aggressively, the production of animal protein by aquaculture could easily be increased many times over within one or two decades. "There are two reasons for this vast potential," Norman Myers has written:

> First, water dwelling creatures enjoy a distinct advantage over their terrestrial relatives in that their body density is almost the same as that of the water they inhabit, so they do not have to direct energy into supporting their body weight; this means, in turn, that they can allocate more food energy to the business of growing than is the case for land animals. Second, fishes, as cold-blooded creatures, do not consume large amounts of energy to keep themselves warm. Carp, for instance, can convert one unit of assimilated food into flesh one and a half times as quickly as can pigs or chickens, and twice as rapidly as cattle or sheep. The tiny shrimplike crustaceans called Daphnia can, when raised in a nutrient-nourished environment, generate almost 20 metric tons of flesh per hectare in just under five weeks, which is ten times the production rate for soybeans—and at one-tenth the cost per unit of protein produced.

Present-day aquaculture resembles the rearing of conventional crops and livestock in utilizing only a small fraction of available diversity. It depends heavily on those species encountered happenstance for the first time by the cultures that invented the practice.

About 300 kinds of fish—finfish, to be exact, as opposed to shellfish—are cultured for food somewhere in the world. But 85 percent of the yield comes from only several carp species, while tilapias contribute a large part of the remainder. There are 18,000 more species known to science, and undoubtedly thousands of others still unknown. In the end only a small minority will prove to be commercially valuable, but even if that figure is only 10 percent, it will vastly increase the utilized diversity

It is within the power of industry to increase productivity while protecting biological diversity, and to proceed in a way that one leads to the other. The forests of the world, for example, are under pressure from a rising demand for paper pulp. Stands of a thousand species in Borneo and ancient forests in North America are converted to pulp at a rising rate, expected to reach 400 million metric tons of annual manufacture by the end of the century. There are better ways to make newspapers and carton boxes than the conversion of wildlands. Kenaf (*Hibiscus cannabinus*), an East African plant related to cotton and okra, is superior to traditional woody plants in almost every respect. Its stands, resembling those of bamboo but bearing white, hibiscus-like flowers, grow to a mature height of 5 meters in just four to five months. In the southern United States, kenaf yields three to five times more pulp than trees, and only minor chemical treatment is needed to whiten the fibers. The young stands can be cut and gathered with a machine similar to a sugarcane harvester.

Pulp and fiber can also be mass-produced from saplings raised in a remarkable form called "wood grass." In a procedure still in the experimental phase, trees are grown in dense stands and mowed like grass while still young and flexible. Their vegetation is then converted into pulp, fuel, or livestock feed. When the right tree species are chosen, the stands are fast-growing, sprout from the root stocks like grass and do not require reseeding. And when predominantly legumes, they nitrify the soil, reducing the need for fertilizer.

Kenaf and wood-grass plantations are among the latest discoveries in a saga that began with the origins of agriculture. The crucial innovations made here and there five to ten thousand years ago were the cultivation of certain food species already harvested in the wild, followed by selection of the best varieties within the species. Hunter-gatherers must have understood for millennia that plants produce seeds, which then grow into plants. It was but a small step for them to plant the seeds in convenient places. When they also learned to cultivate the plants on prepared ground, selecting the best ones to spawn the next generation, they became farmers, and agriculture was born. A chain of events had been set in motion that took the plants and their descendants, in Erich Hoyt's image, on a strange and wondrous ride into modern history.

Today the old geographic localities of neolithic agriculture support not only the domestic varieties on agricultural land but also those original wild species still surviving in the dwindling natural habitats nearby. The combination of domestic and native strains makes these sites the headquarters of genetic diversity. They are called Vavilov centers, in recognition of the pioneering work of the Russian botanist Nikolai Vavilov, who in the 1920s and 1930s traveled through Afghanistan, Ethiopia, Mexico, Central America, and the far reaches of the Soviet Union to collect plants for agricultural use. Our geographical knowledge of diversity centers has been augmented by other botanists in recent decades. There is nothing mysterious about Vavilov centers. They are for the most part simply the places where agriculture began, and hence fall within the ranges of the plant species chosen by the first farmers. In southwest Asia, for example, lived the grasses that became barley and wheat. In Mexico grew wild corn (maize), squashes, and beans, and in Peru the ancestors of potatoes.

With cultivation comes evolution by artificial selection of the succulent foliage, large tubers, and tender fruits favored by human beings. Specialization of this kind means a reduced ability to persist unattended in the original habitats. No domestic strain I know of has reentered the natural habitat of its ancestors and competed successfully there. Domestic strains are also more vulnerable to diseases and plant-eating insects and other pests. Artificial selection has always been a tradeoff between the genetic creation of traits desired by human beings and an unintended but inevitable genetic weakness in the face of natural enemies.

With the Green Revolution of agrotechnology, the tradeoff became more pronounced. Highly productive strains have been bred and mass-cultivated during the past forty years, and domestic species have become even more specialized and homogeneous than before. In India farmers originally grew as many as 30,000 varieties of rice. That diversity is being whittled down so fast that by the year 2005 three quarters of the ricefields may contain no more than ten varieties.

In a world created by natural selection, homogeneity means vulnerability. Purity of stock lowers resistance to disease, while monocultures spread contiguously over vast areas are an invitation to enemies made newly formidable. The consolidated rice paddies of Asia, rendered even more vulnerable by year-round cropping, have been opened to the swift spread of maladies that can threaten the livelihood of millions. During the 1970s the grassy-stunt virus devastated fields from India to Indonesia. Fortunately, enough wild species and varieties of rice existed to handle the problem. The International Rice Institute assayed 6,273 kinds of rice for resistance to grassy stunt. Of this array only one, the relatively feeble Indian species *Oryza nivara* (which has been known to science only since 1966), had genes with the desired qualities. It was bred with the prevailing cultivated type to create a resistant hybrid, which is now grown across 110,000 square kilometers of ricefields in Asia.

Most of the coffee plantations of Brazil are descended from a single tree that originated in East Africa. Plantings were first made in the West Indies, and some of the progeny reared there were transferred to South America. In 1970 coffee rust, a disease that had already destroyed most of the crops of Sri Lanka, appeared in Brazil and spread to Central America, threatening the economies of several countries. It happens that wild varieties of coffee still grow in the Kaffa region of southwestern Ethiopia, the presumed ancestral land of domestic coffee. Genes resistant to coffee rust were found there and bred into the Brazilian and Central American crops just in time to save the industry.

Crop species owe roughly 50 percent of their increased productivity to selective breeding and hybridization, in other words to agricultural programs that deliberately reshuffle genes among species and varieties. The modern tomato *(Lycopersicon esculentum)* is benefited by genes of many related species and races. At least nine stocks, all native to Central and South America, have either contributed valuable traits to this crop or at least possess genes that can make such a contribution:

Lycopersicon cheesmanii. Endemic of the Galápagos Islands; can be irrigated with seawater.

Lycopersicon chilense. Drought resistance.

Lycopersicon chmielewskii. Color intensity, increased sugar content.

Lycopersicon esculentum cerasiforme. Tolerance of high temperature and humidity.

Lycopersicon hirsutum. High-altitude form, resistant to many diseases and pests.

Lycopersicon parviflorum. Color intensity, increased soluble solids.

Lycopersicon pennellii. Drought resistance, increased vitamin C and sugar content.

Lycopersicon peruvianum. Pest resistance, rich source of vitamin C.

Lycopersicon pimpinellifolium. Wide disease resistance, lower acidity, higher vitamin content.

The creation of today's domestic tomato was a skilled feat of plant breeding, but one that requires many generations to accomplish. A wild species or race bred into the domestic stock also carries with it baggage of less desirable genes that reduce yield and quality. Breeders must delete these traits through repeated backcrossing, mating the hybrids back to the domestic strains, in a way that preserves only the desirable genes of both domestic and wild forms in the breeding stock. Finally, conventional hybridization can be accomplished solely among species and strains similar enough to be bred together, as in the case of the multiple parents of *Lycopersicon esculentum.*

Now, however, traditional selective breeding can be short-circuited. New methods of genetic engineering have made it possible to transfer genes directly, excising them from the chromosomes of one species and placing them into the chromosomes of another species without hybridization of the entire genomes. In other words, sex has been bypassed. Furthermore, the exchange can be accomplished among species of plants and animals so different as to make ordinary hybridization impossible. Thomas Eisner has described the possibilities in striking imagery:

> A biological species, nowadays, must be regarded as more than a unique conglomerate of genes. As a consequence of recent advances in genetic engineering, it must be viewed also as a depository of genes that are potentially transferable. A species is not merely a hard-bound volume of the library of nature. It is also a loose-leaf book, whose individual pages, the genes, might be available for selective transfer and modification of other species.

A species in the tomato genus treated as a loose-leaf notebook might share genes with species outside the genus, say plants in the larger nightshade family, or even beyond, with radically different flowering plants—donating or acquiring disease resistance, larger fruit mass, cold hardiness, the ability to grow all year, and so on through the full gamut of desired biological qualities. The possibilities are there, and they raise the potential importance for humanity of every wild species and race.

I do not mean to suggest that every ecosystem now be viewed as a factory of useful products. Wilderness has virtue unto itself and needs no extraneous justification. But every ecosystem, including those in wilderness reserves, can be the source of species to be cultivated elsewhere for practical purposes or of genes for transfer to domesticated species.

The supreme test of the utilitarian principle will be the rain forests. It is now highly profitable in most tropical countries simply to log all the trees from one tract and move on to the next. Land is cheap enough to turn a profit by the destruction of primal forests, allowing more land to be purchased and the cycle continued until the last of the trees are down. The alternative is to use the rain forests as extractive reserves, for the harvesting of "minor" products such as edible fruits, oils, latex, fibers, and medicines.

The key question from an economic viewpoint is whether the income from minor products is high enough to justify preserving rain forests as extractive reserves. The answer turns out to be yes, at least in some localities, even with the limited knowledge at hand. In 1989 Charles Peters, Alwyn Gentry, and Robert Mendelsohn demonstrated that not only are minor products in the Peruvian Amazon potentially more profitable in the long run, but considerably more so than conventional one-time logging. Among 275 kinds of trees they identified in a 1-hectare plot near the town of Mishana, 72 (26 percent) yielded fruits, vegetables, wild chocolate, and latex that could be sold in Peruvian markets. The annual net yield, after deducting costs for collecting and transport, was estimated to be $422. The Mishana plot contains enough timber to generate a net revenue of $1,000 on delivery to the sawmill if cut once, the usual practice. Within a short time, then, sustained harvesting of fruits and latex can be made more profitable than clear-cutting, and the forest is left intact. Even if trees of high profitability were removed at intervals that allowed maximal yield of their timber, the long-term income would still be less than one tenth that from fruit and latex harvesting. These are the unmined riches of the Mishana rain forest (1 hectare):

Product	Number of plants	Annual production per plant	Value (U.S. $)
Palm fruits			
Aguaje	8	195.8 lbs.	$177.60
Aguajillo	25	66.15 lbs.	75.00
Sinamillo	1	3,000 fruits	22.50
Ungurahui	36	80.48 lbs.	115.92
Other edible fruits			
Charichuelo	2	100 fruits	1.50
Lachehuaya	2	1,060 fruits	70.67
Naranjo podrido	3	150 fruits	112.50
Masaranduba	1	800 fruits	3.75
Tamamuri	3	500 fruits	11.25
Other edible products			
Sacha cacao (wild chocolate)	3	50 fruits	22.50
Shimbillo (legume)	9	200 fruits	27.00
Rubber-tree products			
Shiringa (latex)	24	4.41 lbs.	57.60
Total	117		$697.79
Cost of harvesting and transport			$276.00
Net value			$421.79

The computed yield of extractive products is actually the most conservative for the Mishana plot, since it was based exclusively on the inventory of commercially tested materials and a still poorly developed market. Little effort has been made to perform bioeconomic assays of whole ecosystems, identifying the species that can produce food and pharmaceuticals, as well as serve as agents of pest control and as restorers and enrichers of the soil. Almost all of the potent species are destroyed when the forest is clear-cut for timber and agricultural land. The old ways of using the land, chained to markets handed down by the traditions of conquistadors and the vagaries of foreign markets, extract only a small portion of the wealth while discarding the rest. The same is true, to only slightly lesser degree, in the valuation and full use of forests in the temperate zones.

Economists are struggling to fit wilderness and living species into their equations. They have created a new field, ecological economics, devoted to the sustainability of the environment and its long-term productivity. I believe they will succeed in an accurate assessment of the fraction of biodiversity subject to inventory and cost-benefit analyses, as already accomplished for the consumable plant products of the Mishana tract. They will also be able to add the revenues from "ecotourism." More and more people from developed countries are willing to pay to experience, however briefly, the prehuman earth. In 1990 tourism had risen to become the second most important source of outside income in Costa Rica, ahead of bananas and closing fast on coffee. Rain forests used for the purpose have become many times more profitable per hectare than land cleared for pastures and fields. Ecotourism is the third most important source of income in Rwanda, rising fast behind coffee and tea, largely because that tiny, overpopulated East African country is home to the mountain gorilla. As Rwanda protects the gorilla, the gorilla will help to save Rwanda.

Beyond commodity value, the economists fall short. In dimensions other than produce and tourist dollars, their yardsticks are elastic and poorly calibrated. They have no sure way of valuating the ecosystem services that species provide singly and in combination—the soil we plow, the air we breathe, the water we draw. Natural ecosystems regulate the atmospheric gases, which in turn alter temperature, wind patterns, and precipitation. The vast Amazonian rain forests create half their own rainfall. As the forests are cut, the water supply is diminished to corresponding degree. Mathematical models of the cycle of precipitation and evaporation suggest that a critical threshold of green cover exists below which the forests will no longer be able to last, converting much of the great river basin irreversibly into scrubby grassland. The pall might then travel southward to desiccate parts of Brazil's rich agricultural heartland.

When forests are leveled, the elements that composed the wood and tissue are partially converted into greenhouse gases. Then when forests are regrown, an equivalent amount of the elements are recalled into solid matter. The net loss of tropical forest cover worldwide during 1850-1980 contributed between 90 and 120 billion metric tons of carbon dioxide to the earth's atmosphere, not far below the 165 billion metric tons emanating from the burning of coal, oil, and gas. These two processes together have raised the concentration of carbon dioxide in the global atmosphere by more than 25 percent, setting the stage for global warming and a rise in sea level. The second most important greenhouse gas, methane, has doubled at about the same time; and 10 to 15 percent of the increase is thought to be due to tropical deforestation. If 4 million square kilometers of the tropical regions were replanted in forest, in other words an area half the size of Brazil, all of the current buildup of atmospheric carbon dioxide from human agents would be canceled. In addition, the rise in methane and other greenhouse gases would be slowed.

The very soils of the world are created by organisms. Plant roots shatter rocks to form much of the grit and pebbles of the basic substrate. But soils are much more than fragmented rock. They are complex ecosystems with vast arrays of plants, tiny animals, fungi, and microorganisms assembled in delicate balance, circulating nutrients in the form of solutions and tiny particles. A healthy soil literally breathes and moves. Its microscopic equilibrium sustains natural ecosystems and croplands alike.

The mere phrase "ecosystems services" has a mundane ring, rather like waste disposal or water-quality control. But if only a small percentage of the journeyman organisms filling these roles were to disappear, human life would be diminished and strikingly less pleasant. It is a failing of our species that we ignore and even despise the creatures whose lives sustain our own.

What then is biodiversity worth? The traditional econometric approach, weighing market price and tourist dollars, will always underestimate the true value of wild species. None has been totally assayed for all of the commercial profit, scientific knowledge, and aesthetic pleasure it can yield. Furthermore, none exists in the wild all by itself. Every species is part of an ecosystem, an expert specialist of its kind, tested relentlessly as it spreads its influence through the food web. To remove it is to entrain changes in other species, raising the populations of some, reducing or even extinguishing others, risking a downward spiral of the larger assemblage.

Downward by how much? The relation between biodiversity and stability is a gray area of science. From a few key studies of forests we know that diversity enlarges the capacity of the ecosystem to retain and conserve nutrients. With multiple plant species, the leaf area is more evenly and dependably distributed. Then the greater the number of plant species, the broader the array of specialized leaves and roots, and the more nutrients the vegetation as a whole can seize from every nook and cranny at every hour through all seasons. The extreme reach of biodiversity anywhere may be that attained by the orchids and other epiphytes of tropical forests, which harvest soil particles directly from mist and airborne dust otherwise destined to blow away. In short, an ecosystem kept productive by multiple species is an ecosystem less likely to fail.

If species composing a particular ecosystem begin to go extinct, at what point will the whole machine sputter and destabilize? We cannot be sure because the requisite natural history of most kinds of organisms does not exist, and experiments on ecosystem failure have been generally lacking. Yet think of how such an experiment *might* unfold. If we were to dismantle an ecosystem gradually, removing one species after another, the exact consequences at each step would be impossible to predict, but one general result seems certain: at some point the ecosystem would suffer a collapse. Most communities of organisms are held together by redundancies in the system. In many cases two or more ecologically similar species live in the same area, and any one can fill the niches of others extinguished, more or less. But inevitably the resiliency would be sapped, efficiency of the food webs would drop, nutrient flow would decline, and eventually one of the elements deleted would prove to be a keystone species. Its extinction would bring down other species with it, possibly so extensively as to alter the physical structure of the habitat itself. Because ecology is still a primitive science, no one is sure of the identity of most keystone species. We are accustomed to thinking of the organisms in this vital category as being large in size—sea otters, elephants, Douglas firs, coral heads—but they might as easily include any of the tiny invertebrates, algae, and microorganisms that teem in the substratum and that also possess most of its protoplasm and move the mass of nutrients.

Economists speak of the "option value" of a species whose worth is still unmeasured, and no measure in all of economics is more intriguing or more elusive. Its greatest difficulty is that it applies equally to commodity, amenity, and morality, the three standard domains of valuation. "As time passes," Bryan Norton has observed,

we gain knowledge in all these areas, and new knowledge may lead to new commodity uses for a species or to a new level of aesthetic appreciation, or our moral values may change and some species will, in the future, prove to have moral value that we cannot now recognize. If placing a dollar figure on these option values seems a daunting task, the situation is actually far worse than it first seems. Calculations of option value can only be begun after we identify a species, guess what uses that species might have, place some dollar value on those uses, and estimate the likelihood of discoveries occurring at any future date.

The attempt to valuate species has led to two competing guidelines of conservation. The first is cost-benefit (CB) analysis, which singles out each threatened species in turn, weighs visible and possible future benefits against the costs of keeping it alive, and decides whether to invest enough land and time to preserve it. The second guideline is the safe minimum standard (SMS), which treats each as an irreplaceable resource for humanity, to be preserved for posterity unless the costs are unbearably high.

Surely prudence and a decent concern for posterity demand the safe minimum standard. Cost-benefit studies consistently undervalue the net benefits conferrable by species since it is much easier to measure the costs of conservation than the ultimate gains, even in purely monetary units. The riches are there, fallow in the wildlands and waiting to be employed by our hands, our wit, our spirit. It would be folly to let any species die by the sole use of the criterion of economic return, however potent, simply because the name of that species happens to be written in red ink.

CHAPTER FOURTEEN

Resolution

EVERY COUNTRY has three forms of wealth: material, cultural, and biological. The first two we understand well because they are the substance of our everyday lives. The essence of the biodiversity problem is that biological wealth is taken much less seriously. This is a major strategic error, one that will be increasingly regretted as time passes. Diversity is a potential source for immense untapped material wealth in the form of food, medicine, and amenities. The fauna and flora are also part of a country's heritage, the product of millions of years of evolution centered on that time and place and hence as much a reason for national concern as the particularities of language and culture.

The biological wealth of the world is passing through a bottleneck destined to last another fifty years or more. The human population has moved past 5.4 billion, is projected to reach 8.5 billion by 2025, and may level off at 10 to 15 billion by midcentury. With such a phenomenal increase in human biomass, with material and energy demands of the developing countries accelerating at an even faster pace, far less room will be left for most of the species of plants and animals in a short period of time.

The human juggernaut creates a problem of epic dimensions: how to pass through the bottleneck and reach midcentury with the least possible loss of biodiversity and the least possible cost to humanity. In theory at least, the minimization of extinction rates and the minimization of economic costs are compatible:

the more that other forms of life are used and saved, the more productive and secure will our own species be. Future generations will reap the benefit of wise decisions taken on behalf of biological diversity by our generation.

What is urgently needed is knowledge and a practical ethic based on a time scale longer than we are accustomed to apply. An ideal ethic is a set of rules invented to address problems so complex or stretching so far into the future as to place their solution beyond ordinary discourse. Environmental problems are innately ethical. They require vision reaching simultaneously into the short and long reaches of time. What is good for individuals and societies at this moment might easily sour ten years hence, and what seems ideal over the next several decades could ruin future generations. To choose what is best for both the near and distant futures is a hard task, often seemingly contradictory and requiring knowledge and ethical codes which for the most part are still unwritten.

If it is granted that biodiversity is at high risk, what is to be done? Even now, with the problem only beginning to come into focus, there is little doubt about what needs to be done. The solution will require cooperation among professions long separated by academic and practical tradition. Biology, anthropology, economics, agriculture, government, and law will have to find a common voice. Their conjunction has already given rise to a new discipline, biodiversity studies, defined as the systematic study of the full array of organic diversity and the origin of that diversity, together with the methods by which it can be maintained and used for the benefit of humanity. The enterprise of biodiversity studies is thus both scientific, a branch of pure biology, and applied, a branch of biotechnology and the social sciences. It draws from biology at the level of whole organisms and populations in the same way that biomedical studies draw from biology at the level of the cell and molecule. Where biomedical studies are concerned with the health of the individual person, biodiversity studies are concerned with the health of the living part of the planet and its suitability for the human species. What follows, then, is an agenda on which I believe most of those who have focused on biodiversity might agree. All the enterprises I will list are directed at the same goal: to save and use in perpetuity as much of earth's diversity as possible.

1. *Survey the world's fauna and flora.* In approaching diversity, biologists are close to traveling blind. They have only the faintest idea

of how many species there are on earth or where most occur; the biology of more than 99 percent remain unknown. Systematists are aware of the urgency of the problem but far from agreed on the best way to solve it. Some have recommended the initiation of a global survey, aimed at the discovery and classification of all species. Others, sensibly noting the shortage of personnel, funds, and time, think the only realistic hope lies in the rapid recognition of the threatened habitats that contain the largest number of endangered endemic species (the hot spots).

In order to move systematics into the larger role demanded by the extinction crisis, its practitioners have to agree on an explicit mission with a timetable and cost estimates. The strategy most likely to work is mixed, aiming at a complete inventory of the world's species, but across fifty years and at several levels, or scales in time and space, from hot-spot identification to global survey, audited and readjusted at ten-year intervals. As each decade comes to a close, progress to that point could be assessed and new directions identified. Emphasis from the outset would be placed on the hottest spots known or suspected.

Three levels can be envisioned. The first is the RAP approach, from the prototypic Rapid Assessment Program created by Conservation International, a Washington-based group devoted to the preservation of global biodiversity. The purpose is to investigate quickly, within several years, poorly known ecosystems that might be local hot spots, in order to make emergency recommendations for further study and action. The area targeted is limited in extent, such as a single valley or isolated mountain. Because so little is known of classification of the vast majority of organisms and so few specialists are available to conduct further studies, it is nearly impossible to catalog the entire fauna and flora of even a small endangered habitat. Instead a RAP team is formed of experts on what can be called the elite focal groups—organisms, such as flowering plants, reptiles, mammals, birds, fishes, and butterflies, that are well enough known to be inventoried immediately and can thereby serve as proxies for the whole biota around them.

The next level of inventory is the BIOTROP approach, from the Neotropical Biological Diversity Program of the University of Kansas and a consortium of other North American universities formed in the late 1980s. Instead of pinpointing brushfires of extinction at selected localities in the RAP manner, BIOTROP explores more systematically across broad areas believed to be major hot spots or at least to contain multiple hot spots. Examples of such regions include the eastern slopes of the Andes and the scattered forests of Guatemala and southern Mexico. Beyond identifying critical localities, the larger goal is to set up research stations across the area that embrace different latitudes and elevations. The work begins with a few focal organisms. It expands to less familiar groups, such as ants, beetles, and fungi, as enough specimens are collected and experts in the groups are recruited to study them. In time, close studies of rainfall, temperature, and other properties of the environment are added to the species inventory. The most important and best equipped of the stations are likely then to evolve into centers of long-term biological research, with leadership roles taken by scientists from the host countries. They can also be used to train scientists from different parts of the world.

We now come to the third and highest stage of the biodiversity survey. From inventories at the RAP and BIOTROP levels in different parts of the world, accompanied by monographic studies of one group of organisms after another, the description of the living world will gradually coalesce to create a fine-grained image of global biodiversity. The growth of knowledge will inevitably accelerate, even given a constant level of effort, by producing its own economies of scale. Costs per species logged into the inventory fall as new methods of collecting and distributing specimens are devised and procedures for accessing information are improved. Costs are not simply additive when nonelite groups of organisms are included, but instead decline on a per-species basis. Botanists, for example, can collect insects living on the plants they study, while identifying these hosts for the entomologists, and entomologists can run the procedure in reverse, gathering plant specimens in company with the insects they collect. Groups such as reptiles, beetles, and spiders can be sampled across entire habitats, then distributed to specialists on each group in turn.

As biodiversity surveys proceed at the several levels, the knowledge gathered becomes an ever more powerful magnet for other kinds of science. Field guides and illustrated treatises open doors to the imagination, and networks of technical information draw geologists, geneticists, biochemists, and others into the enterprise. It will be logical to gather much of the activity into biodiversity centers, where data are gathered and new inquiries planned. The prototype is Costa Rica's National Institute of Biodiversity (Instituto Nacional de Biodiversidad), INBio for short, established on the outskirts of the capital city of San José in 1989. The aim of INBio is nothing less than to account for all the plants and animals of this small Central American country, over half a million species in number, and to use the information to improve Costa Rica's environment and economy. It is perhaps odd that a developing nation should lead the way in such a concerted scientific enterprise, but others will follow. Detailed distribution maps of plants and many kinds of animals have been drawn up in Great Britain, Sweden, Germany, and other European countries under governmental and private auspices. As I write, plans for a national biodiversity center in the United States have been advanced by the Smithsonian Institution and are under wide discussion. Enabling legislation has been placed before Congress but is not yet passed.

The national center of the United States will not have to start from scratch. Many kinds of organisms have been already carefully studied and mapped. Several of the states, including Massachusetts and Minnesota, have undertaken programs to locate endangered species of plants and vertebrate animals within their borders. For fifteen years the Nature Conservancy, one of the premier private American foundations, has conducted a similar effort across all the states. The operation, setting up Natural Heritage Data Centers, has recently been extended to fourteen Latin American and Caribbean countries.

Another key element of biodiversity studies at all levels will be microgeography, the mapping of the structure of the ecosystem in sufficiently fine detail to estimate the populations of individual species and the conditions under which they grow and reproduce. A working technology already exists in the form of Geographic Information Systems, a collection of layers of data on topography, vegetation, soils, hydrology, and species distributions that are registered electronically to a common coordinate system. When applied to biodiversity and endangered species, the cartography is called *gap analysis*. Even though incomplete, gap analysis can reveal the effectiveness of existing parks and reserves. It can be used to help answer the larger questions of conservation practice. Do protected areas in fact embrace the largest possible number of endemic species? Are the surviving habitat fragments large enough to sustain the populations indefinitely? And what is the most cost-effective plan for further land acquisition?

The same information can be used to zone large regions. Parcels of land will have to be set aside as inviolate preserves. Others will be identified as the best sites for extractive reserves, for buffer zones used in part-time agriculture and restricted hunting, and for land

Geographic Information Systems

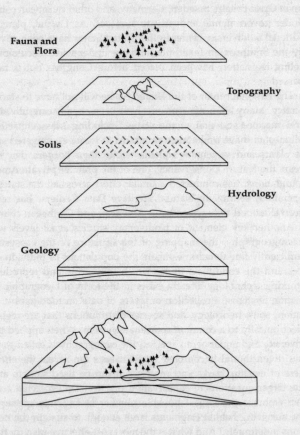

Fauna and
Flora

Topography

Soils

Hydrology

Geology

Geographic Information Systems combine information on physical and
biological environments by joining layered data sets. These can be used
to manage the landscape in a way that protects endangered species and
ecosystems, including the designation of natural reserves.

convertible totally to human use. In the expanded enterprise, landscape design will play a decisive role. Where environments have been mostly humanized, biological diversity can still be sustained at high levels by the ingenious placement of woodlots, hedgerows, watersheds, reservoirs, and artificial ponds and lakes. Master plans will meld not just economic efficiency and beauty but also the preservation of species and races.

The layered data can further aid in defining "bioregions," areas such as watersheds and forest tracts that unite common ecosystems but often extend across the borders of municipalities, states, or even countries. A river may make economic or military sense in dividing two political units, but it makes no sense at all in organizing land-use management. Bioregionalism has had a long but inconclusive history within the United States. It dates back at least as far as John Muir's successful championing of national parks and the establishment of the national forest system in 1891. Since the 1930s it has received increasing governmental sanctions with variable specific agendas, from the Tennessee Valley Authority, which managed land and created hydroelectric power through a large part of the southeast, to the establishment of the Appalachian National Scenic Trail, federal and state management of the south Florida water system and the Everglades, and the multiple regulatory and promotional activities of the New England River Basins Commission during its tenure from 1967 to 1981.

Other examples of bioregionalism abound in the United Sates, but it cannot be said that the movement has coalesced around any single philosophy of land management. Nor has the preservation of biodiversity ranked as more than an auxiliary goal. In fact the great dams built by the Tennessee Valley Authority, while providing cheap electric power to an impoverished part of the nation, inadvertently wiped out a substantial part of the native river fauna. The lower priority given diversity has not been by deliberation but from incomplete knowledge of the faunas and floras of the affected regions.

Systematics, having emerged as a prerequisite for effective long-term zoning and bioregionalism, is a labor-intensive enterprise. Scientists who study the classification of particular organisms, such as centipedes and ferns, are often by default the only authorities on the general biology of those organisms. About 4,000 such specialists in the United States and Canada attempt to manage the classification of the many thousand species of animals, plants, and microorganisms living on the continent. To varying degree they are also responsible for the millions of species occurring elsewhere in the world, since even fewer systematists are active in other countries. Probably a maximum of 1,500 trained professional systematists are competent to deal with tropical organisms, or more than half of the world's biodiversity. A typical case is the shortage of experts on termites, which are premier decomposers of wood, rivals of earthworms as turners of the soil, owners of 10 percent of the animal biomass in the tropics, and among the most destructive of all insect pests. There are exactly three people qualified to deal with termite classification on a worldwide basis. A second revealing case: the oribatid mites, tiny creatures resembling a cross between a spider and a tortoise, are among the most abundant animals of the soil. They are major consumers of humus and fungus spores, and therefore key elements of land ecosystems almost everywhere. In North America only one expert attends to their classification on a full-time basis.

With so few people prepared to launch it, a complete survey of earth's vast reserves of biological diversity may seem beyond reach. But compared with what has been dared and achieved in high-energy physics, molecular genetics, and other branches of big science, the magnitude of its challenge is not all that great. The processing of 10 million species is achievable within fifty years, even with the least efficient, old-fashioned methods. If one systematist proceeded at the cautious pace of ten species per year, including field trips for collecting, analysis of specimens in the laboratory, and publication, taking time out for vacations and family, about one million person-years of work would be required. Given forty years of productive life per scientist, the effort would consume 25,000 professional lifetimes. The number of systematists would still represent less than 10 percent of the current population of scientists active in the United States alone, and it falls well short of the number of enlisted men in the standing armed forces of Mongolia, not to mention the trade and retail personnel of Hinds County, Mississippi. The volumes of published work, one page per species, would fill 12 percent of the shelves of the library of Harvard's Museum of Comparative Zoology, one of the larger institutions devoted to systematics.

I have based these estimates on what is the least efficient procedure imaginable, in order to establish the plausibility of a total inventory of global biodiversity. Systematic work can be speeded up many times over by new techniques now coming into general use. The Statistical Analysis System (SAS), a set of computer programs already running in several thousand institutions worldwide, records taxonomic identifications and localities of individual specimens and automatically integrates data in catalogs and maps. Other computer-aided techniques compare species automatically across large numbers of traits, applying unbiased measures of similarity, the procedure called *phenetics*. Still others assist in deducing the most likely family trees of species, the method called *cladistics*. Scanning-electron microscopy has accelerated the illustration of insects and other small organisms. Computer technology will in time include image scanning that can identify species instantly while flagging specimens that belong to new species. Biologists are also close to electronic publication, which will allow retrieval of descriptions and analyses of particular groups of organisms by desktop personal computers.

Every other form of biological information on species—ecology, physiology, economic uses, status as vectors, parasites, agricultural pests—can be layered in the databases. DNA and RNA sequences and gene maps can be added. GenBank, the genetic-sequence bank, has been chartered to provide a computer database for all known DNA and RNA sequences and related biological information. By 1990 it had accumulated 35 million sequences distributed through 1,200 species of plants, animals, and microorganisms. The rate of data accession is ascending swiftly with the advent of improved sequencing methods.

2. *Create biological wealth.* As species inventories expand, they open the way to bioeconomic analysis, the broad assessment of the economic potential of entire ecosystems. Every community of organisms contains species with potential commodity value—timber and wild-plant products to be harvested on a sustained basis, seeds and cuttings that can be transplanted to grow crops and ornamentals elsewhere, fungi and microorganisms to be cultured as sources of medicinals, organisms of all kinds offering new scientific knowledge that points to still more practical applications. And the wild habitats have recreational value, which will grow as a larger sector of the public travels and learns to enjoy natural history.

The decision to make bioeconomic analysis a routine part of land-management policy will protect ecosystems by assigning them future value. It can buy time against the removal of entire communities of organisms ignorantly assumed to lack such value. When local faunas and floras are better known, the decision can be taken on how to use them optimally—whether to protect them, to extract products from them on a sustainable yield basis, or to destroy their habitat for full human occupation. Destruction is anathema to conservationists, but the fact remains that most people, lacking knowledge, regard it as perfectly acceptable. Somehow knowledge and reason must be made to intrude. I am willing to gamble that familiarity will save ecosystems, because bioeconomic and aesthetic values grow as each constituent species is examined in turn—and so will sentiment in favor of preservation. The wise procedure is for law to delay, science to evaluate, and familiarity to preserve. There is an implicit principle of human behavior important to conservation: *the better an ecosystem is known, the less likely it will be destroyed.* As the Senegalese conservationist Baba Dioum has said, "In the end, we will conserve only what we love, we will love only what we understand, we will understand only what we are taught."

A key enterprise in bioeconomic analysis is what Thomas Eisner has called *chemical prospecting,* the search among wild species for new medicines and other useful chemical products. The logic of prospecting is supported by everything we have learned about organic evolution. Each species has evolved to become a unique chemical factory, producing substances that allow it to survive in an unforgiving world. A newly discovered species of roundworm might produce an antibiotic of extraordinary power, an unnamed moth a substance that blocks viruses in a manner never guessed by molecular biologists. A symbiotic fungus cultured from the rootlets of a nearly extinct tree might yield a novel class of growth promoters for plants. An obscure herb could be the source of a sure-fire blackfly repellent—at last. Millions of years of testing by natural selection have made organisms chemists of superhuman skill, champions at defeating most of the kinds of biological problems that undermine human health.

Because chemical prospecting depends so heavily on classification, it is best conducted in tandem with biodiversity surveys. In order to succeed, investigators must also work in laboratories equipped with advanced facilities, which are usually available only in industrialized countries. In 1991 Merck and Company, the world's largest pharmaceutical firm, agreed to pay Costa Rica's National Institute of Biodiversity $1 million to assist in such a screening effort. The institute will collect and identify the organisms, sending chemical samples from the most promising species to the Merck laboratories for medicinal assay. If natural substances are marketed, the company is committed to pay the Costa Rican government a share of the royalties, which will then be earmarked for conservation programs. Merck has previously marketed four drugs from soil organisms originating from other countries. One, derived from a fungus, is Mevacor, an effective agent for lowering cholesterol levels. In 1990 Merck sold $735 million worth of this substance alone. It follows that a single success in Costa Rica—a commercial product from, say, any one species among the 12,000 plants and 300,000 insects estimated to live in the country—could handsomely repay Merck's entire investment.

There are historical reasons why Merck and other research and commercial organizations are increasingly inclined to take on chemical prospecting. The search for naturally occurring drugs and other chemical products has been cyclical through the years. In the 1960s and 1970s pharmaceutical companies phased out the screening of plants on the grounds that it was too complicated and expensive. With only one in 10,000 species yielding a promising substance (by procedures then in use) and millions of dollars needed to bring a product fully on line, the eventual payoff seemed marginal. The companies turned to new technologies in microbiology and synthetic chemistry, hoping to design the magic bullets of the new medical age with chemicals taken from the shelf. To rely on human ingenuity rather than evolved natural chemistry in distant jungles seemed much more "scientific" and direct, and perhaps less expensive. Yet natural products remained a potential shortcut, a Columbus-like journey west, for those willing to acquire the essential skills. Now the pendulum has begun to swing back, again from advances in technology, because high-volume, robot-controlled biological assays allow larger companies to screen up to 50,000 samples a year using only bits of fresh tissue or extract flown to them from any part of the world.

The path from wild organism to commercial production can sometimes be shortened further by taking clues from the lore and traditional medicine of indigenous peoples. It is a remarkable fact that of the 119 known pure pharmaceutical compounds used somewhere in the world, 88 were discovered through leads from traditional medicine. The knowledge of all the world's indigenous cultures, if gathered and catalogued, would constitute a library of Alexandrian proportions. The Chinese, for example, employ materials from about 6,000 of the 30,000 plant species in their country for medicinal purposes. Among them is artemisinin, a terpene derived from the annual wormwood (*Artemisia annua*), which shows promise as an alternative to quinine in the treatment of malaria. Because the molecular structures of the two substances are entirely different, artemisinin would have been discovered much less quickly if not for its folkloric reputation.

Because the lives of people and the reputations of shamans have depended on it for generations, much of the traditional pharmacopoeia is reliable. Extraction procedures and dosage have been tested by trial and error countless times. But this preliterate knowledge, like so many of the plant and animal species to which it pertains, is disappearing rapidly as tribes move from their homelands onto farms and into cities and villages. When they take up new trades, their languages fall into disuse and the old ways are forgotten. During the 1980s, all but 500 of the 10,000 Penans of Borneo abandoned their centuries-old seminomadic life in the forests and settled in villages. Today their memories are fading quickly. Eugene Linden notes, "Villagers know that their elders used to watch for the appearance of a certain butterfly, which always seemed to herald the arrival of a herd of boar and the promise of good hunting. These days, most of the Penans cannot remember which butterfly to look for." On the other side of the world, 90 of Brazil's 270 Indian tribes have vanished since 1900, and two thirds of those remaining contain populations of less than a thousand. Many have lost their lands and are forgetting their cultures.

Small farms around the world are giving way to the monocultures of agrotechnology. The raised garden squares of the Incas have all but vanished; the densely variegated gardens of Mesoamerica and West Africa are threatened. The revitalization of local farming is another aim of biodiversity studies. The goal is to make the practice more economically practical, while conserving the genetic reserves that will contribute to crops of the future. Species and strains of high economic efficiency, from perennial corn to amaranth and iguanas, can be fed through research centers into the local regions best suited to use them. A successful prototype of such enterprises is the Tropical Agricultural Research and Training Center (CATIE) at Turrialba, Costa Rica. Created by the Organization of American States in 1942, CATIE maintains large samples of plant species, including disease-resistant strains of cacao and other tropical crops. Its staff members experiment with propagation methods for crops and timber, design wildland preservation programs, search for new crop species and varieties, and train students in the new methods of agriculture and conservation. Institutions of the future can be profitably built to include not only these activities but also chemical prospecting and molecular techniques of gene transfer from wild to domestic species.

3. *Promote sustainable development.* The rural poor of the Third World are locked onto a downward spiral of poverty and the destruction of

diversity. To break free they need work that provides the basic food, housing, and health care taken for granted by a great majority of people in the industrialized countries. Without it, lacking access to markets, hammered by exploding populations, they turn increasingly to the last of the wild biological resources. They hunt out the animals within walking distance, cut forests that cannot be regrown, put their herds on any land from which they cannot be driven by force. They use domestic crops ill suited to their environment, for too many years, because they know no alternative. Their governments, lacking an adequate tax base and saddled with huge foreign debts, collaborate in the devastation of the environment. Using an accountant's trick, they record the sale of forests and other irreplaceable natural resources as national income without computing the permanent environmental losses as expense.

The poor are denied an adequate education. They cannot all move into the cities; in most countries, and especially those in the tropics, industrialization will be too slow to absorb more than a small fraction into the labor force. Their striving billions will, for the next century at least, have to be accommodated in rural areas. So the issue comes down to this: how can people in developing countries achieve a decent living from the land without destroying it?

The proving ground of sustainable development will be the tropical rain forests. If the forests can be saved in a manner that improves local economies, the biodiversity crisis will be dramatically eased. Within that "if" are folded technical and social difficulties of the most vexing kind. But many paths to the goal have been suggested, and some have successfully tested.

One of the most encouraging advances to date is the demonstration, cited in the last chapter, that the extraction of nontimber products from Peruvian rain forests can yield similar levels of income as logging and farming, even with the limited outlets available in existing local markets. The practice has been regularized by the rubber tappers of Brazil without a bit of theory or cost-benefit analysis. The tappers, or *seringueiros* as they are locally called, are the descendants of immigrants from northeastern Brazil who colonized portions of the Amazon during the late nineteenth century and found a steady living in latex harvesting. Half a million strong, they draw their principal income today not only from rubber but also from Brazil nuts, palm hearts, tonka beans, and other wild products. Each family owns a house in the midst of harvesting pathways shaped like clover leaves. In addition to harvesting natural products, rubber tappers also hunt, fish, and practice small-scale agriculture in forest clearings.

Because they depend on biological diversity, the tappers are devoted to the preservation of the forests as stable and productive ecosystems. They are in fact full members of the ecosystems. In 1987 the Brazilian government authorized the establishment of *seringueiro* extractive reserves on state land, with thirty-year renewable leases and a prohibition on the clear-cutting of timber.

Extractive reserves represent a major conceptual advance, but they are not enough to save more than a small portion of the rain forests. In 1980 rubber-tapper households occupied 2.7 percent of the area of the North Region of Brazilian Amazonia, including the states of Amazonas and Acre, while farms and ranches occupied 24 percent. Only a small fraction of the flood of new immigrants now pouring into the region can become extractivists. The rest will seek income wherever they find it, primarily by advancing the agricultural frontier. The key to the future of Amazonia and other forested tropical regions is whether employment made available to them saves or destroys the environment. "The real challenge," John Browder writes, "is not where to designate extractive reserves, but rather, how to integrate sustainable extraction and other natural forest management practices into the production strategies of those existing rural properties, small farms and large ranches alike, that are responsible for most of the devastation being visited upon Amazonian rainforests. Fundamentally, the problem is not where to sequester forests, but how to turn people into better forest managers."

It is possible to harvest timber from the Amazonian wilderness and other great remaining rain forests extensively and profitably with little loss of biodiversity. The method of choice, first suggested by Gary Hartshorn in 1979 and extended by other foresters, is strip logging. While lowland forested basins are not rugged in terrain, most are moderately rolling with well-defined slopes and dense systems of drainage streams. Strip logging imitates the natural fall of trees that create linear gaps through the forest, with the artificial gaps being aligned along the contours. The technique is described by Carl Jordan:

> In this scheme, a strip is harvested on the contour of a slope, parallel to the stream. Along the upper edge of the strip is a road used for hauling out the logs. After harvesting, the area is left for a few years until saplings begin to grow in the cut areas. Then the loggers clear-cut another strip, this time above the road. The advantages of this system are that the nutrients from the freshly cut second strip wash downslope

into the rapidly regenerating first strip, where the trees can quickly use the nutrients, and that seeds from the mature forest above the cut area will roll down into the recently cut strip. In contrast, in clear-cutting there are no saplings with well-developed roots capable of retaining nutrients in the system, nor is there a source of seed for regeneration of the forest.

So far so good, but how can governments and local peoples be persuaded to adopt such innovations as extractive reserves and strip logging? The shift to sustainable development will depend as much on education and social change as on science. Around the world modest projects are being advanced with one common result: if procedures tailored to the special case are used, economic development and conservation can both be served. People can be persuaded; they understand their own long-term interest and they can adapt. Here are three successful programs from Latin America.

• By Panama law, the Kuna Indians hold sovereign rights over the San Blas Islands and 300,000 hectares of adjacent mainland forest. The Kuna maintain "spirit sanctuaries," areas of primary forest in which only certain kinds of trees may be cut and no farming is allowed. Local communities depend on the sea for most of their protein, on the forests for wood, game, and medicine, and on limited patches of cleared land for domestic crops. When a spur of the Pan-American Highway was brought to the edge of their land, the Kuna established a forest reserve and guarded it with their own people. Well aware of the outside world, welcoming to visitors, the tribes have nevertheless chosen to discourage immigration and to preserve their own culture within the bountiful natural environment that has sustained them for centuries.

• Most of Central America, unlike the land of the Kuna, is plagued by soil erosion and nutrient loss owing to the excessive cultivation of maize and other crops, leading to the cutting of forests on ever steeper slopes, all driven in turn by overpopulation. As production declines, farmers invade the remaining natural areas in search of more arable land. The process is especially acute in the Güinope region of Honduras. In 1981 two private foundations, one international and one Honduran, commenced a pilot program in some of the Güinope villages under government auspices to raise productivity and restore the land. They introduced drainage ditches, contour furrows, grassy barriers, and intercropping with nitrogen-restoring legumes. The field labor and implementation costs were provided entirely by the farmers. Within several years, yields tripled and emigration nearly ceased. The new agricultural methods began to spread to surrounding areas.

• When a highway, the Carretera Marginal de la Selva, was cut into Peru's Palcazú Valley, 85 percent of the land was still clothed by rain forest. Like most of the eastern tropical slopes of the Andes, the valley is biologically rich, containing for example more than a thousand species of trees. The region also supported about 3,000 Amuesha Indians and an equal number of settlers who had established small landholdings over the previous fifty years. Once opened to outside commerce, the typical fate of a western Amazonian valley is to be clear-cut by new immigrants and logging companies, then used for cattle ranches and small farms. The thin, acidic soil soon loses most of its free phosphates and other nutrients, launching the next phase: erosion, poverty, partial abandonment. For this valley, however, an alternative plan was proposed by the U.S. Agency for International Development and approved by the Peruvian government. It is to extract timber by strip cutting, regulated to allow perpetual regeneration of the forest through thirty- to forty-year rotations. The plan permits limited permanent conversion of the most arable land to agriculture and livestock production. But it also calls for the establishment of a watershed reserve in the adjacent San Matias mountain range and the designation of the neighboring Yana- chaga range as the Yanachaga-Chemillén National Park. With luck, the Palcazú will support a healthy human population and a slice of Peru's biodiversity into the next century.

Wildlands and biological diversity are legally the properties of nations, but they are ethically part of the global commons. The loss of species anywhere diminishes wealth everywhere. Today the poor- est countries are rapidly decapitalizing their natural resources and unintentionally wiping out much of their biodiversity in a scramble to meet foreign debts and raise the standard of living. By perceived necessity they follow environmentally destructive policies that yield the largest short-term profits. The rich debt-holding nations aggra- vate the practice by encouraging a free market in poor countries while providing subsidies to farmers at home.

Consider the infamous "hamburger connection" between the United States and Central America. By 1983, in response to the excellent U.S. market for beef, Costa Rican landowners had acceler- ated the creation of new pastures until only 17 percent of the coun- try's original forest cover was left. For a time it was the world's leading exporter of beef to the United States. When northern tastes

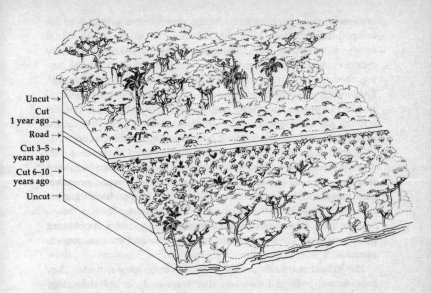

Uncut →
Cut 1 year ago →
Road →
Cut 3–5 years ago →
Cut 6–10 years ago →
Uncut →

Strip logging allows a sustainable timber yield from forests, including the relatively fragile rain forests. A corridor is cleared along the contours of the land, narrow enough to allow natural regeneration within a few years. Another corridor is then cut above the first, and so on, through a cycle lasting many decades.

changed somewhat and the market fell, Costa Rica was left with a denuded landscape and widespread soil erosion. It had also lost part of its biological diversity.

Developing countries competing in an international free market have a strong incentive to transfer capital into single-money crops such as bananas, sugar cane, and cotton. To that end governments often subsidize the clearing of wildlands and the overuse of pesticides and fertilizers. The rush to maximize export income also concentrates ever more acreage in the hands of a relatively few, politically favored landowners. Small farmers are then forced to seek new land of marginal productivity, including natural habitats. Faced with ruin, they have no choice but to press into nutrient-poor tropical forests, steep hillside watersheds, coastal wetlands, and other final refuges of terrestrial diversity.

This journey to the precipice is hastened by the agricultural support systems of the richest nations. At the present time subsidies to developed-world farmers total $300 billion a year, six times the official foreign aid to Third World countries. When European Community countries recently underwrote a large program of feedlot cattle raising, they created a huge artificial market for cassava. Landowners in Thailand responded by clearing more tropical forest to grow cassava, and in the process displaced large numbers of subsistence farmers into the deep forest and up the eroding hillsides. When the United States tightened import quotas of cane sugar to aid domestic growers, U.S. imports from the Caribbean countries dropped 73 percent in ten years, forcing many of the rural poor out of jobs in the plantations and into marginal habitats for subsistence farming. Japan's extravagant subsidy to its own rice farmers, intended to continue an ancient agricultural tradition (the Japanese written character for rice means "root of life"), has a depressing effect on the rice-growing populations of tropical Asia. Once again, the impact on natural environments is increased.

The richest countries set the rules for international trade. They provide the bulk of loans and direct aid and control technology transfer to the poor nations. It is their responsibility to use this power wisely, in a manner that both strengthens these trading partners and protects the global environment. They themselves will suffer if the wildlands and biological diversity are not entered into the calculus of trade agreements and international aid.

The raging monster upon the land is population growth. In its presence, sustainability is but a fragile theoretical construct. To say, as many do, that the difficulties of nations are not due to people but to poor ideology or land-use management is sophistic. If Bangladesh had 10 million inhabitants instead of 115 million, its impoverished people could live on prosperous farms away from the dangerous floodplains midst a natural and stable upland environment. It is also sophistic to point to the Netherlands and Japan, as many commentators incredibly still do, as models of densely populated but prosperous societies. Both are highly specialized industrial nations dependent on massive imports of natural resources from the rest of the world. If all nations held the same number of people per square kilometer, they would converge in quality of life to Bangladesh rather than to the Netherlands and Japan, and their irreplaceable natural resources would soon join the seven wonders of the world as scattered vestiges of an ancient history.

Every nation has an economic policy and a foreign policy. The time has come to speak more openly of a population policy. By this I mean not just the capping of growth when the population hits the wall, as in China and India, but a policy based on a rational solution of this problem: what, in the judgment of its informed citizenry, is the *optimal* population, taken for each country in turn, placed against the backdrop of global demography? The answer will follow from an assessment of the society's self-image, its natural resources, its geography, and the specialized long-term role it can most effectively play in the international community. It can be implemented by encouragement or relaxation of birth control and the regulation of immigration, aimed at a target density and age distribution of the national population. The goal of an optimal population will require addressing, for the first time, the full range of processes that lock together the economy and the environment, the national interest and the global commons, the welfare of the present generation with that of future generations. The matter should be aired not only in think tanks but in public debate. If humanity then chooses to breed itself and the rest of life into impoverishment, at least it will have done so with open eyes.

4. *Save what remains.* Biodiversity can be saved by a mixture of programs, but not all the programs proposed can work. Consider one often raised in discussions by futurists. Suppose that we lost the race to save the environment, that all natural ecosystems were allowed to vanish. Could new species be created in the laboratory, after genetic engineers have learned how to assemble life from raw organic compounds? It is doubtful. There is no assurance that organisms can be generated artificially, at least not any as complex as flowers or butterflies—or amoebae for that matter. Even this godlike power would solve only half the problem, and the easy one at that. The technicians would be working in ignorance of the history of the extinct life they presumed to simulate. No knowledge exists of the endless mutations and episodes of natural selection that inserted billions of nucleotides into the now-vanished genomes, nor can it be deduced in more than tiny fragments. The neospecies would be creations of the human mind—plastic, neither historical nor adaptive, and unfit for existence apart from man. Ecosystems built from them, like zoos and botanical gardens, would require intensive care. But this is not the time for science-fiction dreams.

On then to the next technical remedy that springs up in scientific conferences and corridor arguments. Can extinct species be resurrected from the DNA still preserved in museum specimens and fossils? Again the answer is no. Fractions of genetic codes have been sequenced from a 2400-year-old Egyptian mummy and magnolia leaves preserved as rock fossils 18 million years ago, but they constitute only the smallest portion of the genetic codes. Even that part is hopelessly scrambled. To clone these organisms or a mammoth or a dodo or any other extinct organism would be, as the molecular biologist Russell Higuchi recently said, like taking a large encyclopedia in an unknown language previously ripped into shreds and trying to reassemble it without the use of your hands.

Consider the next possibility raised with regularity: why not just forget the problem and let natural evolution replace the species that are disappearing? It can be done if our descendants are willing to wait several million years. Following the five great extinction episodes of geological history, full recovery of biodiversity required between 10 and 100 million years. Even if *Homo sapiens* lasts that long, the recovery would require returning a large part of the land to its natural state. By appropriating or otherwise disturbing 90 percent of the land surface, humanity has already closed most of the theaters of natural evolution. And even if we did that much and waited that long, the new biota would be very different from the one we destroyed.

Then why not scoop up tissue samples of all living species and freeze them in liquid nitrogen? They could be cloned later to produce whole organisms. The method works for some microorganisms, including viruses, bacteria, and yeasts, as well as the spores of fungi. The American Type Culture Collection, located at Rockville, Maryland, contains over 50,000 species suspended in the deep sleep of absolute biochemical inactivity, ready for warming and reactivation as needed. The cultures are used in research, primarily in molecular biology and medicine. It is possible that many larger organisms could be similarly preserved in nitrogen sleep, at least as fertilized eggs, to be reared later into mature individuals. Even scraps of undifferentiated tissue might be stimulated into normal growth and development. It has been done for organisms as complex as carrots and frogs.

So let us suppose for argument that all kinds of plants and animals are salvageable by such means, that biologists will perfect the techniques of total inactivation and total recovery. The cryotorium in which they would rest, the new Noah's ark, must house tens of millions of species. The preservation of the content of even one endangered habitat (say a mountain-ridge forest in Ecuador) would be an immense operation enveloping thousands of species, most of which are still unknown to science. Even if completed at the species level, only a small fraction of the genetic variability of each species could be practicably included. Unless the samples numbered into the millions, great arrays of naturally occurring genetic strains would be lost. And when the time comes to return the species to the wild, the physical base of the ecosystem, including its soil, its unique nutrient mix, and its patterns of precipitation, will have been altered so as to make restoration doubtful. Cryopreservation is at best a last-ditch operation that might rescue a few select species and strains certain to die otherwise. It is far from the best way to save ecosystems and could easily fail. The need to put an entire community of organisms in liquid nitrogen would be tragic. Its enactment would be, in a particularly piercing sense of the word, obscene.

I have spoken so far of the maintenance of species and genetic stocks away from their natural habitats. Not all such methods are fantastic or repugnant. One that works for many plants is the maintenance of seed banks: seeds are dried and kept in repositories over long periods. The banks are kept in cool temperatures (about −20°C is typical) but not in the suspended animation of liquid nitrogen. Botanists have proved the technique effective for preserving most strains of crop species. About a hundred countries maintain seed banks and are adding to them steadily by exchanges and new collecting expeditions. Their efforts are aided by the "Green Board," the International Board for Plant Genetic Resources (IBPGR), an autonomous scientific organization located in Rome that composes part of the network of the International Agricultural Research Centers. In 1990 over 2 million sets of seeds were on deposit, representing more than 90 percent of the known local geographic varieties—landraces, as they are called—of many of the basic food crops. Especially well represented are wheat, maize, oats, potatoes, rice, and millet. An effort has begun to include the wild relatives of existing crop species, such as the richly promising perennial maize of Mexico. The method can be extended to wild, noncrop floras of the world.

But there are serious problems with seed banks. Up to 20 percent of plant species, some 50,000 in all, possess "recalcitrant" seeds that cannot be stored by conventional means. Even if seed storage were perfected for all kinds of plants, an unlikely prospect for the immediate future, the task of collecting and maintaining many thousands of endangered species and races would be stupendous. All the efforts of the existing seed banks to date have been barely enough to cover a hundred species, and even those are in many cases poorly recorded and of uncertain survival ability. Another difficulty: if reliance were placed entirely on seed banks, and the species then disappeared in the wild, the bank survivors would be stripped of their insect pollinators, root fungi, and other symbiotic partners, which cannot be put in cold storage. Most of the symbionts would go extinct, preventing the salvaged plant species from being replanted in the wild.

Other *ex situ* methods rely more realistically on captive populations that grow and reproduce. There are about 1,300 botanical gardens and arboretums in the world, many harboring plant species that are endangered or extinct in the wild. As of June 1991, twenty such institutions in the United States that subscribe to the registry of the National Collection of Endangered Plants contained seeds, plants, and cuttings of 372 species native to the United States. Some of the gardens in North America and Europe are more global in their reach. Harvard's Arnold Arboretum, for one, is famous for its collection of Asiatic trees and shrubs. England's magnificent Kew Gardens is engaged in a bold attempt to preserve and cultivate the last remnants of the nearly vanished tree flora of St. Helena.

Animals are vastly more difficult than plants and microorganisms to maintain *ex situ*. Zoos and other animal facilities have attempted the task in heroic fashion. By the late 1980s, those around the world whose stocks are known had gathered breeding populations of 540,000 individuals belonging to more than 3,000 species of mammals, birds, reptiles, and amphibians. The collections include roughly 13 percent of the known land-dwelling species of vertebrate animals. The better-financed zoos, including those in London, Frankfurt, Chicago, New York, San Diego, and Washington, D.C., conduct basic and veterinarian research with results that are applied to both captive and wild populations. The rosters of 223 zoos in Europe and North America are tracked by the International Species Inventory System (ISIS), which uses the data to coordinate preservation and cross-breeding. The ISIS zoos and research institutions aim not only to save endangered animals but to reintroduce species into their native habitats when land is made available. They have been successful with three species, the Arabian oryx, the black-footed ferret, and the golden lion tamarin. Attempts are underway or planned for at least four other species, the California condor, the Bali starling, the Guam rail, and the Przewalski horse, the ancestor of all domestic horses. The ISIS facilities are trying to get ready if the giant panda, the

Sumatran rhinoceros, and the Siberian tiger, now on the brink, should go extinct in the wild.

The best efforts by zoos, zooparks, aquariums, and research facilities, however, slow the tide of extinction by a barely perceptible amount. Even the groups of animals most favored by the public cannot be completely served. Conservation biologists estimate that as many as 2,000 species of mammals, birds, and reptiles can only be salvaged if they are bred in captivity, a task beyond reach with the means at hand. William Conway, director of the comprehensive zoo maintained by the New York Zoological Society, believes that existing facilities worldwide can sustain viable populations of no more than 900 species. At best these survivors would contain only a small fraction of their species' original genes. And far worse: no provision at all has been made for the many thousands of species of insects and other invertebrates that are equally at risk.

The dreams of scientists come to this: *ex situ* conservation is not enough and will never be enough. Some of the methods are invaluable as safety nets for the fraction of endangered species that biology best understands and the lay public is willing to support. But even if countries everywhere chose to finance greatly enlarged cryobiological vaults, seed banks, botanical gardens, and zoos, the facilities could not be assembled quickly enough to save a majority of species close to extinction from habitat destruction alone. Biologists are hampered by lack of knowledge of more than 90 percent of the species of fungi, insects, and smaller organisms on earth. They have no way to ensure a reasonable sampling of genetic variation even in the species rescued. They have only the faintest idea of how to reassemble ecosystems from salvaged species, if indeed such a feat is possible. Not least, the entire process would be enormously expensive.

All these considerations converge to the same conclusion: *ex situ* methods will save a few species otherwise beyond hope, but the light and the way for the world's biodiversity is the preservation of natural ecosystems. If that is accepted, we must face two realities squarely. The first is that the habitats are disappearing at an accelerating rate and with them a quarter of the world's biodiversity. The second is that the habitats cannot be saved unless the effort is of immediate economic advantage to the poor people who live in and around them. Eventually idealism and high purpose may prevail around the world. Eventually an economically secure populace will treasure their native biodiversity for its own sake. But at this moment they are not secure and they, and we, have run out of time.

The rescue of biological diversity can only be achieved by a skillful blend of science, capital investment, and government: science to blaze the path by research and development; capital investment to create sustainable markets; and government to promote the marriage of economic growth and conservation.

The primary tactic in conservation must be to locate the world's hot spots and to protect the entire environment they contain. Whole ecosystems are the targets of choice because even the most charismatic species are but the representatives of thousands of lesser-known species that live with them and are also threatened. The most inclusive federal legislation in the United States is the Endangered Species Act of 1973, which throws a protective shield around species of "fish, wildlife, and plants" that are "endangered and threatened" by human activities; as amended in 1978, the act also includes subspecies. A bold and creative advance, the legislation is nevertheless destined to be an arena of rising litigation. As any natural environment is reduced in area, the number of species that can live in it indefinitely is also reduced. In other words, some species are doomed to extinction even if all of the remaining habitat were to be preserved from that time on. One of the principles of ecology, as I have stressed, is that the number of species eventually declines by an amount roughly equal to the sixth to third root of the area already lost. Because the great majority of species of microorganisms, fungi, and insects are not well known, it follows that they have been slipping unnoticed through the cracks in the Endangered Species Act. Conflicts between developers and conservationists over birds, mammals, and fishes are already commonplace. As ecosystems are better explored, less-conspicuous endangered species will come to light and the number of clashes will grow.

There is a way out of the dilemma, other than abandoning legal protection of America's fauna and flora altogether. As biodiversity surveys are improved, the hot spots will come more sharply into focus. Well-documented examples already include the embattled coral reef of the Florida Keys and the rain forests of Hawaii and Puerto Rico. As other local habitats are pinpointed, they can be assigned the highest priority for conservation. This means, in most cases, that they will be set aside as inviolate reserves. Warm spots, areas less threatened or containing fewer species not found elsewhere, can be zoned for partial development, with core preserves centered on endemic species and races and buffer strips around the preserves kept partly wild. Agricultural landscapes and harvested forest tracts can be better designed to harbor rare species and races.

All these actions together, wisely administered, will be effective. But the Endangered Species Act or an equivalent is also needed to serve as a safety net for threatened forms of life in all environments, whether harbored in reserves or not. Finally, in those rare cases where the costs are perceived as intolerable by the electorate, a compromise can be sought by means of population management. This means transplantation of the species to suitable habitats nearby, or restoring its environment in places where it was previously extinguished outside the zone of conflict, or—when all else fails—exile to botanical gardens, zoos, or other *ex situ* preserves.

The area-species relation governing biodiversity shows that maintenance of existing parks and reserves will not be enough to save all the species living within them. Only 4.3 percent of the earth's land surface is currently under legal protection, divided among national parks, scientific stations, and other classes of reserves. These fragments represent recently shrunken habitat islands, whose faunas and floras will continue to dwindle until a new, often lower equilibrium is reached. Over 90 percent of the remaining land surface, including most of the surviving high-diversity habitats, has been altered. If the disturbance continues until most of the natural outside reserves are swept away, a majority of the world's terrestrial species will be either extinguished or put at extreme risk. And more: even the existing reserves are in harm's way. Poachers and illegal miners invade them, timber thieves work their margins, developers find ways to convert them in part. During recent civil wars in Ethiopia, Sudan, Angola, Uganda, and other African countries, many of the national parks were left to ruin.

So we should try to expand reserves from 4.3 percent to 10 percent of the land surface, to include as many of the undisturbed habitats as possible with priority given to the world's hot spots. One of the more promising means to attain this goal is by debt-for-nature swaps. As currently practiced, conservation organizations such as Conservation International, the Nature Conservancy, and the World Wildlife Fund (U.S.) raise funds to purchase a portion of a country's commercial debt at a discount, or else they persuade creditor banks to donate some of it. This first step is easier than it sounds because so many developing countries are close to default. The debts are then exchanged in local currency or bonds set at favorable rates. The enlarged equity is used to promote conservation, especially by the purchase of land, environmental education, and the improvement of land management. By early 1992 a total of twenty such agreements totaling $110 million had been arranged in nine countries, including Bolivia, Costa Rica, Dominican Republic, Ecuador, Mexico, Madagascar, Zambia, the Philippines, and Poland.

In February 1991, to take one example, Conservation International was authorized to buy $4 million in debt from Mexico's creditors. After discounting on secondary markets, the actual cost is expected to be as little as $1.8 million. The conservation organization has agreed to forgive the full amount in return for the expenditure of $2.6 million by the Mexican government on a broad range of conservation projects. The most important initiative will be to preserve the Lacandan tract in the extreme south of Mexico, the largest rain forest in North America.

The debt of Third World countries has been reduced so far by only one part in 10,000 through debt-for-nature swaps. Nor are the arrangements without risk for the receiving country, notably in the crowding out of domestic expenditures and a sparking of local inflation. But these temporary effects are offset by the immense gain, dollar for dollar, in the stabilization of the environment.

More potent still are unencumbered contributions from wealthier nations channeled and carefully targeted through international assistance organizations. The most important enterprise of this kind is the Global Environment Facility (GEF), established in 1990 by the World Bank, the United Nations Environmental Program, and the United Nations Development Program. At this writing, $450 million has been committed to set up national parks, promote sustainable forestry, and establish conservation trust funds in developing countries. Under consideration or already approved are proposals from Bhutan, Indonesia, Papua New Guinea, the Philippines, Vietnam, and the Central African Republic. Two principal difficulties have appeared within the GEF agenda. One is the limited absorptive power of the recipient nations. With limited trained personnel and expert knowledge, national leaders find it difficult to select the best projects and initiate them effectively. Of much greater significance, the brief terms of funding leave little prospect for the proper management and protection of reserves when the money runs out. Fearing loss of employment, the brightest professionals are likely to look to other activities to ensure their futures. The solution to both problems may lie in the establishment of national trust funds, producing income that can be fed into the conservation programs gradually and over a period of many years. One such fund has recently been established for Bhutan with the help of the World Wildlife Fund.

We come then to the design of the reserves themselves. As land is set aside, the primary goal is to place the reserves in the regions of highest diversity and to make them as large as possible. Another goal is to design their shape and spacing for retaining efficiency. In approaching that secondary end, a debate has arisen in conservation circles on the so-called SLOSS problem: whether to invest allotted land into a Single Large reserve Or into Several Small reserves. A single large reserve, to put the matter as simply as possible, possesses larger populations of each species, but they all fit into one basket. A single catastrophic fire or flood could extinguish a large part of the diversity of the region. Breaking the reserve into several pieces reduces that problem, but it also diminishes the size of the constituent populations and hence threatens each with extinction. All might easily decline in the face of widespread stress, such as drought or unseasonable cold.

Some biologists have suggested a compromise solution to the SLOSS problem, which is to create small reserves connected by corridors of natural habitat. For example, several forest patches (say 10 kilometers square each) might be joined by strips of forest 100 meters across. Then if a species vanishes from one of the patches, it can be replaced by colonists immigrating along the forest corridor from another patch. The disadvantage that critics of the compromise have been quick to identify is that disease, predators, and exotic competitors can also use corridors to move through the network. Since populations in the patches are small and vulnerable, all might fall like a row of dominoes. I doubt that any general principle of population dynamics exists that can resolve the SLOSS controversy, at least not in the clean manner suggested by its simple geometric imagery. Instead each ecosystem must be studied in turn to decide the best design, which will depend on the species the system contains and the year-by-year fluctuation of its physical environment. For the time being, conservation biologists will agree on the cardinal rule: to save the most biodiversity, make the reserves as large as possible.

5. *Restore the wildlands.* The grim signature of our time has been the reduction of natural habitats until a substantial portion of the kinds of plants and animals, certainly more than 10 percent, have already vanished or else are consigned to early extinction. The toll of genetic races has never been estimated, but it is almost certainly much higher than that of species. Yet there is still time to save many

of the "living dead"—those so close to the brink that they will disappear soon even if merely left alone. The rescue can be accomplished if natural habitats are not only preserved but enlarged, sliding the numbers of survivable species back up the logarithmic curve that connects quantity of biodiversity to amount of area. Here is the means to end the great extinction spasm. The next century will, I believe, be the era of restoration in ecology.

In haphazard manner, largely through the abandonment of small farms, the area of coniferous and hardwood forests in the eastern United States has increased during the past hundred years. Deliberate efforts to enlarge wild areas are also underway. In 1935 a pioneering effort resulted in the planting of 24 hectares of tall-grass prairie at the University of Wisconsin Arboretum. The arboretum has also served as the headquarters of the Center for Restoration Ecology, devoted to research and the collation of information from projects in other parts of the country. Elsewhere in the United States, small restoration projects by the hundreds have been initiated, all devoted to the increase in area of natural habitats and the return of degraded ecosystems to full health. They range broadly in ecosystem types, from the ironwood groves of Santa Catalina Island to the Tobosa grassland of Arizona, the oakland understory of California's Santa Monica Mountains, the magnificent open mountain woodlands of Colorado, and last savanna remnants of Illinois. They include fragments of salt and freshwater wetlands from California to Florida and Massachusetts.

In Costa Rica an audacious effort by the American ecologist Daniel Janzen and local conservation leaders has led to the establishment of Guanacaste National Park, a 50,000-hectare reserve in the northwestern corner of the country. The park will be created—literally created—by the regrowth of dry tropical forest planted on cattle ranches. The Guanacaste dream was born of recognition that in Central America dry forest is even more threatened than rain forest, down to only 2 percent of its original cover. The plan is to use existing patches of the original forest to seed a steadily growing area of ranchland. The conversion will be made easier by the low density of the human population in the area. The regenerating woodland will provide a protected watershed, an income from tourism expected to reach $1 million or more annually, and a net increase in employment of the area's residents. Most important in the long run, it will save a significant part of Costa Rica's natural heritage.

I have spoken of the salvage and regeneration of existing ecosystems. There will come a time when even more is possible with the aid of scientific knowledge. The return to biology's Eden might also include the creation of synthetic faunas and floras, assemblages of species carefully selected from different parts of the world and introduced into impoverished habitats. The idea struck home for me one late afternoon as I sat at the edge of the artificial lake near the center of the University of Miami campus, surrounded by the densely urbanized community of Coral Gables. At least six species of fishes swarmed in the clear brackish water within 2 meters of shore, some as solitary foragers, others in schools. Most were exotics. Their unusual diversity and beauty reminded me of a newly created coral reef. As the sun set and the water darkened, a large predator fish, probably a gar, broke the surface in the middle of the lake. A small alligator glided out from reeds across the way and cruised into open water. Well beyond the far shore, a flock of parrots returned noisily to their palm-top evening roost. They belonged to one of more than twenty exotic species that breed or occur in the Miami area, all originating from individuals that escaped or were deliberately released from captivity. Thus has the parrot family, the Psittacidae, returned to Florida with a vengeance, only decades after the extermination of the Carolina parakeet, last of the endemic North American species. With flashing wings they salute the vanished native.

It is dangerous, I must quickly add, to think too freely of introducing exotics anywhere. They might or might not take to the new environment—between 10 and 50 percent of bird species have succeeded, depending on the part of the world and the number of attempts made to introduce them. Exotics might become economic pests or force out native species. A few, like rabbits, goats, pigs, and the notorious Nile perch are capable not only of extinguishing individual species but of degrading entire habitats. Ecology is still too primitive a science to predict the outcome of the synthesis of predesigned biotas. No responsible person will risk dumping destroyers into the midst of already diminished communities. Nor should we delude ourselves into thinking that synthetic biotas increase global diversity. They only increase local diversity by expanding the ranges and population sizes of selected species.

Yet the search for the safe rules of biotic synthesis is an enterprise of high intellectual daring. If the effort is successful, regions already stripped of their native biotas can be restored to places of diversity and environmental stability. A wilderness of sorts can be reborn in the wasteland. Species already extinct in the wild, those now maintained in zoos and gardens, deserve high priority. Transplanted into impoverished or synthetic biotas, they can endure as orphan species in foster ecosystems. Even though their original home has been closed to them, they will regain security and independence. They will repay us by attaining one criterion of wilderness—that we are allowed to lay down the burden of their care and visit them as equal partners, on our own time. A few species will be prosthetic. As keystone elements, such as a tree able to grow rapidly and shelter many other plant and animal species, they will play a disproportionate role in holding the new communities together.

Finally, the question of central interest is how much of the world's biodiversity we can expect to carry with us out of the bottleneck fifty or a hundred years hence. Let me venture a guess. If the biodiversity crisis remains largely ignored and natural habitats continue to decline, we will lose at least one quarter of the earth's species. If we respond with the knowledge and technology already possessed, we may hold the loss to 10 percent. At first glance the difference may seem bearable. It is not; it amounts to millions of species.

I feel no hesitance in urging the strong hand of protective law and international protocols in the preservation of biological wealth, as opposed to tax incentives and marketable pollution permits. In democratic societies people may think that their government is bound by an ecological version of the Hippocratic oath, to take no action that knowingly endangers biodiversity. But that is not enough. The commitment must be much deeper—to let no species knowingly die, to take all reasonable action to protect every species and race in perpetuity. The government's moral responsibility in the conservation of biodiversity is similar to that in public health and military defense. The preservation of species across generations is beyond the capacity of individuals or even powerful private institutions. Insofar as biodiversity is deemed an irreplaceable public resource, its protection should be bound into the legal canon.

The Environmental Ethic

THE SIXTH GREAT extinction spasm of geological time is upon us, grace of mankind. Earth has at last acquired a force that can break the crucible of biodiversity. I sensed it with special poignancy that stormy night at Fazenda Dimona, when lightning flashes revealed the rain forest cut open like a cat's eye for laboratory investigation. An undisturbed forest rarely discloses its internal anatomy with such clarity. Its edge is shielded by thick secondary growth or else, along the river bank, the canopy spills down to ground level. The nighttime vision was a dying artifact, a last glimpse of savage beauty.

A few days later I got ready to leave Fazenda Dimona: gathered my muddied clothes in a bundle, gave my imitation Swiss army knife to the cook as a farewell gift, watched an overflight of Amazonian green parrots one more time, labeled and stored my specimen vials in reinforced boxes, and packed my field notebook next to a dog-eared copy of Ed McBain's police novel *Ice*, which, because I had neglected to bring any other reading matter, was now burned into my memory.

Grinding gears announced the approach of the truck sent to take me and two of the forest workers back to Manaus. In bright sunlight we watched it cross the pastureland, a terrain strewn with fire-blackened stumps and logs, the battlefield my forest had finally lost. On the ride back I tried not to look at the bare fields. Then, abandoning my tourist Portuguese, I turned inward and daydreamed. Four splendid lines of Virgil came to mind, the only ones I ever memorized, where the Sibyl warns

Aeneas of the Underworld:

> The way downward is easy from Avernus.
> Black Dis's door stands open night and day.
> But to retrace your steps to heaven's air,
> There is the trouble, there is the toil . . .

For the green prehuman earth is the mystery we were chosen to solve, a guide to the birthplace of our spirit, but it is slipping away. The way back seems harder every year. If there is danger in the human trajectory, it is not so much in the survival of our own species as in the fulfillment of the ultimate irony of organic evolution: that in the instant of achieving self-understanding through the mind of man, life has doomed its most beautiful creations. And thus humanity closes the door to its past.

The creation of that diversity came slow and hard: 3 billion years of evolution to start the profusion of animals that occupy the seas, another 350 million years to assemble the rain forests in which half or more of the species on earth now live. There was a succession of dynasties. Some species split into two or several daughter species, and their daughters split yet again to create swarms of descendants that deployed as plant feeders, carnivores, free swimmers, gliders, sprinters, and burrowers, in countless motley combinations. These ensembles then gave way by partial or total extinction to newer dynasties, and so on to form a gentle upward swell that carried biodiversity to a peak—just before the arrival of humans. Life had stalled on plateaus along the way, and on five occasions it suffered extinction spasms that took 10 million years to repair. But the thrust was upward. Today the diversity of life is greater than it was a 100 million years ago—and far greater than 500 million years before that.

Most dynasties contained a few species that expanded disproportionately to create satrapies of lesser rank. Each species and its descendants, a sliver of the whole, lived an average of hundreds of thousands to millions of years. Longevity varied according to taxonomic group. Echinoderm lineages, for example, persisted longer than those of flowering plants, and both endured longer than those of mammals.

Ninety-nine percent of all the species that ever lived are now extinct. The modern fauna and flora are composed of survivors that somehow managed to dodge and weave through all the radiations and extinctions of geological history. Many contemporary world-dominant groups, such as rats, ranid frogs, nymphalid butterflies, and plants of the aster family Compositae, attained their status not long before the Age of Man. Young or old, all living species are direct descendants of the organisms that lived 3.8 billion years ago. They are living genetic libraries, composed of nucleotide sequences, the equivalent of words and sentences, which record evolutionary events all across that immense span of time. Organisms more complex than bacteria—protists, fungi, plants, animals—contain between 1 and 10 billion nucleotide letters, more than enough in pure information to compose an equivalent of the *Encyclopaedia Britannica*. Each species is the product of mutations and recombinations too complex to be grasped by unaided intuition. It was sculpted and burnished by an astronomical number of events in natural selection, which killed off or otherwise blocked from reproduction the vast majority of its member organisms before they completed their lifespans. Viewed from the perspective of evolutionary time, all other species are our distant kin because we share a remote ancestry. We still use a common vocabulary, the nucleic-acid code, even though it has been sorted into radically different hereditary languages.

Such is the ultimate and cryptic truth of every kind of organism, large and small, every bug and weed. The flower in the crannied wall—it *is* a miracle. If not in the way Tennyson, the Victorian romantic, bespoke the portent of full knowledge (by which "I should know what God and man is"), then certainly a consequence of all we understand from modern biology. Every kind of organism has reached this moment in time by threading one needle after another, throwing up brilliant artifices to survive and reproduce against nearly impossible odds.

Organisms are all the more remarkable in combination. Pull out the flower from its crannied retreat, shake the soil from the roots into the cupped hand, magnify it for close examination. The black earth is alive with a riot of algae, fungi, nematodes, mites, springtails, enchytraeid worms, thousands of species of bacteria. The handful may be only a tiny fragment of one ecosystem, but because of the genetic codes of its residents it holds more order than can be found on the surfaces of all the planets combined. It is a sample of the living force that runs the earth—and will continue to do so with or without us.

We may think that the world has been completely explored. Almost all the mountains and rivers, it is true, have been named, the coast and geodetic surveys completed, the ocean floor mapped to the deepest trenches, the atmosphere transected and chemically analyzed. The planet is now continuously monitored from space by satellites; and, not least, Antarctica, the last virgin continent, has become a research station and expensive tourist stop. The biosphere, however, remains obscure. Even though some 1.4 million species of organisms have been discovered (in the minimal sense of having specimens collected and formal scientific names attached), the total number alive on earth is somewhere between 10 and 100 million. No one can say with confidence which of these figures is the closer. Of the species given scientific names, fewer than 10 percent have been studied at a level deeper than gross anatomy. The revolution in molecular biology and medicine was achieved with a still smaller fraction, including colon bacteria, corn, fruit flies, Norway rats, rhesus monkeys, and human beings, altogether comprising no more than a hundred species.

Enchanted by the continuous emergence of new technologies and supported by generous funding for medical research, biologists have probed deeply along a narrow sector of the front. Now it is time to expand laterally, to get on with the great Linnean enterprise and finish mapping the biosphere. The most compelling reason for the broadening of goals is that, unlike the rest of science, the study of biodiversity has a time limit. Species are disappearing at an accelerating rate through human action, primarily habitat destruction but also pollution and the introduction of exotic species into residual natural environments. I have said that a fifth or more of the species of plants and animals could vanish or be doomed to early extinction by the year 2020 unless better efforts are made to save them. This estimate comes from the known quantitative relation between the area of habitats and the diversity that habitats can sustain. These area-biodiversity curves are supported by the general but not universal principle that when certain groups of organisms are studied closely, such as snails and fishes and flowering plants, extinction is determined to be widespread. And the corollary: among plant and animal remains in archaeological deposits, we usually find extinct species and races. As the last forests are felled in forest strongholds like the Philippines and Ecuador, the decline of species will accelerate even more. In the world as a whole, extinction rates are already hundreds or thousands of times higher than before the coming of man. They cannot be balanced by new evolution in any period of time that has meaning for the human race.

Why should we care? What difference does it make if some species are extinguished, if even half of all the species on earth disappear? Let me count the ways. New sources of scientific information will be lost. Vast potential biological wealth will be destroyed. Still undeveloped medicines, crops, pharmaceuticals, timber, fibers, pulp, soil-restoring vegetation, petroleum substitutes, and other products and amenities will never come to light. It is fashionable in some quarters to wave aside the small and obscure, the bugs and weeds, forgetting that an obscure moth from Latin America saved Australia's pasture-land from overgrowth by cactus, that the rosy periwinkle provided the cure for Hodgkin's disease and childhood lymphocytic leukemia, that the bark of the Pacific yew offers hope for victims of ovarian and breast cancer, that a chemical from the saliva of leeches dissolves blood clots during surgery, and so on down a roster already grown long and illustrious despite the limited research addressed to it.

In amnesiac revery it is also easy to overlook the services that ecosystems provide humanity. They enrich the soil and create the very air we breathe. Without these amenities, the remaining tenure of the human race would be nasty and brief. The life-sustaining matrix is built of green plants with legions of microorganisms and mostly small, obscure animals—in other words, weeds and bugs. Such organisms support the world with efficiency because they are so diverse, allowing them to divide labor and swarm over every square meter of the earth's surface. They run the world precisely as we would wish it to be run, because humanity evolved within living communities and our bodily functions are finely adjusted to the idiosyncratic environment already created. Mother Earth, lately called Gaia, is no more than the commonality of organisms and the physical environment they maintain with each passing moment, an environment that will destabilize and turn lethal if the organisms are disturbed too much. A near infinity of other mother planets can be envisioned, each with its own fauna and flora, all producing physical environments uncongenial to human life. To disregard the diversity of life is to risk catapulting ourselves into an alien environment. We will have become like the pilot whales that inexplicably beach themselves on New England shores.

Humanity coevolved with the rest of life on this particular planet; other worlds are not in our genes. Because scientists have yet to put names on most kinds of organisms, and because they entertain only a vague idea of how ecosystems work, it is reckless to suppose that biodiversity can be diminished indefinitely without threatening humanity itself. Field studies show that as biodiversity is reduced, so is the quality of the services provided by ecosystems. Records of stressed ecosystems also demonstrate that the descent can be unpredictably abrupt. As extinction spreads, some of the lost forms prove to be keystone species, whose disappearance brings down other species and triggers a ripple effect through the demographies of the survivors. The loss of a keystone species is like a drill accidentally striking a powerline. It causes lights to go out all over.

These services are important to human welfare. But they cannot form the whole foundation of an enduring environmental ethic. If a price can be put on something, that something can be devalued, sold, and discarded. It is also possible for some to dream that people will go on living comfortably in a biologically impoverished world. They suppose that a prosthetic environment is within the power of technology, that human life can still flourish in a completely humanized world, where medicines would all be synthesized from chemicals off the shelf, food grown from a few dozen domestic crop species, the atmosphere and climate regulated by computer-driven fusion energy, and the earth made over until it becomes a literal spaceship rather than a metaphorical one, with people reading displays and touching buttons on the bridge. Such is the terminus of the philosophy of exemptionalism: do not weep for the past, humanity is a new order of life, let species die if they block progress, scientific and technological genius will find another way. Look up and see the stars awaiting us.

But consider: human advance is determined not by reason alone but by emotions peculiar to our species, aided and tempered by reason. What makes us people and not computers is emotion. We have little grasp of our true nature, of what it is to be human and therefore where our descendants might someday wish we had directed Spaceship Earth. Our troubles, as Vercors said in *You Shall Know Them*, arise from the fact that we do not know what we are and cannot agree on what we want to be. The primary cause of this intellectual failure is ignorance of our origins. We did not arrive on this planet as aliens. Humanity is part of nature, a species that evolved among other species. The more closely we identify ourselves with the rest of life, the more quickly we will be able to discover the sources of human sensibility and acquire the knowledge on which an enduring ethic, a sense of preferred direction, can be built.

The human heritage does not go back only for the conventionally recognized 8,000 years or so of recorded history, but for at least 2 million years, to the appearance of the first "true" human beings, the earliest species composing the genus *Homo*. Across thousands of generations, the emergence of culture must have been profoundly influenced by simultaneous events in genetic evolution, especially those occurring in the anatomy and physiology of the brain. Conversely, genetic evolution must have been guided forcefully by the kinds of selection rising within culture.

Only in the last moment of human history has the delusion arisen that people can flourish apart from the rest of the living world. Preliterate societies were in intimate contact with a bewildering array of life forms. Their minds could only partly adapt to that challenge. But they struggled to understand the most relevant parts, aware that the right responses gave life and fulfillment, the wrong ones sickness, hunger, and death. The imprint of that effort cannot have been erased in a few generations of urban existence. I suggest that it is to be found among the particularities of human nature, among which are these:

• People acquire phobias, abrupt and intractable aversions, to the objects and circumstances that threaten humanity in natural environments: heights, closed spaces, open spaces, running water, wolves, spiders, snakes. They rarely form phobias to the recently invented contrivances that are far more dangerous, such as guns, knives, automobiles, and electric sockets.

• People are both repelled and fascinated by snakes, even when they have never seen one in nature. In most cultures the serpent is the dominant wild animal of mythical and religious symbolism. Manhattanites dream of them with the same frequency as Zulus. This response appears to be Darwinian in origin. Poisonous snakes have been an important cause of mortality almost everywhere, from Finland to Tasmania, Canada to Patagonia; an untutored alertness in their presence saves lives. We note a kindred response in many primates, including Old World monkeys and chimpanzees: the animals pull back, alert others, watch closely, and follow each potentially dangerous snake until it moves away. For human beings, in a larger metaphorical sense, the mythic, transformed serpent has come to possess both constructive and destructive powers: Ashtoreth of the Canaanites, the demons Fu-Hsi and Nu-kua of the Han Chinese, Mudamma and Manasa of Hindu India, the triple-headed giant Nehebkau of the ancient Egyptians, the serpent of Genesis conferring knowledge and death, and, among the Aztecs, Cihuacoatl, goddess of childbirth and mother of the human race, the rain god Tlaloc, and Quetzalcoatl, the plumed serpent with a human head who reigned as lord of the morning and evening star. Ophidian power spills over into modern life: two serpents entwine the caduceus, first the winged staff of Mercury as messenger of the gods, then the safe-conduct pass of ambassadors and heralds, and today the universal emblem of the medical profession.

• The favored living place of most peoples is a prominence near water from which parkland can be viewed. On such heights are found the abodes of the powerful and rich, tombs of the great, temples, parliaments, and monuments commemorating tribal glory. The location is today an aesthetic choice and, by the implied freedom to settle there, a symbol of status. In ancient, more practical times the topography provided a place to retreat and a sweeping prospect from which to spot the distant approach of storms and enemy forces. Every animal species selects a habitat in which its members gain a favorable mix of security and food. For most of deep history, human beings lived in tropical and subtropical savanna in East Africa, open country sprinkled with streams and lakes, trees and copses. In similar topography modern peoples choose their residences and design their parks and gardens, if given a free choice. They simulate neither dense jungles, toward which gibbons are drawn, nor dry grasslands, preferred by hamadryas baboons. In their gardens they plant trees that resemble the acacias, sterculias, and other native trees of the African savannas. The ideal tree crown sought is consistently wider than tall, with spreading lowermost branches close enough to the ground to touch and climb, clothed with compound or needle-shaped leaves.

• Given the means and sufficient leisure, a large portion of the populace backpacks, hunts, fishes, birdwatches, and gardens. In the United States and Canada more people visit zoos and aquariums than attend all professional athletic events combined. They crowd the national parks to view natural landscapes, looking from the tops of prominences out across rugged terrain for a glimpse of tumbling water and animals living free. They travel long distances to stroll along the seashore, for reasons they can't put into words.

These are examples of what I have called *biophilia*, the connections that human beings subconsciously seek with the rest of life. To biophilia can be added the idea of wilderness, all the land and communities of plants and animals still unsullied by human occupation. Into wilderness people travel in search of new life and wonder, and from wilderness they return to the parts of the earth that have been humanized and made physically secure. Wilderness settles peace on the soul because it needs no help; it is beyond human contrivance.

Wilderness is a metaphor of unlimited opportunity, rising from the tribal memory of a time when humanity spread across the world, valley to valley, island to island, godstruck, firm in the belief that virgin land went on forever past the horizon.

I cite these common preferences of mind not as proof of an innate human nature but rather to suggest that we think more carefully and turn philosophy to the central questions of human origins in the wild environment. We do not understand ourselves yet and descend farther from heaven's air if we forget how much the natural world means to us. Signals abound that the loss of life's diversity endangers not just the body but the spirit. If that much is true, the changes occurring now will visit harm on all generations to come.

The ethical imperative should therefore be, first of all, prudence. We should judge every scrap of biodiversity as priceless while we learn to use it and come to understand what it means to humanity. We should not knowingly allow any species or race to go extinct. And let us go beyond mere salvage to begin the restoration of natural environments, in order to enlarge wild populations and stanch the hemorrhaging of biological wealth. There can be no purpose more enspiriting than to begin the age of restoration, reweaving the wondrous diversity of life that still surrounds us.

The evidence of swift environmental change calls for an ethic uncoupled from other systems of belief. Those committed by religion to believe that life was put on earth in one divine stroke will recognize that we are destroying the Creation, and those who perceive biodiversity to be the product of blind evolution will agree. Across the other great philosophical divide, it does not matter whether species have independent rights or, conversely, that moral reasoning is uniquely a human concern. Defenders of both premises seem destined to gravitate toward the same position on conservation.

The stewardship of environment is a domain on the near side of metaphysics where all reflective persons can surely find common ground. For what, in the final analysis, is morality but the command of conscience seasoned by a rational examination of consequences? And what is a fundamental precept but one that serves all generations? An enduring environmental ethic will aim to preserve not only the health and freedom of our species, but access to the world in which the human spirit was born.

Notes

Glossary

Acknowledgments

Credits

Index

Notes

1. Storm over the Amazon

Parts of this chapter were modified from my earlier articles, "Storm over the Amazon," in Daniel Halpern, ed., *On Nature: Nature, Landscape, and Natural History* (San Francisco: North Point Press, 1987), pp. 157–159; and "Rain Forest Canopy: The High Frontier," *National Geographic*, 180:78–107 (December 1991). I covered the subject on which I reflected that night in a technical monograph, *Success and Dominance in Ecosystems: The Case of the Social Insects* (Oldendorf/Luhe, Germany: Ecological Institute, 1990). The precise call of the red howler monkey is based on the description by Louise H. Emmons in *Neotropical Rainforest Mammals: A Field Guide* (University of Chicago Press, 1990).

The reflections of **Jöns Jacob Berzelius** are from his *Manual of Chemistry* (vol. 3, 1818), as quoted by Carl Gustaf Bernhard, "Berzelius, Creator of the Chemical Language," reprinted from the *Saab-Scania Griffin 1989/90* by the Royal Swedish Academy of Sciences.

The antifreeze technique of the **notothenioid fishes** is described in Joseph T. Eastman and Arthur L. DeVries, "Antarctic Fishes," *Scientific American*, 255:106–114 (November 1986).

The **archaebacteria** or archaeotida, some of which exist in the most hostile environments on earth, have been most thoroughly documented by Carl R. Woese and his coresearchers. See Robert Pool, "Pushing the Envelope of Life," *Science*, 247:158–247 (1990). Some biologists, including Woese, consider these organisms to constitute a separate kingdom of life apart from true bacteria and other prokaryotic organisms composing the kingdom Monera.

2. Krakatau

The definitive account of the 1883 **Krakatau eruptions,** including personal accounts and research reports of the time, is provided by Tom Simkin and Richard S. Fiske, *Krakatau 1883: The Volcanic Eruption and Its Effects* (Washington, D.C.:

Smithsonian Institution Press, 1983). Additional details on the tidal waves are given by Susanna Van Rose and Ian F. Mercer, *Volcanoes* (Cambridge: Harvard University Press, 1991). Analyses of the recolonization of Rakata are summarized in Robert H. MacArthur and Edward O. Wilson, *The Theory of Island Biogeography* (Princeton: Princeton University Press, 1967); Ian W. B. Thornton et al., "Colonization of the Krakatau Islands by Vertebrates: Equilibrium, Succession, and Possible Delayed Extinction," *Proceedings of the National Academy of Sciences*, 85:515–518 (1988); I. W. B. Thornton and T. R. New, "Krakatau Invertebrates: The 1980s Fauna in the Context of a Century of Recolonization," *Philosophical Transactions of the Royal Society of London*, ser. B, 322:493–522 (1988); and P. A. Rawlinson, A. H. T. Widjoya, M. N. Hutchinson, and G. W. Brown, "The Terrestrial Vertebrate Fauna of the Krakatau Islands, Sunda Strait, 1883–1986," *Philosophical Transactions of the Royal Society of London*, ser. B, 328:3–28 (1990). The estimate of the pumice temperatures following the explosion are by Noboru Oba of Kagoshima University, cited from personal communication by Thornton and New, "Krakatau Invertebrates."

3. The Great Extinctions

I have drawn the details of the **Cretaceous extinction episode** and the meteorite-vulcanism debate from many sources, but especially Matthew H. Nitecki, ed., *Extinctions* (Chicago: University of Chicago Press, 1984); Steven M. Stanley, *Extinction* (New York: Scientific American Books, 1987), and "Periodic Mass Extinctions of the Earth's Species," *Bulletin of the American Academy of Arts and Sciences*, 40(8):29–48 (1987); David M. Raup, *Extinction: Bad Genes or Bad Luck?* (New York: Norton, 1991); Paul Whalley, "Insects and Cretaceous Mass Extinction," *Nature*, 327:562 (1987); Carl O. Moses, "A Geochemical Perspective on the Causes and Periodicity of Mass Extinctions," *Ecology*, 70(4):812–823 (1989); and William Glen, "What Killed the Dinosaurs?," *American Scientist*, 78(4):354–370 (1990). Stanley, for example, has argued persuasively for the role of long-term cooling of the earth's climate as a principal factor in mass extinctions, including the Cretaceous. Whalley details the survival of the insects. The fate of the flowering plants is described by Andrew H. Knoll, in Nitecki, *Extinctions*, pp. 21–68, and R. A. Spicer, "Plants at the Cretaceous-Tertiary Boundary," *Philosophical Transactions of the Royal Society of London*, ser. B, 325:291–305 (1989). The newest fossil evidence from the K-T boundary for many groups of organisms is summarized in *Evolution and Extinction*, special issue of the *Philosophical Transactions of the Royal Society of London*, ser. B, 325:239–488 (1989), ed. W. G. Chaloner and A. Hallam. Additional, unpublished details are reported by Richard A. Kerr in "Dinosaurs and Friends Snuffed Out?," *Science*, 251:160–162 (1991).

The **certainty principle** of opinion was stated by Robert H. Thouless, "The Tendency to Certainty in Religious Belief," *British Journal of Psychology*, 26(1):16–31 (1935). In both religious and secular opinion, Thouless wrote, "there is a real tendency amongst people for degree of belief to approach to certainty. Doubt and skepticism are for most people unusual and, I think, generally unstable states of mind."

Evidence for a **giant meteorite strike** in the Caribbean area at the end of the Cretaceous is given by J.-M. Florentin, R. Maurrasse, and Gautam Sen, "Impacts, Tsunamis, and the Haitian Cretaceous-Tertiary Boundary Layer," *Science*, 252:1690–1693 (1991). They conclude that an immense meteorite impact near the Beloc formation of Haiti "produced microtektites that settled to form a nearly pure layer at the base. Vaporized materials with anomalously high extraterrestrial components settled last, along with carbonate sediments. The entire bed became sparsely consolidated. Subsequently, another major disruptive event, perhaps a giant tsunami, partly reworked the initial deposit . . . This process also may have caused further mixing of Cretaceous and Tertiary microfossils, as observed at Beloc and elsewhere."

The **extinction rates** of families and species during the Permian crisis, based on rarefaction analysis, is given by David M. Raup, "Size of the Permo-Triassic Bottleneck and Its Evolutionary Implications," *Science*, 206:217–218 (1979). His statement on the near extinction of all higher life is given in a later review, "Diversity Crises in the Geological Past," in E. O. Wilson and Frances M. Peter, eds., *Biodiversity* (Washington, D.C.: National Academy Press, 1988), pp. 51–57. A separate examination of the evidence, which considers not only climatic cooling but also a regression (shrinking) of the shallow seas, is provided by Douglas H. Erwin in "The End-Permian Mass Extinction: What Really Happened and Did It Matter?," *Trends in Ecology and Evolution*, 4(8):225–229 (1989).

Evidence of massive volcanic eruptions at the time of the **Permian extinction** spasm was presented by Paul R. Renne and Asish R. Basu, "Rapid Eruption of the Siberian Traps Flood Basalts at the Permo-Triassic Boundary," *Science*, 253:176–179 (1991).

In 1984 David Raup and J. John Sepkoski Jr. proposed that **large-scale extinctions have been periodic,** occurring at intervals of about 26 million years. Their analysis was based on data from the families of marine animals. The Raup-Sepkoski hypothesis set off a round of speculation on possible extraterrestrial causes, such as meteorite or comet showers induced by cyclic approach or realignment of undiscovered celestial bodies. Most intriguing was the postulation of the role of a solar companion star, variously called "Nemesis" or the "Death Star." But the whole idea has been effectively challenged by a combination of critiques of geological dating, statistical analysis, and taxonomic interpretation. The jury is still out on the matter, but an eventual negative verdict seems likely. For reviews of the subject see D. M. Raup, *The Nemesis Affair* (New York: Norton, 1986) and *Extinction: Bad Genes or Bad Luck?* (New York: Norton, 1991); S. M. Stanley, *Extinction* (New York: Scientific American Books, 1987); and a series of articles by paleobiologists in *Ecology*, 70(4):801–834 (1989), ed. Edward F. Connor. In 1991 Raup estimated that half of paleontologists best informed on the subject believed in periodicity; half did not.

4. The Fundamental Unit

A thoroughgoing description and analysis of **species concepts** are given in Douglas J. Futuyma, *Evolutionary Biology,* 2nd ed. (Sunderland, Mass.: Sinauer, 1986); Alan R. Templeton, "The Meaning of Species and Speciation: A Genetic Perspective," in Daniel Otte and John A. Endler, eds., *Speciation and Its Consequences* (Sunderland: Sinauer, 1989), pp. 3–27; and Ernst Mayr and Peter D. Ashlock, *Principles of Systematic Zoology,* 2nd ed. (New York: McGraw-Hill, 1991). The current status of wild tiger populations is documented in Lynn A. Maguire and Robert C. Lacy, "Allocating Scarce Resources for Conservation of Endangered Subspecies: Partitioning Zoo Space for Tigers," *Conservation Biology,* 4(2):157–166 (1990).

The history of the link between malaria and the **sibling species** of *Anopheles maculipennis* is reviewed by Ernst Mayr, *Systematics and the Origin of Species* (New York: Columbia University Press, 1942).

A **history of the biological-species concept** is given by Ernst Mayr, one of its later principal architects, in *Evolution and the Diversity of Life: Selected Essays* (Cambridge: Harvard University Press, 1976).

The idea of a **species as an individual** unto itself has been argued forcefully by Michael T. Ghiselin, "Categories, Life, and Thinking," *Behavioral and Brain Sciences,* 4(2):269–313 (1981).

The **ions chorus** of the Cavendish group, sung to the tune of "My Darling Clementine," is reported by Rutherford's student Samuel Devons in "Rutherford and the Science of His Day," *Notes and Records of the Royal Society of London,* 45(2):221–242 (1991).

Hybridization in oaks and the nature of **semispecies** are discussed by Alan T. Whittemore and Barbara A. Schaal, "Interspecific Gene Flow in Sympatric Oaks," *Proceedings of the National Academy of Sciences,* 88:2540–44 (1991).

The greater frequency of full reproductive isolation among **tropical plant species** has been remarked by Alwyn H. Gentry, "Speciation in Tropical Forests," in L. B. Holm-Nielsen, I. C. Nielsen, and H. Balslev, eds., *Tropical Forests: Botanical Dynamics, Speciation, and Diversity* (New York: Academic Press, 1989), pp. 113–134.

5. New Species

General aspects of **species formation** are well reviewed by Douglas J. Futuyma, *Evolutionary Biology,* 2nd ed. (Sunderland, Mass.: Sinauer, 1986). Special topics are treated at a more advanced level in Daniel Otte and John A. Endler, eds., *Speciation and Its consequences* (Sunderland: Sinauer, 1989).

The gradual evolution of the intermediate **human species** *Homo erectus* is described by Wu Rukang and Lin Shenglong, "Peking Man," *Scientific American*, 248:86–94 (June 1983).

The data on mating times of **giant silkworm moths** are given by Phil and Nellie Rau, "The Sex Attraction and Rhythmic Periodicity in the Giant Saturniid Moths," *Transactions of the Academy of Science of St. Louis*, 26:83–221 (1929).

Details of the courtship displays of **jumping spiders** are given by Jocelyn Crane, "Comparative Biology of Salticid Spiders at Rancho Grande, Venezuela. Part 4: An Analysis of Display," *Zoologica* (New York), 34(4):159–214 (1949).

The definition of formal **subspecies** for use in policy making was proposed by Stephen J. O'Brien and Ernst Mayr, "Bureaucratic Mischief: Recognizing Endangered Species and Subspecies," *Science*, 251:1187–88 (1991). They also reviewed the status of the Florida panther.

The ultrasimple isolating mechanisms of **leafroller moths** is described by Wendell L. Roelofs and Richard L. Brown, "Pheromones and Evolutionary Relationships of Tortricidae," *Annual Review of Ecology and Systematics*, 13:395–422 (1982).

Key review articles on various modes of **sympatric speciation** are Guy L. Bush, "Modes of Animal Speciation," *Annual Review of Ecology and Systematics*, 6:339–364 (1975); Scott R. Diehl and G. L. Bush, "The Role of Habitat Preference in Adaptation and Speciation," in Daniel Otte and John A. Endler, eds., *Speciation and Its Consequences* (Sunderland: Sinauer, 1989), pp. 345–365; and Catherine A. and Maurice J. Tauber, "Sympatric Speciation in Insects: Perception and Perspective," in Otte and Endler, *Speciation*, pp. 307–344. The theory of sympatric speciation by host races was developed principally by Guy Bush. A skeptical analysis is provided by Douglas J. Futuyma and Gregory C. Mayer, "Non-Allopatric Speciation in Animals," *Systematic Zoology*, 29(3):254–271 (1980). Futuyma and Mayer conclude that "the conditions under which host-associated sympatric speciation might occur are so exacting as to be met by very few species."

6. The Forces of Evolution

The data on **number of genes** responsible for variation in simple traits are provided by Russell Lande, "The Minimum Number of Genes Contributing to Quantitative Variation Between and Within Populations," *Genetics*, 99(3,4):541–553 (1981).

The **rates of evolution** due to change in frequencies of single genes are given by Daniel L. Hartl and Andrew G. Clark, *Principles of Population Genetics*, 2nd ed. (Sunderland: Sinauer, 1989).

Allometric variation in the mandibles and horns of male beetles is reviewed by J. T. Clark, "Aspects of Variation in the Stag Beetle *Lucanus cervus* (L.) (Coleoptera:

Lucanidae)," *Systematic Entomology*, 2(1):9–16 (1977). Allometry as the basis of caste differences in ants is presented in detail by Bert Hölldobler and Edward O. Wilson, *The Ants* (Cambridge: Harvard University Press, 1990).

For examples of close linkage between **microevolution and macroevolution**, see: for the adaptive radiation of the Hawaii honeycreepers, Walter J. Bock, "Microevolutionary Sequences as a Fundamental Concept in Macroevolutionary Models," *Evolution*, 24(4):704–722 (1970), and reviewed here in Chapter 7; for the origin of a new vertebrate jaw type in bolyerine snakes on Round Island, Thomas H. Frazzetta, *Complex Adaptations in Evolving Populations* (Sunderland: Sinauer, 1975); and for the origin of new chromosome races and adaptive types in mole rats of the Middle East, Eviatar Nevo, "Speciation in Action and Adaptation in Subterranean Mole Rats: Patterns and Theory," *Bolletino Zoologia*, 52(1–2):65–95 (1985).

The **punctuated equilibrium** thesis was first presented by Niles Eldredge and Stephen J. Gould, "Punctuated Equilibria: An Alternative to Phyletic Gradualism," in T. J. M. Schopf, ed., *Models in Paleobiology* (San Francisco: Freeman, Cooper, 1972), pp. 82–115; and elaborated by Gould, "Is a New and General Theory of Evolution Emerging?," *Paleobiology*, 6(1):119–130 (1980), and by Eldredge, *Time Frames: The Rethinking of Darwinian Evolution and the Theory of Punctuated Equilibria* (New York: Simon and Schuster, 1985). Among the more definitive critiques are Richard Dawkins, *The Blind Watchmaker* (New York: Norton, 1986); Max K. Hecht and Antoni Hoffman, "Why Not Neo-Darwinism? A Critique of Paleobiological Challenges," *Oxford Surveys in Evolutionary Biology*, 3:1–47 (1986); and Jeffrey Levinton, *Genetics, Paleontology, and Macroevolution*, (New York: Cambridge University Press, 1988). The evidence originally presented by Eldredge and Gould has been found not to conform to the punctuated-equilibrium pattern; see William L. Brown Jr., "Punctuated Equilibrium Excused: The Original Examples Fail To Support It," *Biological Journal of the Linnean Society*, 31:383–404 (1987).

The idea of **species selection** in the fossil record was first developed from fossil evidence by Steven M. Stanley, "A Theory of Evolution above the Species Level," *Proceedings of the National Academy of Sciences*, 72:646–650 (1975). The basic argument was expanded by Elisabeth S. Vrba and Stephen J. Gould, "The Hierarchical Expansion of Sorting and Selection," *Paleobiology*, 12(2):217–228 (1986). The basic genetic theory, however, had previously been worked out at the level of multiple competing populations of the same species, following essentially the same model as species selection. The key papers are Richard Levins, "Extinction," in M. Gerstenhaber, ed., *Some Mathematical Questions in Biology* (Providence: American Mathematical Society, 1970), pp. 77–107; and Scott A. Boorman and Paul R. Levitt, "Group Selection on the Boundary of a Stable Population," *Proceedings of the National Academy of Sciences*, 69(9):2711–13 (1972). The enhancement of species proliferation among insects by plant feeding was documented by Charles Mitter, Brian Farrell, and Brian Wiegmann, "The Phylogenetic Study of Adaptive Zones: Has Phytophagy Promoted Insect Diversification?," *American Naturalist*,

132(1):107–128. The connection between species ranges and extinction rates among mollusks is from David Jablonski, "Heritability at the Species Level: Analysis of Geographic Ranges of Cretaceous Mollusks," *Science*, 288:360–363 (1987), and "Estimates of Species Duration: Response," *Science*, 240:969 (1988). The taxon cycle was introduced by Edward O. Wilson, "The Nature of the Taxon Cycle in the Melanesian Ant Fauna," *American Naturalist*, 95:169–193 (1961); and most recently evaluated by James K. Liebherr and Ann E. Hajek, "A Cladistic Test of the Taxon Cycle and Taxon Pulse Hypotheses," *Cladistics*, 6:39–59 (1990). Elisabeth Vrba has documented the turnover rates of African antelopes and other bovid mammals in "African Bovidae: Evolutionary Events since the Miocene," *South African Journal of Science*, 81:263–266 (1985), and "Mammals as a Key to Evolutionary Theory," *Journal of Mammalogy*, 73(1):1–28 (1992). Finally, the example of countervailing organismic and species selection in desert plants was suggested by data from Delbert C. Wiens et al., "Developmental Failure and Loss of Reproductive Capacity in the Rare Paleoendemic Shrub *Dedeckera eurekensis*," *Nature*, 338:65–67 (1989); alternative hypotheses to explain the data are reviewed by Deborah Charlesworth, "Evolution of Low Female Fertility in Plants: Pollen Limitation, Resource Allocation and Genetic Load," *Trends in Ecology and Evolution*, 4(10):289–292 (1989).

7. Adaptive Radiation

The estimate of the number of endemic **Hawaiian insect species** is given by F. G. Howarth, S. H. Sohmer, and W. D. Duckworth, "Hawaiian Natural History and Conservation Efforts," *BioScience*, 38(4):232–238 (1988).

The **honeycreepers of Hawaii** are reviewed by Walter J. Bock, "Microevolutionary Sequences as a Fundamental Concept in Macroevolutionary Models," *Evolution*, 24(4):704–722 (1970), and J. Michael Scott et al., "Conservation of Hawaii's Vanishing Avifauna," *BioScience*, 38(4):238–253 (1988). I have included additional information from Storrs L. Olson (personal communication), who with Helen F. James has pioneered in the study of the subfossil species extinguished by the original Polynesian settlers of Hawaii.

The details of the force of the **woodpecker's strike** are from Philip R. A. May et al., "Woodpeckers and Head Injury," *Lancet*, February 28, 1976, pp. 454–455; and "Woodpecker Drilling Behavior: An Endorsement of the Rotational Theory of Impact Brain Injury," *Archives of Neurology*, 36:370–373 (1979).

The definitive work on the **geospizine finches** is Peter R. Grant, *Ecology and Evolution of Darwin's Finches* (Princeton: Princeton University Press, 1986). Another valuable, more popular account is Sherwin Carlquist, *Island Life: A Natural History of the Islands of the World* (Garden City: Natural History Press, 1965).

The best account of the evolution of **composite herbs on islands** is provided by Sherwin Carlquist in *Island Life* and in *Island Biology* (New York: Columbia University Press, 1974). More recent information on St. Helena flora is given in Mark Williamson, "St. Helena Ebony Tree Saved," *Nature*, 309:581 (1984). The data on St. Helena's mostly extinct beetles are from T. Vernon Wollaston's classic study, *Coleoptera Sanctae-Helenae* (London: John Van Voorst, 1877), updated by P. Basilewski and J. Decelle in their introduction to "La faune terrestre de l'Ile de Sainte-Hélène," *Annales, Musée Royale de l'Afrique Centrale, Tervuren, Belgium, Sciences Zoologiques*, 192:1–9 (1972).

The protean food habits of the **Cocos Island finch** were discovered by Tracey K. Werner and Thomas W. Sherry and reported in "Behavioral Feeding Specialization in *Pinaroloxias inornata*, the 'Darwin's Finch' of Cocos Island, Costa Rica," *Proceedings of the National Academy of Sciences*, 84:5506–10 (1987).

Definitive reviews of **cichlid fish evolution** are presented in *Evolution of Fish Species Flocks*, ed. Anthony A. Echelle and Irv Kornfield (Orono: University of Maine Press, 1984), in articles by Wallace J. Dominey, P. Humphry Greenwood, Leslie S. Kaufman, Karel F. Liem, Kenneth R. McKaye, Richard E. Strauss, and Frans Witte. Molecular evidence of the origin of Lake Victoria cichlids is given by Axel Meyer et al., "Monophyletic Origin of Lake Victoria Cichlid Fishes Suggested by Mitochondrial DNA Sequences," *Nature*, 347:550–553 (1990). A recent analysis of the Lake Victoria species is given by F. Witte and M. J. P. van Oijen, "Taxonomy, Ecology and Fishery of Lake Victoria Haplochromine Trophic Groups," *Zoologische Verhandelingen*, 262:1–47 (1991). Estimates of extinction rates due to the Nile perch are made by C. D. N. Barel et al., "The Haplochromine Cichlids in Lake Victoria: An Assessment of Biological and Fisheries Interests," in M. H. A. Keenleyside, ed., *Cichlid Fishes: Behaviour, Ecology and Evolution* (London: Chapman and Hall, 1991), pp. 258–279.

The role of **plasticity of anatomy and behavior** in macroevolution is emphasized by Mary Jane West-Eberhard, "Phenotypic Plasticity and the Origins of Diversity," *Annual Review of Ecology and Systematics*, 20:249–278 (1989). She provides many interesting examples to show how species formation can proceed rapidly during brief episodes of geographic isolation. A similar effect is posited by Wallace J. Dominey, "Effects of Sexual Selection and Life History on Speciation: Species Flocks in African Cichlids and Hawaiian *Drosophila*," in Echelle and Kornfield, *Evolution of Fish Species Flocks*.

The case of multiple forms in the **arctic char** is described by Skúli Skúlason, David L. G. Noakes, and Sigurdur S. Snorrason, "Ontogeny of Trophic Morphology in Four Sympatric Morphs of Arctic Charr *Salvelinus alpinus* in Thingvallavatn, Iceland," *Biological Journal of the Linnean Society*, 38:281–301 (1989).

Parts of the description of **adaptive radiation in sharks** are adopted from my "In Praise of Sharks," *Discover*, 6(7):40–42, 48, 50–53 (1985).

The description of the **great white shark** by Hugh Edwards is in *Sharks*, ed. J. D. Stevens (New York: Facts on File, 1987), p. 212. Authoritative reviews of shark natural history are given by other authors in the same volume and by Victor G.

Springer and Joy P. Gold, *Sharks in Question* (Washington, D.C.: Smithsonian Institution Press, 1989).

An account of the attacks by **cookie-cutter sharks** on nuclear submarines is provided by C. Scott Johnson, "Sea Creatures and the Problem of Equipment Damage," *U.S. Naval Institute Proceedings*, August 1978, pp. 106–107.

The rapidly developing story of the **megamouth shark** is summarized in Springer and Gold, *Sharks in Question*. An account of the vertical migration is given in the March 1991 *National Geographic*. I have benefited from background information provided in conversation with Robert J. Lavenberg of the Los Angeles County Natural History Museum, who studied the second California specimen live in its natural habitat.

A **fourth adaptive radiation of mammals** occurred on the great island of Madagascar, producing a wide array of lemurs, which are primitive primates resembling monkeys, and tenrecs, which are insectivores variously resembling shrews, moles, and hedgehogs. But the spread of major adaptive types in the fauna as a whole still fell far short of that on Australia, South America, and the World Continent.

The classic survey of **Australian mammals** is Ellis Troughton, *Furred Animals of Australia* (London: Angus and Robertson, 1941). A more recent monograph on Australian mammals, with special attention to conservation, is Tim Flannery's beautifully illustrated *Australia's Vanishing Mammals* (Surry Hills, New South Wales: RD Press, 1990).

An excellent description of the **Great American Interchange** is given by George Gaylord Simpson, *Splendid Isolation: The Curious History of South American Mammals* (New Haven: Yale University Press, 1980). The most recent summary of the fossil and biogeographic evidence, on which my own account is largely based, is Larry G. Marshall et al., "Mammalian Evolution and the Great American Interchange," *Science*, 215:1351–1357 (1982); and L. G. Marshall, "Land Mammals and the Great American Interchange," *American Scientist*, 76:380–388 (1988). A detailed description of the species that were extinguished is provided by Elaine Anderson, "Who's Who in the Pleistocene: A Mammalian Bestiary," in Paul S. Martin and Richard G. Klein, eds., *Quaternary Extinctions*, (Tucson: University of Arizona Press, 1984), pp. 40–89.

I have treated the relation of diversity to **longevity and dominance** more formally in *Success and Dominance in Ecosystems: The Case of the Social Insects* (Oldendorf/ Luhe: Ecology Institute, 1990). Longevity is defined as the duration through geological time of a species and all its descendants. But I will add a note now to say, more precisely, that the longevity of interest is the set of traits by which the species and its descendants are diagnosed, such as possession of a particular gland or bone structure or horn shape. The end of the group of species can then come either by absolute extinction, the death of all the populations, or by "chronotaxon extinction," in which populations of the species group evolve a new set of traits sufficiently different to rank the populations as a different genus or even higher ranking taxon.

8. The Unexplored Biosphere

An incisive review of the **phyla of organisms** is Lynn Margulis and Karlene V. Schwartz, *Five Kingdoms: An Illustrated Guide to the Phyla of Life on Earth* (San Francisco: Freeman, 1982). That book just missed the description of loriciferans by R. M. Kristensen, "Loricifera, a New Phylum with Aschelminthes Characters from the Meiobenthos," *Zeitschrift für Zoologische Systematik und Evolutionsforschung*, 21(3):163–108 (1983). An updated account of loriciferans has been given by Richard C. and Gary J. Brusca, *Invertebrates* (Sunderland: Sinauer, 1990).

The estimates of **numbers of described species** according to group come from my "The Current State of Biological Diversity," in E. O. Wilson and F. M. Peter, eds., *Biodiversity* (Washington, D.C.: National Academy Press, 1988), pp. 3–18. Detailed accounts of diversity and the degree of confidence in estimates of species numbers are provided for many groups individually in Sybil P. Parker, ed., *Synopsis and Classification of Living Organisms*, vols. 1 and 2 (New York: McGraw-Hill, 1982). That work supplied most of my estimates used to reach the world total of 1.4 million. In 1978 T. R. E. Southwood estimated 1.4 million for all described species except fungi, algae, and bacteria and other monerans, in Laurence A. Mound and Nadia Waloff, eds., *Diversity of Insect Faunas* (London: Blackwell, 1978), pp. 19–40. When the missing groups are added, Southwood's world total comes to 1.5 million. Nigel E. Stork, "Insect Diversity: Facts, Fiction and Speculation," *Biological Journal of the Linnean Society*, 35:321–337 (1988), cites an unpublished estimate by N. M. Collins of 1.8 million species of plants and animals alone; if fungi and monerans are added, his figure would rise to about 1.9 million. I am inclined to believe this last number too high, while my own may be too low.

The speculative account of the fate of **a world without insects** was modified from my lecture, "The Little Things That Run the World," given at the National Zoological Park, Washington, D.C., May 7, 1987, and subsequently published in *Conservation Biology*, 1(4):344–346 (1987).

Terry Erwin's estimate of the **diversity of rain-forest arthropods** was first presented in "Tropical Forests: Their Richness in Coleoptera and Other Arthropod Species," *Coleopterists' Bulletin*, 36(1):74–75 (1982), and "Beetles and Other Insects of Tropical Forest Canopies at Manaus, Brazil, Sampled by Insecticidal Fogging," in S. L. Sutton, T. C. Whitmore, and A. C. Chadwick, eds., *Tropical Rain Forest: Ecology and Management* (London: Blackwell, 1983), pp. 59–75. Evaluations of the estimate, with new analyses, are given by Robert M. May, "How Many Species Are There on Earth?," *Science*, 241:1441–49 (1988), and "How Many Species?," *Philosophical Transactions of the Royal Society of London*, ser. B, 330:293–304 (1990); Nigel Stork, "Insect Diversity," and personal communication; and Kevin J. Gaston, "The Magnitude of Global Insect Species Richness," *Conservation Biology*, 5(3):283–296 (1991). The account here is modified from my "Rain Forest Canopy: The High Frontier," *National Geographic*, 180:78–107 (December 1991).

C. William Beebe wrote on the unexplored canopy of the tropical rain forest in *Tropical Wild Life in British Guiana*, by Beebe, G. Inness Hartley, and Paul G. Howes (New York: New York Zoological Society, 1917).

Our knowledge of biological diversity on the **floor of the deep sea** is summarized by J. Frederick Grassle, "Deep-Sea Benthic Biodiversity," *BioScience*, 41(7):464–469 (1991).

The estimation of diversity in **soil bacteria** by means of DNA strand matching is described in two articles by Jostein Goksøyr, Vigdis Torsvik, and their coworkers in *Applied and Environmental Microbiology*, 56(3):776–781, 782–787 (1990). I have benefited from additional unpublished manuscripts kindly provided by Jostein Goksøyr.

The 70 percent **DNA matching criterion** was proposed by the Ad Hoc Committee on Reconciliation of Approaches to Bacterial Systematics in *International Journal of Systematic Bacteriology*, 37:463–464 (1987).

The **new bacterial floras** discovered by deep drilling are characterized by Carl B. Fliermans and David L. Balkwill, "Microbial Life in Deep Terrestrial Subsurfaces," *BioScience*, 39(6):370–377 (1989).

The **diversity of fungi** is another great unknown, its magnitude possibly approaching that of bacteria. In a recent assessment, David L. Hawksworth places the number of known species at 69,000 but the actual number on the earth as very likely 1.5 million. "The Fungal Dimension of Biodiversity: Magnitude, Significance, and Conservation," *Mycological Research*, 95(6):641–655 (1991).

The **symbiosis** of scale insects, yeasts, and bacteria is described by Paul Buchner, *Endosymbiosis of Animals with Plant Microorganisms* (New York: Interscience Publishers, Wiley, 1965), pp. 271–272.

The information on the discovery of new species of **whales and porpoises** is drawn from W. F. J. Mörzer Bruyns, *Field Guide of Whales and Dolphins* (Amsterdam: C. A. Mees, 1971), and Katherine Ralls and Robert L. Brownell Jr., "A Whale of a New Species," *Nature*, 350:560 (1991).

A technical but clear account of equitability and other **diversity measures** of local faunas and floras is given by Anne E. Magurran, *Ecological Diversity and Its Measurement* (Princeton: Princeton University Press, 1988).

The counts of **kingdoms and phyla** of living organisms are based on Lynn Margulis and Karlene V. Schwartz, *Five Kingdoms: An Illustrated Guide to the Phyla of Life on Earth*, (San Francisco: Freeman, 1982). A case can be made for distinguishing the Archaebacteria as a sixth kingdom, but no consensus has been reached by systematists.

The natural history of the **Black Forest hawks** was drawn from Roger Tory Peterson, Guy Montfort, and P. A. D. Hollom, *A Field Guide to the Birds of Britain and Europe*, 2nd ed. (Boston: Houghton Mifflin, 1967). According to Hans Löhrl (personal communication through Ernst Mayr), the goshawk, a threatened species in parts of North America, not only continues to survive in the Black Forest but has increased its numbers, to the point of threatening the capercaillie, the huge grouse-like game bird of the region.

On the **measurement of genetic diversity:** the allozyme estimates are from Robert K. Selander, "Genic Variation in Natural Populations," in F. J. Ayala, ed., *Molecular Evolution* (Sunderland: Sinauer, 1976), pp. 21–45; and "Genetic Variation in Natural Populations: Patterns and Theory," *Theoretical Population Biology*, 13(1):121–177 (1978). Other aspects of allozyme research and the newer measures of nucleotide diversity are given by Wen-Hsiung Li and Dan Graur, *Fundamentals of Molecular Evolution* (Sunderland: Sinauer, 1991), and R. K. Selander, Andrew G. Clark, and Thomas S. Whittam, *Evolution at the Molecular Level* (Sunderland: Sinauer, 1991). I am grateful to Russell Lande for providing valuable advice on the estimation of total genetic diversity based on this research.

9. The Creation of Ecosystems

The role of **sea otters as keystone species** is detailed in David O. Duggins, "Kelp Beds and Sea Otters: An Experimental Approach," *Ecology*, 61(3):447–453 (1980).

The case for **jaguars and pumas as keystone species** is made by John Terborgh, "The Big Things That Run the World—A Sequel to E. O. Wilson," *Conservation Biology*, 2(4):402–403 (1988). The estimate of a tenfold increase of coatis and rodents in the absence of jaguars and pumas on Barro Colorado Island is based on a comparison with the fauna at Cocha Cashu, Peru, where big cats still live.

The **keystone role of large African mammals** is documented by Norman Owen-Smith, "Megafaunal Extinctions: The Conservation Message from 11,000 years B.P.," in *Conservation Biology*, 3(4):405–412 (1989).

The account of **African driver ants** is adapted from my *Success and Dominance in Ecosystems: The Case of the Social Insects* (Oldendorf/Luhe: Ecology Institute, 1990).

The **assembly rules** of community formation were inferred for the birds of New Guinea by Jared M. Diamond, "Assembly of Species Communities," in *Ecology and Evolution of Communities*, ed. M. L. Cody and J. M. Diamond (Cambridge: Harvard University Press, 1975), pp. 342–444. A critique of Diamond's approach based on statistical analysis is provided by Daniel Simberloff, "Using Island Biogeographic Distributions To Determine If Colonization Is Stochastic," *American Naturalist*, 112:713–726 (1978); "Competition Theory, Hypothesis Testing, and Other Community-Ecology Buzzwords," *American Naturalist*, 122:626–635 (1983). A general review across many groups of microorganisms and animals was recently provided by James A. Drake, "Communities as Assembled Structures: Do

Rules Govern Pattern?," *Trends in Ecology and Evolution*, 5(5):159–164 (1990). Drake, who concludes that competition-based assembly rules do in fact exist, also employs the jigsaw-puzzle analogy to describe colonization sequences. Other, generally favorable approaches to the role of competition are provided by the authors of *Community Ecology*, ed. Jared Diamond and Ted J. Case (New York: Harper and Row, 1986).

The **fire-ant competition** story is told in Hölldobler and Wilson, *The Ants*.

Compression and release among competing Darwin's finches is described by Peter R. Grant, *Ecology and Evolution of Darwin's Finches* (Princeton: Princeton University Press, 1986).

Character displacement in Darwin's finches was first suggested by David Lack in his classic *Darwin's Finches* (Cambridge: Cambridge University Press, 1947). It was then documented in convincing detail by Peter Grant in his own 1986 classic, *Ecology and Evolution of Darwin's Finches*. The plier analogy to bill structure was introduced by Robert I. Bowman as part of a detailed analysis in "Morphological Differentiation and Adaptation in the Galápagos Finches," *University of California Publications in Zoology*, 58:1–302 (1961).

The relation between **predation and species numbers** in intertidal mollusks is reported by Robert T. Paine, "Food Web Complexity and Species Diversity," *American Naturalist*, 100:65–75 (1966).

For showing me how to extract and examine **forehead mites,** I am grateful to Michael Huben.

For authoritative reviews of **food webs** see Joel E. Cohen, *Food Webs and Niche Space* (Princeton: Princeton University Press, 1978); Joel E. Cohen, Frédéric Briand, and Charles M. Newman, eds., *Community Food Webs* (New York: Springer, 1990); and Stuart L. Pimm, John H. Lawton, and Joel E. Cohen, "Food Web Patterns and Their Consequences," *Nature*, 350:669–674 (1991).

The strange reciprocal predation of **mosquito larvae and protozoans** in the genus *Lambornella* was reported by Jan O. Washburn et al., "Predator-Induced Trophic Shift of a Free-Living Ciliate: Parasitism of Mosquito Larvae by Their Prey," *Science*, 240:1193–95 (1988).

10. Biodiversity Reaches the Peak

Details of **microbial mats and stromatolites** are given by David J. Des Marais, "Microbial Mats and the Early Evolution of Life," *Trends in Ecology and Evolution*, 5(5):140–144 (1990); and lecture materials provided by J. William Schopf in Steve Olson, *Shaping the Future: Biology and Human Values* (Washington, D.C.: National Academy Press, 1989).

The details of the **history of diversity** have been taken from many sources, especially Andrew H. Knoll and John Bauld, "The Evolution of Ecological Tolerance in Prokaryotes," *Transactions of the Royal Society of Edinburgh, Earth Sciences,* 80:209–223 (1989); lecture materials from J. William Schopf in Olson, *Shaping the Future;* Philip W. Signor, "The Geologic History of Diversity," *Annual Review of Ecology and Systematics,* 21:509–539 (1990); and Mark A. S. McMenamin, "The Emergence of Animals," *Scientific American,* 256:94–102 (April 1987). Reports of spores of earliest land plants and burrows of invertebrate animals are by Gregory J. Retallack and Carolyn R. Feakes, "Trace Fossil Evidence for Late Ordovician Animals on Land," *Science,* 235:61–63 (1987). I have also benefited from an unpublished manuscript by A. H. Knoll and Heinrich D. Holland, "Oxygen and Proterozoic Evolution: An Update."

The concept of **evolutionary progress** given here was first presented in my *Success and Dominance in Ecosystems: The Case of the Social Insects* (Oldendorf/Luhe: Ecology Institute, 1990).

The idea of a moving average toward **larger and more complex animals through geological time** is documented by Geerat J. Vermeij, *Evolution and Escalation: An Ecological History of Life* (Princeton: Princeton University Press, 1987); and by John Tyler Bonner, *The Evolution of Complexity, by Means of Natural Selection* (Princeton: Princeton University Press, 1988).

The age of the **Cambrian period** and hence of the entire Phanerozoic eon, 550 million years, is the consensus among geologists, according to Simon Conway Morris (personal communication). The connection between the rise of atmospheric oxygen and the origin of macroscopic animals in late Precambrian and early Cambrian times was first proposed as a theoretical model by Preston Cloud.

The estimates of **extinction rates** among marine organisms are based on many studies of the Permian and Triassic fossil record, which are reviewed in D. H. Erwin, "The End-Permian Mass Extinction," *Annual Review of Ecology and Systematics,* 21:69–91 (1990).

The **Field of Bullets** scenario is given by David M. Raup, *Extinction: Bad Genes or Bad Luck?* (New York: Norton, 1991). It is derived from his techniques for estimating extinction rates according to taxonomic rank in "Taxonomic Diversity Estimation Using Rarefaction," *Paleobiology,* 1(4):333–342 (1975). I have given the scenario a military twist.

An authoritative analysis of the **Cambrian explosion** in the evolution of marine animals is offered by S. Conway Morris, "Burgess Shale Faunas and the Cambrian Explosion," *Science,* 246:339–346 (1989). The affinities of several problematica, including *Hallucigenia,* to the living phylum Onychophora are suggested by L. Ramsköld and Hou Xianguang, "New Early Cambrian Animal and Onychophoran Affinities of Enigmatic Metazoans," *Nature,* 351:225–228 (1991). The diagnosis of the bizarre *Wiwaxia corrugata* was made by Nicholas J. Butterfield, "A Reassessment of the Enigmatic Burgess Shale Fossil *Wiwaxia corrugata* (Matthew) and Its Relationship to the Polychaete *Canadia spinosa* Walcott," *Paleobiology*, 16(3):287–303 (1990).

For Signor's summary of the correlation of **continental geography and global biodiversity,** see his "Geologic History of Diversity."

The trend toward **enrichment of local faunas and floras** is documented by J. John Sepkoski Jr. et al., "Phanerozoic Marine Diversity and the Fossil Record," *Nature*, 293:435–437 (1981); and Andrew H. Knoll, "Patterns of Change in Plant Communities through Geological Time," in Jared M. Diamond and Ted J. Case, eds., *Community Ecology* (New York: Harper and Row, 1986), pp. 126–141.

Latitudinal species gradients. The numbers of breeding bird species are from Adrian Forsyth, *Portraits of the Rainforest* (Ontario: Camden House, Camden East, 1990). Raymond A. Paynter supplied me with the number of Colombian species. A list of publications documenting the latitudinal diversity gradient in a wide range of plant and animals is given by George C. Stevens, "The Latitudinal Gradient in Geographical Range: How So Many Species Coexist in the Tropics," *American Naturalist*, 133(2):240–256 (1989).

Estimates of tropical versus temperate **plant diversity** are given by Peter H. Raven, "The Scope of the Plant Conservation Problem World-Wide," in David Bramwell, Ole Hamann, V. H. Heywood, and Hugh Synge, eds., *Botanic Gardens and the World Conservation Strategy* (New York: Academic Press, 1987), pp. 19–29. Alwyn H. Gentry's count of tree species in Peru, the world record, is given in "Tree Species Richness of Upper Amazonian Forests," *Proceedings of the National Academy of Sciences*, 85:156–159 (1988). Peter S. Ashton's unpublished estimates of Bornean tree diversity were provided in a personal communication.

The **butterfly diversity data** from Peru and Brazil are cited by Gerardo Lamas, Robert K. Robbins, and Donald J. Harvey, "A Preliminary Survey of the Butterfly Fauna of Pakitza, Parque Nacional del Manu, Peru, with an Estimate of Its Species Richness," *Publicaciones del Museo de Historia Natural, Universidad Nacional Mayor de San Marcos, serie A Zoologia*, 40:1–19 (1991); and Thomas C. Emmel and George T. Austin, "The Tropical Rain Forest Butterfly Fauna of Rondonia, Brazil: Species Diversity and Conservation," *Tropical Lepidoptera*, 1(1):1–12 (1990).

The **ants of a single tree** in Peruvian rain forest were analyzed by me in "The Arboreal Ant Fauna of Peruvian Amazon Forests: A First Assessment," *Biotropica*, 19(3):245–251 (1987). Terry L. Erwin estimated the **number of beetle species** in a Panamanian rain forest in "Tropical Forests: Their Richness in Coleoptera and Other Arthropod Species," *Coleopterist's Bulletin*, 36(1):74–75 (1982). Estimates of beetle diversity in North America and the world are given in Ross H. Arnett Jr., *American Insects: A Handbook of the Insects of America North of Mexico* (New York: Van Nostrand Reinhold, 1985).

David J. Currie's correlation of **species richness of vertebrates and trees** in North America with environmental variables is presented in "Energy and Large-Scale

Patterns of Animal- and Plant-Species Richness," *American Naturalist*, 137(1):27–49 (1991).

Rapoport's rule, as George Stevens has called it, was proposed by the Argentine ecologist Eduardo H. Rapoport in *Aerography: Geographical Strategies of Species*, English translation from the 1975 Spanish original (New York: Pergamon, 1982). Stevens himself, however, compiled the published data from many sources that sealed the point. He also made the connection between Rapoport's rule—that temperate species have wider latitudinal distributions—and the need for temperate species to occupy more variable local environments. The narrowing of altitudinal ranges on the sides of tropical mountains by the same effect, an idea essentially the same as Rapoport's rule, was introduced in 1967 by Daniel H. Janzen, "Why Mountain Passes Are Higher in the Tropics," *American Naturalist*, 101:233–249.

The **New Guinea weevils** with gardens of algae, lichens, and mosses on their backs were discovered by J. Linsley Gressitt, "Epizoic Symbiosis," *Entomological News*, 80(1):1–5 (1969).

Dynamine hoppi and many other rare and beautiful **butterfly species** are described by Philip J. DeVries, *The Butterflies of Costa Rica and Their Natural History: Papilionidae, Pieridae, Nymphalidae* (Princeton: Princeton University Press, 1987).

The **source-sink model** is evaluated by H. Ronald Pulliam, "Sources, Sinks, and Population Regulation," *American Naturalist*, 132(5):652–661 (1988). It has been especially well documented in the exhaustive study of tree diversity in Panama by Stephen Hubbell and Robin Foster: "Commonness and rarity in a Neotropical Forest: Implications for Tropical Tree Conservation," in Michael E. Soulé, ed., *Conservation Biology: The Science of Scarcity and Diversity* (Sunderland: Sinauer, 1986), pp. 205–231.

The account of **epiphytes** is modified from my "Rain Forest Canopy: The High Frontier," *National Geographic*, 180:78–107 (December 1991).

The importance of **deep-sea animals** as support for environmental stability was first pointed out by Howard L. Sanders, "Marine Benthic Diversity: A Comparative Study," *American Naturalist*, 102:243–282 (1968).

Analyses of the **effect of the size of organisms** on biological diversity are given by D. R. Morse et al, "Fractal Dimension of Vegetation and the Distribution of Arthropod Body Lengths," *Nature*, 314:731–733 (1985); and Robert M. May, "How Many Species Are There on Earth?," *Science*, 241:1441–49 (1988).

G. Evelyn Hutchinson and Robert H. MacArthur proposed the **logarithmic rule of increasing biodiversity** with decreasing organism size in "A Theoretical Ecological Model of Size Distributions among Species of Animals," *American Naturalist*, 93:117–125 (1959).

The **fractal analysis of niche size** as a determinant of biodiversity was introduced by Morse et al., "Fractal Dimension." These ecologists measured actual vegetation surfaces to get the differences perceived by organisms of different sizes.

The **feather-mite world** in the plumage of parrots is described by Tila M. Pérez and Warren T. Atyeo, "Site Selection of the Feather and Quill Mites of Mexican Parrots," in D. A. Griffiths and C. E. Bowman, eds., *Acarology VI* (Chichester, Eng.: Ellis Horwood, 1984), pp. 563–570. Additional details were kindly provided to me by Tila Pérez in personal communication.

The final days of the **Carolina parakeet** are described by Doreen Buscami, "The Last American Parakeet," *Natural History,* 87(4):10–12 (1978).

The most thorough statistical studies of the **factors affecting the numbers of animal species** were recently conducted by Kenneth P. Dial and John M. Marzluff. See "Are the Smallest Organisms the Most Diverse?," *Ecology,* 69(5):1620–24 (1988); "Nonrandom Diversification within Taxonomic Assemblages," *Systematic Zoology,* 38(1):26–37 (1989); and "Life History Correlates of Taxonomic Diversity," *Ecology,* 72(2):428–439 (1990).

The account of **insect diversity and dominance** given here is based on my "First Word," *Omni,* 12:6 (September 1990).

The reasons for the great variety and **ecological importance of insects** are assessed by T. R. E. Southwood, "The Components of Diversity," in Laurence A. Mound and Nadia Waloff, eds., *Diversity of Insect Faunas* (London: Blackwell, 1978), pp. 19–40.

The description of **adaptive radiation in African mammals** is modified from Charles J. Lumsden and Edward O. Wilson, *Promethean Fire* (Cambridge: Harvard University Press, 1983).

11. The Life and Death of Species

The account of the extinction of the **New Zealand mistletoe** is based on David A. Norton, "*Trilepidea adamsii:* An Obituary for a Species," *Conservation Biology,* 5(1):52–57 (1991).

The data on **extinction rates of marine organisms** are from David M. Raup, "Extinction: Bad Genes or Bad Luck?," *Acta geològica hispànica,* 16(1–2):25–33 (1981); and "Evolutionary Radiations and Extinction," in H. D. Holland and A. F. Trandall, eds., *Patterns of Change in Evolution* (Berlin: Dahlem Konferenzen, Abakon Verlagsgesellschaft, 1984), pp. 5–14.

The approximate **constancy of species extinction** within a clade—and of clades within larger clades—was documented by Leigh van Valen, "A New Evolutionary Law," *Evolutionary Theory,* 1:1–30 (1973). An updated evaluation of longevity, confirming constancy but with a great many caveats, is provided by Jeffrey

Levinton, *Genetics, Paleontology, and Macroevolution* (New York: Cambridge University Press, 1988).

The recent **history of African buffalos and antelopes,** including a mass extinction episode 2.5 million years ago, is detailed by Elisabeth S. Vrba, "African Bovidae: Evolutionary Events since the Miocene," *South African Journal of Science,* 81:263–266 (1985).

Rapid species formation in Andean plants, especially orchids, is argued by Alwyn H. Gentry and Calaway H. Dodson, "Diversity and Biogeography of Neotropical Vascular Epiphytes," *Annals of the Missouri Botanical Garden,* 74:205–233 (1987).

The **birth of the island Surtsey** on November 14, 1963, was followed by the colonization of plants and animals in a manner paralleling that of Krakatau (Chapter 2), though with many fewer species. The history of the island is detailed by Sturla Fridriksson, *Surtsey: Evolution of Life on a Volcanic Island* (New York: Halsted Press, Wiley, 1975). The Icelandic people have witnessed similar episodes many times. The tenth-century poem *Völuspá* transforms the eruptions into the rampages of the fire-giant Surtur the Black: "The hot stars down / from Heaven are whirled. / Fierce grows the steam / and the life-feeding flame. / Till fire leaps high / about Heaven itself." The name Surtsey means island of Surtur.

The theory of island biogeography was presented in 1963 by Robert H. MacArthur and Edward O. Wilson, "An Equilibrium Theory of Insular Zoogeography," *Evolution,* 17(4):373–387, and elaborated in our *The Theory of Island Biogeography* (Princeton: Princeton University Press, 1967). There have been many discussions and improvements of the idea, perhaps best presented by Mark Williamson in *Island Populations* (Oxford: Oxford University Press, 1981) and "Natural Extinction on Islands," *Philosophical Transactions of the Royal Society of London,* ser. B, 325:457–468 (1989). The rule that increasing the area of an island tenfold doubles the number of species was first suggested by Philip J. Darlington, *Zoogeography: The Geographical Distribution of Animals* (New York: Wiley, 1957).

The biogeographic experiment on the **Florida Keys** is reported in Daniel S. Simberloff and Edward O. Wilson, "Experimental Zoogeography of Islands: Defaunation and Monitoring Techniques," *Ecology,* 50(2):267–278 (1969); and "Experimental Zoogeography of Islands: A Two-Year Record of Colonization," *Ecology,* 51(5):934–937 (1970). The theory of island biogeography, especially its central proposition of a dynamic equilibrium in species numbers, has been tested by many other experiments using miniature systems, including diatoms suspended on slides in freshwater streams and microorganisms in bottles of water. Studies of turnover in patches of islands of varying areas have also contributed, as well as analyses of the postcatastrophe histories of Krakatau and Surtsey.

The early results of the **Forest Fragments Project** in Brazil are reported in Thomas E. Lovejoy et al., "Ecosystem Decay of Amazon Forest Remnants," in Matthew H. Nitecki, ed., *Extinction* (Chicago: University of Chicago Press, 1984), pp. 295–

325; and Lovejoy et al., "Edge and Other Effects of Isolation on Amazon Forest Fragments," in Michael E. Soulé, ed., *Conservation Biology: The Science of Scarcity and Diversity* (Sunderland: Sinauer, 1986), pp. 257–285. The loss of beetle diversity was demonstrated by Bert C. Klein, "Effects of Forest Fragmentation on Dung and Carrion Beetle Communities in Central Amazonia," *Ecology*, 70(6):1715–25 (1989).

The theory of **extinction probability,** along with data from small British islands testing the theory, is presented in Stuart L. Pimm, H. Lee Jones, and Jared Diamond, "On the Risk of Extinction," *American Naturalist*, 132(6):757–785 (1988). In *The Theory of Island Biogeography* (1967), MacArthur and Wilson provide equations measuring the heavy dependence of the longevity of populations on population size and the birth and death rates of the member organisms.

Details on **endangered North American bird species** have been drawn from John W. Terborgh, "Preservation of Natural Diversity: The Problem of Extinction Prone Species," *BioScience*, 24(12):715–722 (1974); and *Where Have All the Birds Gone? Essays on the Biology and Conservation of Birds That Migrate to the American Tropics* (Princeton: Princeton University Press, 1989); David S. Wilcove and J. W. Terborgh, "Patterns of Population Decline in Birds," *American Birds*, 38(1):10–13 (1984); and Russell Lande, "Genetics and Demography in Biological Conservation," *Science*, 241:1455–60 (1988). The diverse properties of rareness in organisms are classified by Deborah Rabinowitz, Sara Cairns, and Theresa Dillon, "Seven Forms of Rarity and Their Frequency in the Flora of the British Isles," in Michael E. Soulé, ed., *Conservation Biology: The Science of Scarcity and Diversity*, (Sunderland: Sinauer, 1986), pp. 182–204.

The account of **Paleozoic snails** dwelling on the anuses of sea lilies is from Steven M. Stanley, "Periodic Mass Extinctions of the Earth's Species," *Bulletin of the American Academy of Arts and Sciences*, 40(8):29–48 (1987).

My study of extinction in **ants of the West Indies** was presented in "Invasion and Extinction in the West Indian Ant Fauna: Evidence from the Dominican Amber," *Science*, 229:265–267 (1985).

Steven Stanley on the greater **longevity of abundant mollusks** in the fossil record: "Periodic Mass Extinctions," pp. 34–36.

The **50–500 rule** of minimum population size was introduced by Ian Robert Franklin, "Evolutionary Changes in Small Populations," in Michael E. Soulé and Bruce A. Wilcox, eds., *Conservation Biology: An Evolutionary-Ecological Perspective* (Sunderland: Sinauer, 1980), pp. 135–149. The lethal equivalents in the genetic makeup of zoo-animals are analyzed by John W. Senner, "Inbreeding Depression and the Survival of Zoo Populations," in Soulé and Wilcox, *Conservation Biology*, pp. 209–224; and by Katherine Ralls, Jonathan D. Ballou, and Alan Templeton, "Estimates of Lethal Equivalents and the Cost of Inbreeding in Mammals," *Conservation Biology*, 2(2):185–193 (1988). The 50–500 rule is reexamined by Otto Frankel and Michael E. Soulé, *Conservation and Evolution* (Cambridge: Cambridge

University Press, 1981), and more critically by Russell Lande, "Genetics and Demography in Biological Conservation," *Science*, 241:1455–60 (1988).

The **tiny populations** of the Frigate Island darkling beetle and Socorro sowbug are described in *The IUCN Invertebrate Red Data Book* (Old Woking: Unwin Brothers, 1983) and that of the hau kuahiwi tree of Kauai in *Plant Conservation* (Center for Plant Conservation), 3(4):1–8 (1988).

The **metapopulation concept,** originated by Richard Levins in 1970, has been most recently explored by Isabelle Olivieri et al., "The Genetics of Transient Populations: Research at the Metapopulation Level," *Trends in Ecology and Evolution*, 5(7):207–210 (1990); and in fine detail by authors in *Metapopulation Dynamics: Empirical and Theoretical Investigations*, ed. Michael Gilpin and Ilkka Hanski (New York: Academic Press, 1991), a book reprinted from the *Biological Journal of the Linnean Society*, 42(1–2) (1991).

Information on the **Karner blue butterfly** comes from "Minimum Area Requirements for Long-Term Conservation of the Albany Pine Bush and Karner Blue Butterfly: An Assessment," an unpublished report for the state of New York by Thomas J. Givnish, Eric S. Menges, and Dale F. Schweitzer, August 9, 1988; cited by permission of the authors. The Karner blue is one of a few scattered metapopulations classified as the eastern race of the Melissa blue, *Lycaeides melissa*. It was formally described by Vladimir Nabokov, the novelist and distinguished aurelian.

The final days of **Spix's macaw** in the wild were reported by Jorgen B. Thomsen and Charles A. Munn, "*Cyanopsitta spixii:* A Non-Recovery Report," *Parrotletter*, 1(1):6–7 (1987) and in a news report, "Lone Macaw Makes a Vain Bid for Survival," *New Scientist*, August 18, 1990. I am indebted to Jorgen Thomsen for additional details on the status of the last surviving male.

12. Biodiversity Threatened

I am grateful to Alwyn H. Gentry for supplying me with the **history of Centinela.** Some of the characteristics of the flora are provided by Gentry in "Endemism in Tropical versus Temperate Plant Communities," in Michael E. Soulé, ed., *Conservation Biology: The Science of Scarcity and Diversity* (Sunderland: Sinauer, 1986), pp. 153–181. A history of deforestation in Ecuador is traced in Calaway Dodson and Gentry, "Biological Extinction in Western Ecuador," *Annals of the Missouri Botanical Gardens*, 78(2):273–295 (1991).

Mass extinction of Polynesian birds. The extinction of Hawaiian landbirds by the Polynesian colonists is described in Storrs L. Olson and Helen F. James, "Descriptions of Thirty-Two New Species of Birds from the Hawaiian Islands, Part 1: Non-Passeriformes," *Ornithological Monographs*, 45:1–88 (1991); and "Descriptions of Thirty-Two New Species of Birds from the Hawaiian Islands, Part 2: Passeriformes," *Ornithological Monographs*, 46:1–88 (1991). The destruction of the faunas in other parts of Polynesia are documented in David W. Steadman, "Extinction of Birds in Eastern Polynesia: A Review of the Record and Comparisons with Other Pacific Island Groups," *Journal of Archaeological Science*, 16:177–205 (1989); and Tom Dye and D. W. Steadman, "Polynesian Ancestors and Their Animal World," *American Scientist*, 78:207–215 (1990). The Henderson story is told by Steadman and Olson, "Bird Remains from an Archaeological Site on Henderson Island, South Pacific: Man-Caused Extinctions on an 'Uninhabited' Island," *Proceedings of the National Academy of Sciences*, 82:6191–95 (1985).

Ice Age extinctions. The definitive work on extinctions at the end of the last Ice Age, about 11,000 years ago, is the multiauthored *Quaternary Extinctions: A Prehistoric Revolution*, ed. Paul S. Martin and Richard G. Klein (Tucson: University of Arizona Press, 1984). The authors consulted here in order of appearance are David W. Steadman and Paul S. Martin (North American Pleistocene extinctions and late Pleistocene birds), Leslie F. Marcus and Rainer Berger (late Pleistocene megafauna as disclosed at Rancho La Brea), Larry D. Agenbroad (mammoths), Arthur M. Phillips III (ground sloths), C. Vance Haynes (Clovis culture and megafauna extinction), Jared M. Diamond (Iceland's bird fauna), James E. King and Jeffrey J. Saunders (mastodons), S. David Webb (North American mammalian extinctions for the past 10 million years), and Donald K. Grayson (history of nineteenth-century explanations of Pleistocene extinctions).

The **extinction of the moas** and other endemic birds on New Zealand is a story told by Michael M. Trotter and Beverley McCulloch, Atholl Anderson, and Richard Cassels, in Martin and Klein, *Quaternary Extinctions;* and more recently again by Anderson in *Prodigious Birds: Moas and Moa-hunting in Prehistoric New Zealand* (New York: Cambridge University Press, 1990).

The fates of the **Madagascan and Australian faunas** are described by Robert E. Dewar, Peter Murray, Duncan Merrilees, and D. R. Horton in Martin and Klein, *Quaternary Extinctions.*

Jared Diamond's case identifying **prehistoric man as destroyer** of the world's megafauna is an improvement on that developed by Paul Martin and others, with important additions from Diamond's own research on birds of the Pacific region. It is presented in "Quaternary Megafaunal Extinctions: Variations on a Theme by Paganini," *Journal of Archaeological Science*, 16:167–175 (1989).

The demise of the **imperial woodpecker** in Mexico was reported by George Plimpton, "Un gran pedazo de carne," *Audubon Magazine*, 79(6):10–25 (1977).

The origin and impact of **exotic species** are treated in Harold A. Mooney and James A. Drake, eds., *Ecology of Biological Invasions of North America and Hawaii* (New York: Springer, 1986).

The status of **extinct and vulnerable fish species** in North America is reviewed by Jack E. Williams et al., "Fishes of North America. Endangered, Threatened, or of Special Concern: 1989," *Fisheries* (American Fisheries Society), 14(6):2–20

(1989); R. R. Miller et al., "Extinctions of North American Fishes During the Past Century," *Fisheries*, 14(6):22–38 (1989); and Jack E. Williams and Robert R. Miller, "Conservation Status of the North American Fish Fauna in Fresh Water," *Journal of Fish Biology*, 37(A):79–85 (1990). I am grateful to Karsten E. Hartel for sharing his unpublished analysis of data pertaining to species decline.

The anecdotes of **extinction of birds** are based on Jared M. Diamond, "The Present, Past and Future of Human-Caused Extinction," *Philosophical Transactions of the Royal Society of London*, ser. B, 325:469–477 (1989); and John Terborgh, *Where Have All the Birds Gone? Essays on the Biology and Conservation of Birds That Migrate to the American Tropics* (Princeton: Princeton University Press, 1989).

On the high rate of **extinction of freshwater fishes**, see Diamond, and Walter R. Courtenay Jr. and Peter B. Moyle, "Introduced Fishes, Aquaculture, and the Biodiversity Crisis," *Abstracts, 71st Annual Meeting, American Society of Ichthyologists and Herpetologists*, no pp.; and Irv Kornfield and Kent E. Carpenter, "Cyprinids of Lake Lanao, Philippines: Taxonomic Validity, Evolutionary Rates and Speciation Scenarios," in Anthony A. Echelle and Irv Kornfield, eds., *Evolution of Fish Species Flocks* (Orono: University of Maine Press, 1984). The total of 18 species accepted in the classical accounts of the Lake Lanao cyprinid species flock may be excessive, even though the Maranao people of the region recognize all of them. Some of the species may instead be morphs of very plastic species, as I described for the Mexican cichlid and arctic char in Chapter 7. However the matter is judged taxonomically, the adaptive radiation of the Lanao cyprinids is extreme for a single lake, and it has been almost completely erased during the past fifty years. The fate of the Lake Victoria fishes is described by Christopher G. Barlow and Allan Lisle, "Biology of the Nile Perch *Lates niloticus* (Pisces: Centropomidae) with Reference to Its Proposed Role as a Sport Fish in Australia," *Biological Conservation*, 39(4):269–289 (1987); Daniel J. Miller, "Introductions and Extinction of Fish in the African Great Lakes," *Trends in Ecology and Evolution*, 4(2):56–59 (1989); and C. D. N. Barel et al., "The Haplochromine Cichlids in Lake Victoria: An Assessment of Biological and Fisheries Interests," in M. H. A. Keenleyside, ed., *Cichlid Fishes: Behaviour, Ecology and Evolution* (London: Chapman and Hall, 1991), pp. 258–279.

The decline of **freshwater mollusks** is documented in *The IUCN Invertebrate Red Data Book* (Gland, Switzerland: International Union for Conservation of Nature and Natural Resources, 1983).

The **Moorean tree snails** have been the subject of classic studies of microevolution by Henry E. Crampton and Bryan C. Clarke. The snails' total destruction in the wild is described by James Murray, Elizabeth Murray, Michael S. Johnson, and Bryan Clarke, "The Extinction of *Partula* on Moorea," *Pacific Science*, 42(3,4):150–153 (1988); I am grateful to Bryan Clarke for supplying additional unpublished details on the episode. The loss of the Hawaiian tree snails is documented in *The ICUN Invertebrate Red Data Book* (1983).

The **threatened plant species of the United States** are counted by Linda R. McMahan, "CPC Survey Reveals 680 Native U.S. Plants May Become Extinct within 10 Years," *Plant Conservation* (Center for Plant Conservation), 3(4):1–2 (1988). The species already extinct were tabulated by Michael O'Neal and other CPC staff members in 1992 (personal communication). The account of the Puerto Rican endemic *Banara vanderbiltii* is based on John Popenoe, "One of the World's Rarest Species," *Plant Conservation*, 3(4):6 (1988).

The numbers of threatened and **endangered invertebrate species of Europe** were reported by Eladio Fernandez-Galiano in *IUCN Special Report Bulletin* (International Union for Conservation of Nature and Natural Resources), 18(7–9):7 (1987). In 1989, 501 insect species were listed as threatened under provisions of the U.S. Endangered Species Act. This represents only about 1 percent of the total known fauna, but it is also a gross underestimate owing to the poor state of taxonomic knowledge in all but a few groups.

The **decline of European fungi** is reviewed by John Jaenike, "Mass Extinction of European Fungi," *Trends in Ecology and Evolution*, 6(6):174–175 (1991). Similar studies have not yet been undertaken in North America.

The case of the **northern spotted owl** is discussed by Russell Lande, "Demographic Models of the Northern Spotted Owl (*Strix occidentalis caurina*)," *Oecologia*, 75(4):601–607 (1988), and "Genetics and Demography in Biological Conservation," *Science*, 241:1455–60 (1988).

Rare frogs and salamanders of the Pacific Northwest forests are described by Hartwell H. Welsh Jr., "Relictual Amphibians and Old-Growth Forests," *Conservation Biology*, 4(3):308–319 (1990).

A catalogue of **threatened and endangered habitats** is provided in *The IUCN Invertebrate Red Data Book* (1983).

Norman Myers' eighteen **hot spots** were identified in two articles, "Threatened Biotas: 'Hot spots' in Tropical Forests," *Environmentalist*, 8(3):187–208 (1988); and "The Biodiversity Challenge: Expanded Hot-Spots Analysis," *Environmentalist*, 10(4):243–256 (1990).

The present condition of the **Brazilian Atlantic forest** is detailed in Mark Collins, ed., *The Last Rain Forests: A World Conservation Atlas* (New York: Oxford University Press, 1990). This beautifully illustrated book, containing maps of the former and present extent of all the major tropical forests, is the best popular reference work of its kind.

Among the concerns of ecologists, **tropical deciduous forests** have stood in the shadow of the rain forests, but they are in even greater peril. Because they occupy potentially prime agricultural and cattle land and are easily cleared, they are among the most heavily exploited of the world's land environments. In Central America they have been reduced to less than 10 percent of the original cover. Tropical deciduous forests are intermediate between rain forests and temperate deciduous forests in amount of diversity. A review is presented by Manuel Lerdau, Julie Whitbeck, and N. Michele Holbrook, "Tropical Deciduous Forest: Death of a Biome," *Trends in Ecology and Evolution*, 6(7):201–233 (1991).

The **reduction of coral reefs** by both natural and human-caused stress is reported in "Coral Reefs off 20 Countries Face Assaults from Man and Nature," *New York Times*, March 27, 1990; Peter W. Glynn, "Coral Reef Bleaching in the 1980s and Possible Connections with Global Warming," *Trends in Ecology and Evolution*, 6(6):175–179 (1991); and Leslie Roberts, "Greenhouse Role in Reef Stress Unproven," *Science*, 253:258–259 (1991).

The effects of **climatic warming** on biodiversity are predicted by Robert L. Peters and Joan D. S. Darling, "The Greenhouse Effect and Nature Reserves," *BioScience*, 35(11):707–717 (1985); Andy Dobson, Alison Jolly, and Dan Rubenstein, "The Greenhouse Effect and Biological Diversity," *Trends in Ecology and Evolution*, 4(3):64–68 (1989); and Robert L. Peters and Thomas E. Lovejoy, eds., *Global Warming and Biological Diversity* (New Haven: Yale University Press, 1992). The account given here is drawn from these sources and from my "Threats to Biodiversity," *Scientific American*, 260(9):108–116 (1989).

The expected impact of the **rise in sea level** on biodiversity is examined by Walter V. Reid and Mark C. Trexler, *Drowning the National Heritage: Climate Change and U.S. Coastal Biodiversity* (Washington, D.C.: World Resources Institution, 1991).

The estimate of **energy appropriated** on the land by people was made by Peter M. Vitousek, Paul R. Ehrlich, Anne H. Ehrlich, and Pamela A. Matson, "Human Appropriation of the Products of Photosynthesis," *BioScience*, 36(6):368–373 (1986). The measure used by these authors was net primary production, the amount of energy left after subtracting the respiration of primary producers (mostly plants) from the total amount of energy (mostly solar) that is fixed biologically. The appropriation includes consumption of food, fiber, and timber; the productivity of all the land devoted exclusively to human needs, such as croplands (in addition to crops actually eaten); land burned over for clearing; and land devoted to dwellings or reduced to unproductive wastelands by overuse. The human appropriation of marine production remains relatively small. The relation of body size to population density and energy consumption among animal species is analyzed by James H. Brown and Brian A. Maurer, "Macroecology: The Division of Food and Space among Species on Continents," *Science*, 243:1145–50 (1989).

Global population trends were taken from *The Economist Book of Vital World Statistics* (New York: Times Books, 1990).

The account of the **fragility of tropical rain forests** is drawn from my "The Current State of Biological Diversity," in E. O. Wilson and F. M. Peter, eds., *Biodiversity* (Washington, D.C.: National Academy Press, 1988), pp. 3–18; from Christopher Uhl, "Restoration of Degraded Lands in the Amazon Basin," ibid., pp. 326–332; and from T. C. Whitmore, "Tropical Forest Nutrients: Where Do

We Stand? A *Tour de Horizon*," in J. Proctor, ed., *Mineral Nutrients in Tropical Forest and Savanna Ecosystems* (Boston: Blackwell Scientific Publications, 1990), pp. 1–13.

Accounts of the record 1987 **destruction of Amazonian forest** are given by Mac Margolis, "Thousands of Amazon Acres Burning," *Washington Post*, September 8, 1988; Marlise Simons, "Vast Amazon Fires, Man-Made, Linked to Global Warming," *New York Times*, August 12, 1988; and "Amazon Holocaust: Forest Destruction in Brazil, 1987–88," *Briefing Paper*, Friends of the Earth (London, 1988).

The estimates of **annual tropical deforestation rates** in 1989 were taken from the report by Norman Myers, *Deforestation Rates in Tropical Forests and Their Climatic Implications* (London: Friends of the Earth, 1989). They are based on data assembled country by country. Myers provides a summary of his study in "Tropical Deforestation: The Latest Situation," *BioScience*, 41(5):282 (1991). He defines tropical moist forests, roughly equated with tropical rain forests, as "evergreen or partly evergreen forests, in areas receiving not less than 100 mm of precipitation in any month for two out of three years, with mean annual temperature of 24-plus degrees Celsius, and essentially frost-free; in these forests some trees may be deciduous; the forests usually occur at altitudes below 1300 metres (though often in Amazonia up to 1,800 metres and generally in South-east Asia up to only 750 metres); and in mature examples of these forests, there are several more or less distinctive strata." In late 1991 the Food and Agriculture Organization of the United Nations released a preliminary report ("Second Interim Report on the State of Tropical Forests") that independently conforms to the assessment by Myers. The authors estimate that in 1981–1990 tropical forests were being removed at a rate of 170,000 square kilometers per year. The figure is 20 percent higher than Myers', but the FAO measurements included removal of thinner forests than those considered by Myers, as well as high bamboo stands. More precisely, forests were defined as collections of trees or bamboos with a minimum of 10 percent crown cover associated with wild floras and faunas and relatively undisturbed soil conditions. The extent of prehistoric forest cover is reviewed in Peter H. Raven, "The Scope of the Plant Conservation Problem World-Wide," in David Bramwell, Ole Hamann, V. H. Heywood, and Hugh Synge, eds., *Botanic Gardens and the World Conservation Strategy* (New York: Academic Press, 1987), pp. 20–29. The history of estimation of tropical deforestation rates from the 1970s to Myers' 1989 report is evaluated by J. A. Sayer and T. C. Whitmore, "Tropical Moist Forests: Destruction and Species Extinction," *Biological Conservation*, 55(2):199–213 (1991). They conclude that deforestation grew worse during the 1980s. They doubt that extinction was greatly increased as a result, but they make no reference to many of the data and models in the literature.

A comprehensive review of the large number of **z values** collected from faunas and floras around the world is provided in Mark Williamson, *Island Populations* (New York: Oxford University Press, 1981).

Species extinction from loss of rain forest: projections similar to the ones I have made globally were obtained independently by Daniel S. Simberloff for plants and birds in the American tropics, "Are We on the Verge of a Mass Extinction in Tropical Rain Forests?," in David K. Elliott, ed., *Dynamics of Extinction* (New York: Wiley, 1986), pp. 165–180. Simberloff projects that with a halving of the original rain forest, expected by the end of this century (parallel to but not the same as cutting in half the amount left at this moment), 15 percent of the plant species—about 13,600 in all—will become extinct. If forests are saved only in existing parks and reserves, 66 percent will suffer extinction. For birds of the Amazon Basin, the figures are 12 and 70 percent respectively.

The **extinction of the birds of Cebu** is cited by Jared Diamond, "Playing Dice with Megadeath," *Discover*, April 1990, pp. 55–59.

The use of **land-bridge islands** to estimate rates of species extinction was introduced by Jared Diamond, "Biogeographic Kinetics: Estimation of Relaxation Times for Avifaunas of Southwest Pacific Islands," *Proceedings of the National Academy of Sciences*, 69:3199–03 (1972), and "'Normal' Extinctions of Isolated Populations," in Matthew H. Nitecki, ed., *Extinction* (Chicago: University of Chicago Press, 1984), pp. 191–246; and by John Terborgh, "Preservation of Natural Diversity: The Problem of Extinction-Prone Species," *BioScience*, 24(12):715–722 (1974). The exponential-decay function in the decline of species is an assumption not yet proved, since extinction rates are difficult to track on single islands: Stanley H. Faeth and Edward F. Connor, "Supersaturated and Relaxing Island Faunas: A Critique of the Species-Age Relationship," *Journal of Biogeography*, 6(4):311–316 (1979).

Bird extinction in isolated patches of Brazilian subtropical forest was reported by Edwin O. Willis, "The Composition of Avian Communities in Remanescent Woodlots in Southern Brazil," *Papéis avulsos de zoologia*, 33(1):1–25 (1979). The parallel study in the Bogor Botanical Garden was described by Jared M. Diamond, K. David Bishop, and S. Van Balen, "Bird Survival in an Isolated Javan Woodland: Island or Mirror?," *Conservation Biology*, 1(2):132–142 (1987). The decline of the bird fauna of southwestern Australia's wheatland was reported by D. A. Saunders, "Changes in the Avifauna of a Region, District and Remnant as a Result of Fragmentation of Native Vegetation: The Wheatbelt of Western Australia," *Biological Conservation*, 50(1–4):99–135 (1989).

13. Unmined Riches

The discovery of a new species of **perennial maize** is reported by Hugh H. Iltis, John F. Doebley, Rafael Guzmán, and Batia Pazy, "*Zea diploperennis* (Gramineae): A New Teosinte from Mexico," *Science*, 203:186–188 (1979). The site of the wild population of perennial maize, together with surrounding land, totalling 139,000 hectares, has been set aside as the Sierra de Manantlán Biosphere Reserve by the Mexican government, specifically to protect the maize and other wild-crop relatives. It will also save many other native plant species as well as animals including ocelots and jaguars.

The status of the **Catharanthus periwinkles** of Madagascar is described in Mark Plotkin et al., *Ethnobotany in Madagascar: Overview, Action Plan, Database* (Gland: International Union for Conservation of Nature and Natural Resources and World Wide Fund for Nature, 1985). Other details, including a discussion of the general promise of medicinal alkaloids, are provided in Thomas Eisner, "Prospecting for Nature's Chemical Riches," *Issues in Science and Technology*, 6(2):31–34 (1990). The alkaloid products of the rosy periwinkle have the following clinical record: vinblastine increases the ten-year survival rate for Hodgkin's disease from 2 percent to 58 percent, and vincristine increases the ten-year survival rate from 20 to 80 percent. The drugs are also effective against some other cancers, including Wilms' tumor, primary brain tumors, and testicular, cervical, and breast cancers. See Margery L. Oldfield, *The Value of Conserving Genetic Resources* (Sunderland: Sinauer, 1989).

Information on the natural origins of **medicines used in the United States** is provided in Chris Hails, *The Importance of Biological Diversity* (Gland: World Wide Fund for Nature, 1989).

An authoritative account of **pharmaceuticals harvested from plants,** including a complete list of the 119 substances used in pure form, is provided by Norman R. Farnsworth, "Screening Plants for New Medicines," in E. O. Wilson and F. M. Peter, eds., *Biodiversity* (Washington, D.C.: National Academy Press, 1988), pp. 83–97. Additional perspectives are provided by D. D. Soejarto and N. R. Farnsworth, "Tropical Rain Forests: Potential Source of New Drugs?," *Perspectives in Biology and Medicine*, 32(2):244–256 (1989).

The properties of the **neem tree** are described in Noel D. Vietmeyer, ed., *Neem: A Tree for Solving Global Problems* (Washington, D.C.: National Academy Press, 1992).

An account of **leeches** and the anticoagulant they produce is given by Paul S. Wachtel, "Return of the Bloodsucker," *International Wildlife*, September 1987, pp. 44–46. A news report of the new anticoagulants from **vampire bats** and **pit vipers** was published in *Science*, 253:621 (1991).

The list of **pharmaceuticals derived from plants and fungi** is drawn from Hails, *The Importance of Biological Diversity*; D. D. Soejarto and N. R. Farnsworth, "Tropical Rain Forests: Potential Source of New Drugs?," *Perspectives in Biology and Medicine*, 32(2):244–256 (1989); and Margery L. Oldfield, *The Value of Conserving Genetic Resources* (Sunderland: Sinauer, 1989). An impressive number of Amerindian natural products, few of which have been investigated to date, are described by Richard E. Schultes and Robert F. Raffauf, *The Healing Forest: Medicinal and Toxic Plants of the Northwest Amazonia* (Portland: Dioscorides Press, 1990).

The examples of **food and forage plant species** in early stages of economic development are taken in part from the much-esteemed "green book," *Underexploited Tropical Plants with Promising Economic Value*, published by the National

Academy Press in 1975. This work is part of a series sponsored by the U.S. National Academy of Sciences under the direction of the Board on Science and Technology for International Development (BOSTID). Other studies in the series are *Tropical Legumes: Resources for the Future* (1979), *The Winged Bean: A High-protein Crop for the Tropics*, 2nd ed. (1981), *Amaranth: Modern Prospects for an Ancient Crop* (1983), and *Lost Crops of the Incas* (1989). Equally useful semitechnical reviews are found in Margery L. Oldfield, *The Value of Conserving Genetic Resources* (Sunderland: Sinauer, 1989), and Noel D. Vietmeyer, "Lesser-Known Plants of Potential Use in Agriculture and Forestry," *Science*, 232:1379–84 (1986). The best popular introductions, both influential in the development of this important subject, are Norman Myers, *A Wealth of Wild Species: Storehouse for Human Welfare* (Boulder: Westview Press, 1983), and the booklet compiled by Myers, *The Wild Supermarket* (Gland: World Wide Fund for Nature, 1990).

The **potential of wild plant and animal species** is detailed in the previously cited studies by Margery Oldfield, Norman Myers, and the authors in *Biodiversity*, as well as in Hails, *The Importance of Biological Diversity*. Inca agriculture is described in Hugh Popenoe, Noel D. Vietmeyer, and a panel of coauthors, *Lost Crops of the Incas* (Washington, D.C.: National Academy Press, 1989).

The history of **amaranth** as an Amerindian crop is told by Jean L. Marx, "Amaranth: A Comeback for the Food of the Americas?," *Science*, 198:40 (1977).

The stellar qualities of the **babassu palm** are detailed by Anthony B. Anderson, Peter H. May, and Michael J. Balick, *The Subsidy from Nature: Palm Forests, Peasantry, and Development on an Amazon Frontier* (New York: Columbia University Press, 1991).

The promise of **salt-tolerant plants** is explored in two publications of the National Academy Press, prepared under the direction of the Board on Science and Technology for International Development: *Underexploited Tropical Plants with Promising Economic Value* (1975) and *Saline Agriculture: Salt-tolerant Plants for Developing Countries* (1990). An evaluation of the latter is provided by Susan Turner-Lewis, *National Research Council News Report*, May 1990, pp. 2–4.

The status and economic potential of the **Podocnemis river turtles** is described by Russell A. Mittermeier, "South American River Turtles: Saving Their Future," *Oryx*, 14(3):222–230 (1978).

The descriptions of **wild animal species as potential food sources** are based on *Little-known Asian Animals with a Promising Economic Future*, ed. Noel D. Vietmeyer (Washington, D.C.: National Academy Press, 1983); Oldfield, *The Value of Conserving Genetic Resources; Neotropical Wildlife Use and Conservation*, eds. John G. Robinson and Kent H. Redford (Chicago: University of Chicago Press, 1991); and *Microlivestock*, ed. Noel D. Vietmeyer (Washington, D.C.: National Academy Press, 1991).

Chris Wille and Diane Jukofsky wrote on the **green iguana** in "Savory 'Chicken of the Trees' Could Play a Role in Saving Forests," *Canopy* (Rainforest Alliance),

Summer 1991, p. 7. Dagmar Werner, who cheerfully calls herself the Iguana Mama, has provided a technical report on breeding and marketing the species, "The Rational Use of Green Iguanas," in J. G. Robinson and K. H. Redford, eds. *Neotropical Wildlife Use and Conservation* (Chicago: University of Chicago Press, 1991), pp. 181–201.

The account of **aquaculture** is based on Myers, *A Wealth of Wild Species*.

New sources of pulp are recounted by Myers, *A Wealth of Wild Species*.

Wood grass is described by Sinyan Shen in "Biological Engineering for Sustainable Biomass Production," in Wilson and Peter, *Biodiversity*, pp. 377–389.

The history of **wild relatives and genetic diversity** of crop plants is based principally on Erich Hoyt, *Conserving the Wild Relatives of Crops* (Rome and Gland: International Board for Plant Genetic Resources, International Union for Conservation of Nature and Natural Resources, and World Wide Fund for Nature, 1988); Hails, *The Importance of Biological Diversity*; Cary Fowler and Pat Mooney, *Shattering: Food, Politics, and the Loss of Genetic Diversity* (Tucson: University of Arizona Press, 1990); and "Bad Seed," a review of the Fowler-Mooney book by Ann Misch in *World-Watch*, 4(4):39–40 (1991).

The **metaphor of a species as a loose-leaf book** was used by Thomas Eisner, "Chemical Ecology and Genetic Engineering: The Prospects for Plant Protection and the Need for Plant Habitat Conservation," *Symposium on Tropical Biology and Agriculture* (St. Louis: Monsanto Company, July 15, 1985).

The **potential economic yield of Amazon rain forests** is provided by Charles M. Peters, Alwyn H. Gentry, and Robert O. Mendelsohn, "Valuation of an Amazonian Rainforest," *Nature*, 339:655–656 (1989). The detailed ledger is from Charles M. Peters as quoted in the *New York Times*, July 4, 1989.

Key contributions to the new interdisciplinary field of **ecological economics** include Herman E. Daly, *Steady-State Economics* (San Francisco: Freeman, 1977), and most recently, Robert Constanza, ed., *Ecological Economics: The Science and Management of Sustainability* (New York: Columbia University Press, 1991). An evaluation of the field from an environmentalist's perspective is provided by David W. Orr, "The Economics of Conservation," *Conservation Biology*, 5(4):439–441 (1991). A new journal devoted to the subject, *Ecological Economics*, was started by Elsevier (New York) in 1989. A related journal, *Ecological Engineering*, was inaugurated by the same publishers in 1992.

Ecotourism is analyzed by Elizabeth Boo, *Ecotourism: The Potentials and Pitfalls* (Washington, D.C.: World Wildlife Fund, 1990). I am grateful to Gary Hartshorn and James Hirsch for information on ecotourism income in Costa Rica and to Elizabeth Boo for the most recent report from Rwanda. According to Hirsch, countryside ecotourism accounted for 7 percent, or $20 million, of the $275 million spent by visitors to Costa Rica in 1990.

The possible consequences of deforestation of the **Amazon forest** on the region's climate was examined by J. Shukla, C. Nobre, and P. Sellers, "Amazon Deforestation and Climate Change," *Science*, 247:1322–25 (1990).

The role of tropical deforestation in the **buildup of atmospheric carbon dioxide** has been analyzed by many authors; the sources used here are Richard A. Houghton and George M. Woodwell, "Global Climatic Change," *Scientific American*, 260(4):36–44 (April 1989), and R. A. Houghton, "Emission of Greenhouse Gases," in Myers, *Deforestation Rates in Tropical Forests*, pp. 53–62.

The **genesis of soils** by living organisms is described in Paul R. and Anne H. Ehrlich, *Healing the Planet: Strategies for Resolving the Environmental Crisis* (Reading: Addison-Wesley, 1991).

The evidence for the role of biodiversity in the **conservation and circulation of nutrients** in forests is reviewed by Ariel E. Lugo, "Diversity of Tropical Species: Questions That Elude Answers," *Biology International* (International Union of Biological Sciences, Paris), special issue no. 19, 37 pp. (1988).

Bryan G. Norton's assessment of the **option value of species** is given in "Commodity, Amenity, and Morality: the Limits of Quantification in Valuing Biodiversity," in Wilson and Peter eds., *Biodiversity*, pp. 200–205. General aspects of economic analysis are explained by other authors in the same volume, including Nyle C. Brady, J. William Burley, Robert J. A. Goodland, and John Spears. They are also treated by Harold J. Morowitz, "Balancing Species Preservation and Economic Considerations," *Science*, 253:752–754 (1991).

In thinking about **economic and moral foundations of conservation,** I have been informed by the writings of ethical philosophers, including David Ehrenfeld, *The Arrogance of Humanism* (New York: Oxford University Press, 1978); Bryan Norton, "Commodity," and *Why Preserve Natural Variety?* (Princeton: Princeton University Press, 1987); Peter Singer, *The Expanding Circle: Ethics and Sociobiology* (New York: Farrar, Straus, and Giroux, 1981); Holmes Rolston III, *Philosophy Gone Wild: Essays in Environmental Ethics* (Buffalo: Prometheus Books, 1986), and *Environmental Ethics: Duties to and Values in the Natural World* (Philadelphia: Temple University Press, 1988); Alan Randall, "The Value of Biodiversity," *Ambio*, 20(2):64–68 (1991); and the authors of *The Preservation of Species: The Value of Biological Diversity*, ed. Bryan G. Norton (Princeton: Princeton University Press, 1986).

14. Resolution

The discussion of the **conservation ethic** is based in part on my *Biophilia* (Cambridge: Harvard University Press, 1984). The general definition of *ethic* comes from Aldo Leopold, *A Sand County Almanac and Sketches Here and There* (New York: Oxford University Press, 1949).

The **definition of biodiversity studies** given here and a discussion of its ramifications were presented in Paul R. Ehrlich and Edward O. Wilson, "Biodiversity Studies: Science and Policy," *Science*, 253:758–762 (1991).

The **three-level approach to surveying global biodiversity** was developed in collaboration with Peter H. Raven.

The **RAP search** for hot spots is described by Sarah Pollock, "Biological SWAT Team Ranks for Diversity, Endemism," *Pacific Discovery*, 44(3):6–7 (1991).

An account of INBio, **Costa Rica's National Institute of Biodiversity,** is provided by Laura Tangley, "Cataloging Costa Rica's Diversity," *BioScience*, 40(9):633–636 (1990); and by Daniel H. Janzen, one of INBio's architects, in "How To Save Tropical Biodiversity," *American Entomologist*, 37(3):159–171 (1991). An equivalent institute for the United States is included in the National Biological Diversity Conservation and Environmental Research Act, which as of February 1992 remains to be passed by Congress.

The use of **Geographic Information Systems** to map ecosystems is described by J. Michael Scott et al., "Species Richness: A Geographic Approach to Protecting Future Biological Diversity," *BioScience*, 37(11):782–788 (1987). On a much broader scale, essentially the same method has been applied by Eric Dinerstein and Eric D. Wikramanayake to assess reserves and parks in Asia and the western Pacific, in "Beyond 'Hotspots': How to Prioritize Investments in Biodiversity in the Indo-Pacific Region," *Conservation Biology*, 7(1): 53–65 (1993). Techniques for mapping endangered species are given by many authors in Larry E. Morse and Mary Sue Henifin, eds., *Rare Plant Conservation: Geographical Data Organization* (New York: New York Botanical Garden, 1981).

The employment of **landscape design** to enhance biodiversity has been widely discussed. Summaries of key topics are provided in separate chapters by Bryn H. Green, Larry D. Harris (with John F. Eisenberg), and David Western, in Western and Mary C. Pearl, eds., *Conservation for the Twenty-First Century* (New York: Oxford University Press, 1989).

The concept of **bioregions,** dating back to the 1800s and developed in modern form by Raymond F. Dasmann, Peter Berg, Charles H. W. Foster, and others, is reviewed in C. H. W. Foster, *Experiments in Bioregionalism: The New England River Basins Story* (Hanover: University Press of New England, 1984), and "Bioregionalism," *Renewable Resources Journal*, 4(3):12–14 (1986).

The **shortage of systematists** is cited in my "The Biological Diversity Crisis: A Challenge to Science," *Issues in Science and Technology*, 2(1):20–29 (1985), and "Time to Revive Systematics," *Science*, 230:1227 (1985).

The progress of **GenBank** in recording DNA and RNA sequences is described by Christian Burks et al., in Russell F. Doolittle, ed., *Molecular Evolution: Computer Analysis of Protein and Nucleic Acid Sequences* (New York: Academic Press, 1990), pp. 3–22.

Baba Dioum on knowledge and conservation is quoted by John Hopkins, "Preserving Native Biodiversity," Sierra Club special publication (San Francisco, 1991).

The concept of **chemical prospecting** was developed by Thomas Eisner during the late 1980s and presented in "Prospecting for Nature's Chemical Riches," *Issues in Science and Technology*, 6(2):31–34 (1990); and "Chemical Prospecting: A Proposal for Action," in F. Herbert Bormann and Stephen R. Kellert, eds., *Ecology, Economics, Ethics: The Broken Circle* (New Haven: Yale University Press, 1991), pp. 196–202.

The 1991 agreement between **Merck and Costa Rica's National Institute of Biodiversity** was reported by William Booth, "U.S. Drug Firm Signs Up To Farm Tropical Forests," *Washington Post*, September 21, 1991. The cyclical nature of investment in natural products is described by Deborah Hay, "Pharmaceutical Industry's Renewed Interest in Plants Could Sow Seeds of Rainforest Protection," *The Canopy* (Rainforest Alliance), Spring 1991, pp. 1, 7. The use of wild species as sources of medicine is reviewed by Norman R. Farnsworth, "Screening Plants for New Medicines," in E. O. Wilson and F. M. Peter, eds., *Biodiversity* (Washington, D.C.: National Academy Press, 1988), pp. 83–97.

The data on pharmaceuticals discovered from **folkloric medicine** are reported in Farnsworth, "Screening Plants." Excellent brief accounts of traditional knowledge and the endangered status of indigenous people who possess it are given by Mark J. Plotkin, "The Outlook for New Agricultural and Industrial Products from the Tropics," in Wilson and Peter, *Biodiversity*, pp. 106–116, and by Eugene Linden, "Lost Tribes, Lost Knowledge," *Time*, September 23, 1991, pp. 46–56. The citation of Chinese traditional medicine was provided by Peter H. Raven (personal communication) and, for artemisinin, by Daniel L. Klayman, "*Qinghaosu* (Artemisinin): An Antimalarial Drug from China," *Science*, 228:1049–55 (1985), and by Xuan-De Luo and Chia-Chiang Shen, "The Chemistry, Pharmacology, and Clinical Applications of Qinghaosu (Artemisinin) and Its Derivatives," *Medicinal Research Reviews*, 7(1):29–52 (1987).

The operations of the **Tropical Agricultural Research and Training Center** in Costa Rica is described by Laura Tangley in "Fighting Central America's Other War," *BioScience*, 37(11):772–777 (1987).

The failure of governments, as well as of the United Nations Statistical Office and World Bank, to include deforestation and use of other natural resources in **national depletion accounts** is reported by Malcolm Gillis, "Economics, Ecology, and Ethics: Mending the Broken Circle for Tropical Forests," in Bormann and Kellert, *Ecology, Economics, Ethics*, pp. 155–179.

The **extractive reserves of the Amazon region** are described by Walter V. Reid, James N. Barnes, and Brent Blackwelder, *Bankrolling Successes: A Portfolio of Sustainable Development Projects* (Washington, D.C.: Environmental Policy Institute and National Wildlife Federation, 1989), and Philip M. Fearnside, "Extractive Reserves in Brazilian Amazonia," *BioScience*, 39(6):387–393 (1989). A critique of extractive reserves is presented by John O. Browder, "Extractive Reserves Will Not Save Tropics," *BioScience*, 40(9):626 (1990).

The Brazilian **rubber-tappers movement** of the 1980s was bitterly opposed by some of the wealthy landowners of the western Amazon. On December 22, 1988, its leader Chico Mendes was killed by gunmen. The murder and the circumstances surrounding the struggle for control of the Amazonian environment are chronicled by Andrew Revkin, *The Burning Season* (Boston: Houghton Mifflin, 1990).

Strip logging as a sustainable industry is described by Carl F. Jordan, in "Amazon Rain Forests," *American Scientist*, 70:394–401 (1982), and by Gary S. Hartshorn, "Natural Forest Management by the Yanesha Forestry Cooperative in Peruvian Amazonia," in A. B. Anderson, ed., *Alternatives to Deforestation: Steps Toward Sustainable Use of the Amazon Rain Forest* (New York: Columbia University Press, 1990), pp. 128–137.

The examples of successful **local sustainable development** in Latin America are taken from Reid, Barnes, and Blackwelder, *Bankrolling Successes*. An account of local planning for sustainable tropical forest extraction is detailed in Leonard Berry et al., *Technologies To Sustain Tropical Forest Resources* (Washington, D.C.: Office of Technology Assessment, U.S. Congress, 1984).

The impact of **trade and subsidy policies** of rich nations is described by Roger D. Stone and Eve Hamilton, *Global Economics and the Environment: Toward Sustainable Rural Development in the Third World* (New York: Council on Foreign Relations, 1991).

The present status of research on **DNA in fossils and archaeological remains** is reviewed by Jeremy Cherfas, "Ancient DNA: Still Busy after Death," *Science*, 253:1354–56 (1991).

The status of **microbial preservation** is described in "American Type Culture Collection Seeks To Expand Research Effort," *Scientist*, 4(16):1–7 (1990).

Seed banks are reviewed by Erich Hoyt, *Conserving the Wild Relatives of Crops* (Rome and Gland: International Board for Plant Genetic Resources, etc., 1988); Jeffrey A. McNeely et al., *Conserving the World's Biological Diversity* (Gland and Washington, D.C.: International Union for Conservation of Nature and Natural Resources, World Resources Institute, etc., 1990); Joel I. Cohen et al., "Ex Situ Conservation of Plant Genetic Resources: Global Development and Environmental Concerns," *Science*, 253:866–872 (1991).

The **National Collection of Endangered Plants** is the subject of a report in *Plant Conservation*, 6(1):6–7 (1991).

The performance of **zoos** and other captive-animal facilities in maintaining diversity is described by William Conway, "Can Technology Aid Species Preservation?"

in Wilson and Peter, *Biodiversity*, pp. 263–268; and by Colin Tudge, *Last Animals at the Zoo* (London: Hutchinson Radius, 1991).

The number of species of **mammals facing extinction** and requiring rescue is from Michael E. Soulé et al., "The Millennium Ark: How Long a Voyage, How Many Staterooms, How Many Passengers?," *Zoo Biology*, 5:101–114 (1986). William Conway is quoted on the limits of zoos by Edward C. Wolf, *On the Brink of Extinction: Conserving the Diversity of Life* (Washington, D.C.: Worldwatch Institute, 1987).

A pioneering set of **recommendations to save tropical ecosystems** was advanced in 1980 by Peter H. Raven et al., *Research Priorities in Tropical Biology* (Washington, D.C.: National Academy Press, 1980). A review of ongoing efforts is provided by the authors in Wilson and Peter, *Biodiversity;* by McNeely et al., *Conserving;* by Janet N. Abramovitz, *Investing in Biological Diversity: U.S. Research and Conservation Efforts in Developing Countries* (Washington, D.C.: World Resources Institute, 1991); and by Kathleen Courrier, ed., *Global Biodiversity Strategy* (Washington, D.C.: World Resources Institute; Gland: World Conservation Union; New York: United Nations Environment Program, 1992).

The provisions of the **Endangered Species Act** of the United States, as well as those of international regulatory protocols, are reviewed by Robert Boardman, *International Organization and the Conservation of Nature* (Bloomington: Indiana University Press, 1981); by Michael J. Bean, *The Evolution of National Wildlife Law* (New York: Praeger, 1983); and by Simon Lyster, *International Wildlife Law* (Cambridge, Eng.: Grotius, 1985).

The **status of national parks** and other reserves is documented by Walter V. Reid and Kenton R. Miller, *Keeping Options Alive: The Scientific Basis for Conserving Biodiversity* (Washington, D.C.: World Resources Institute, 1989); and by Michael E. Soulé, "Conservation: Tactics for a Constant Crisis," *Science*, 253:744–750 (1991). The percentage of the world's land surface under legal protection is from *1990 United Nations List of National Parks and Protected Areas*.

Debt-for-nature exchange schemes have been well explained by José Márcio Ayres, "Debt-for-Equity Swaps and the Conservation of Tropical Rain Forests," *Trends in Ecology and Evolution*, 4(11):331–332 (1989); and by Roger D. Stone and Eve Hamilton, *Global Economics and the Environment* (New York: Council on Foreign Relations, 1991). I have also used a master's thesis from University College, London, by Victoria C. Drake, "Debt-for-Nature Swaps: An Economic Appraisal." The Mexican trade was reported by Mark A. Uhlig, "Mexican Debt Deal May Save Jungle," *New York Times*, February 26, 1991. The idea of debt-for-nature was first proposed by Thomas Lovejoy of the Smithsonian Institution.

The **SLOSS controversy** is examined, with different conclusions, by James F. Quinn and Alan Hastings, "Extinction in Subdivided Habitats," *Conservation Biology*, 1(3):198–208 (1987); and by Michael E. Gilpin, "A Comment on Quinn and Hastings: Extinction in Subdivided Habitats," *Conservation Biology*, 2(3):290–292

(1988). The advantages and disadvantages of corridors between small reserves are reviewed by William Stolzenburg, "The Fragment Connection," *Nature Conservancy*, July-August 1991, pp. 18–25.

The progress of **ecosystem restoration** in the United States can be followed in issues of *Restoration and Management Notes*, published by the University of Wisconsin Press since 1982. A recent account of prairie renewal and the general hopes and misgivings of restorationists is provided by William K. Stevens, "Green-Thumbed Ecologists Resurrect Vanished Habitats," *New York Times*, March 19, 1991. The creation of new dry tropical forest in Costa Rica's Guanacaste National Park is described by Reid et al., *Bankrolling Successes*.

The history of **animal-species introduction** into new environments is reviewed by Paul R. Ehrlich, "Which Animal Will Invade?," in Harold A. Mooney and James A. Drake, eds., *Ecology of Biological Invasions of North America and Hawaii* (New York: Springer, 1986), pp. 79–95.

Ariel E. Lugo has spoken on behalf of **exotic species** in expanding local biodiversity. While conceding the high risk of introductions and the need to remove elements that endanger native fauna and flora, he notes that most such species are naturalized without creating ecological problems. "Exotics appear to do best in human-disturbed environments. Exotics can provide food and fiber without causing ecological havoc. For example, when managed properly, certain exotic trees grow well in highly degraded lands where they contribute to soil rehabilitation and reestablishment of native species." "Removal of Exotic Organisms," *Conservation Biology*, 4(4):345 (1990).

15. The Environmental Ethic

The **Sibyl's advice to Aeneas** is from the translation by Robert Fitzgerald, *The Aeneid: Virgil* (New York: Random House, 1983), book 6, p. 164.

The greater **attendance of people at zoos and aquariums** than at professional sporting events (football, baseball, basketball, ice hockey) is cited in *Directory of the American Association of Zoological Parks and Aquaria*, ed. Linda Boyd (Wheeling, West Virginia: Ogle Bay Park, 1990–91).

The innate **affiliation of human beings with the natural world** is elaborated in my *Biophilia* (Cambridge: Harvard University Press, 1984). The imagery of the serpent was drawn from Balaji Mundkur's masterful *The Cult of the Serpent: An Interdisciplinary Survey of Its Manifestations and Origins* (Albany: State University of New York Press, 1983). The concept of the idealized living place as a biological adaptation was developed by Gordon H. Orians, "Habitat Selection: General Theory and Applications to Human Behavior," in Joan S. Lockard, ed., *The Evolution of Human Social Behavior* (New York: Elsevier North Holland, 1980), pp. 46–66; and "An Ecological and Evolutionary Approach to Landscape Aesthetics," in Edmund C. Penning-Rowsell and David Lowenthal, eds., *Landscape Meanings and Values* (London: Allen and Unwin, 1986), pp. 3–22.

Excellent histories of **wilderness in the human imagination,** especially in Europe and America, have been presented by Roderick Nash, *Wilderness and the American Mind* (New Haven: Yale University Press, 1967); and by Max Oelschlaeger, *The Idea of Wilderness: From Prehistory to the Age of Ecology* (New Haven: Yale University Press, 1991).

Glossary

Included in this list of terms is biographical information for those scientists and other principal contributors to biodiversity studies mentioned in the text.

abyssal benthos   The community of organisms living on or close to the floor of the deep sea.

adaptation   A particular part of the anatomy (such as color), a physiological process (such as respiration rate), or behavior pattern (such as a mating dance) that improves an organism's chances to survive and reproduce. Also the evolution that creates such a trait.

adaptive radiation   The evolution of a single species into many species that occupy diverse ways of life within the same geographical range. Example: the origin of kangaroos, koalas, and other present-day Australian marsupials from a single distant ancestor.

allele   A particular form of a gene, where multiple such forms occur. Sickle-cell anemia is caused by one such variant of a gene; another variant of the same gene contributes to normal hemoglobin.

allometry   The condition in which one part of the body grows faster relative to another part, so that the larger the organism, the greater the disproportion; the large males of many kinds of beetles and deer, for example, develop horns that are enormous in comparison with the rest of the body.

allopatric   Occupying different geographical ranges.

allopatric speciation   The same as geographic speciation: the splitting of a population into two or more subpopulations by a geographic barrier, followed by the evolutionary divergence of the population until it attains the status of full species.

alternation of generations   The alternation of haploid organisms (possessing one chromosome of each kind per cell) in one generation with diploid organisms (two chromosomes of each kind per cell) in the next generation, then back to haploid organisms, and so on. In most species the haploid generation consists of only the eggs and sperm, which fuse to create the diploid generation, which in time produces more eggs and sperm, the next haploid generation.

Alvarez, Luis W. (1911–1988)   Particle physicist at the University of California, Berkeley; led the team that discovered a high level of iridium at the Cretaceous-Tertiary boundary and interpreted it as the result of massive meteorite strikes.

amphibian   A member of the vertebrate class Amphibia, such as a frog or salamander.

analogy   In biology, a resemblance in appearance and function between structures that occur in two kinds of organisms but not because of common ancestry. The wings of birds and insects are analogues, they are analogous to each other, and the similarity constitutes an analogy; they were not evolved from the same organ in a common ancestor. Cf. *homology.*

angiosperm   A flowering plant, member of the plant phylum that dominates land vegetation, characterized by seeds from fruits.

annelid   A worm of the phylum Annelida, such as an earthworm, leech, or nereid.

area-species curve   The relationship between the area of an island or some other discrete geographic region and the number of species living there. Approximated by the equation $S = CA^z$, where A is the area, S is the number of species, and C and z are constants that depend on the place and group of organisms (such as birds or trees) being considered. Also called the species-area curve.

arthropod   A member of the phylum Arthropoda, such as an insect, spider, or crustacean, bearing an articulated, external skeleton.

asexual species   Populations of organisms that are different enough to be conveniently distinguished as species, even though they do not reproduce sexually and the criterion of reproductive isolation cannot be applied to them.

assembly rules   The combinations of species that can live together in a community of animals and plants and the sequences in which they can invade and persist in the community.

autochthon   A species that originated in a certain place, such as New Zealand or Lake Victoria, and is found only there. Cf. *endemic.*

bacteria   Microscopic single-celled organisms that are prokaryotic, or lacking nuclear membranes around the genes.

base pair   A pair of organic bases constituting a letter of the genetic code; usually adenine (A) paired with thymine (T), or cytosine (C) paired with guanine (G). Each base is sited on one strand of the DNA double helix and opposes the other base at the same position on the second strand. The code is then read off as a sequence of four possible letters on the double helix, AT, TA, CG, and GC. Versions of the same gene differ by the sequence of these four letters.

biodiversity The variety of organisms considered at all levels, from genetic variants belonging to the same species through arrays of species to arrays of genera, families, and still higher taxonomic levels; includes the variety of ecosystems, which comprise both the communities of organisms within particular habitats and the physical conditions under which they live.

biodiversity studies The systematic examination of the full array of different kinds of organisms, together with a consideration of the technology by which the diversity can be maintained and used for the benefit of humanity.

bioeconomic analysis The assessment of the potential economic value of all the organisms in an ecosystem, from their natural products to their use in ecotourism.

biogeography The scientific study of the geographic distribution of organisms.

biological diversity See *biodiversity*.

biomass The total weight (usually, dry weight) of a designated group of organisms in a particular area, as of all the birds living in a woodlot or all the algae in a pond or all the organisms in the world.

biome A major category of habitat in a particular region of the world, such as the tundra of northern Canada or the rain forest of the Amazon basin.

bioregion A continuous natural area, such as a river system or mountain range, large enough to extend beyond political boundaries.

biota The combined flora, fauna, and microorganisms of a given region. Microorganisms are often referred to as fauna or flora depending on the group to which they belong, such as the bacterial flora.

biotic Biological, especially referring to the characteristics of faunas, floras, and ecosystems.

Bush, Guy L. (1929–) Entomologist and evolutionary biologist at Michigan State University; the foremost researcher on host races and their role in the formation of species.

Cambrian Referring to the earliest period of the Paleozoic era, extending 550–500 million years ago, when larger marine animals vastly increased in both numbers and diversity—the Cambrian explosion of animal evolution.

centimeter One-hundredth of a meter; equal to 0.39 inch (2.5 centimeters is approximately 1 inch).

centinelan extinctions An expression (proposed in this book) to designate extinctions of species unknown before their demise and hence unrecorded.

character A trait that varies so as to be useful in classification, such as a flower part that varies among plants or a dental formula that varies among mammals. The differences from one species to the next are called character states.

character displacement The process by which two species evolve away from one another, acquiring greater differences, as a result of competition or the risk of lowered survival and fertility caused by hybridization.

chemical prospecting The screening of wild species of plants, animals, and microorganisms for natural substances of practical use, especially in medicine.

chromosome A structure, visible under the light microscope and usually rod-shaped, that carries genes. Chromosomes are made up of DNA, which compose the genes, and a supporting matrix of proteins.

chronospecies A population that evolves so much that it is regarded as a different species, even though the population has not divided into multiple coexisting species; its division into two species across time is based on subjective judgment about the degree of change.

clade A group of species all descended from a common ancestor. Cats of the genus *Felis*, living and extinct, are a clade descended from a single ancestor that lived in the geological past. The clade includes the ancestor.

class In classification, a group of species of common ancestry ranked below the phylum and above the order; hence one or more orders.

coevolution The evolution of two or more species due to mutual influence; for example, many species of flowering plants and their insect pollinators have coevolved in a way that makes the relationship more effective.

Cohen, Joel E. (1944–) Professor of population biology at Rockefeller University; a major contributor to the interpretation of food webs in ecosystems.

commensalism A form of symbiosis (intimate coexistence) in which one species profits from the association without harming or benefiting the other.

community All the organisms—plants, animals, and microorganisms—that live in a particular habitat and affect one another as part of the food web or through their various influences on the physical environment.

competitive exclusion The extinction of one species by another in a habitat through competition.

conservation biology The relatively new discipline that treats the content of biodiversity, the natural processes that produce it, and the techniques used to sustain it in the face of human-caused environmental disturbance.

continental drift The gradual breaking up of the continents that has occurred steadily over the past 200 million years.

convergence In evolutionary biology, the same as convergent evolution, the increasing similarity during evolution of two or more unrelated species. Example: the placental wolf of the northern hemisphere and its remarkable look-alike, the marsupial "wolf" of Australia.

Conway Morris, Simon (1951–) A paleontologist at the University of Cambridge; a leading scholar on the Cambrian explosion of invertebrate animals and early evolution of arthropods.

cryopreservation The storage of organisms and tissue samples at extremely cold temperatures, usually in liquid nitrogen.

cyanobacteria Formerly called blue-green algae, these organisms are not true algae but single-celled prokaryotes resembling bacteria. They were dominant elements during the early history of life and are still ecologically prominent.

Darwin, Charles Robert (1809–1882) Originator with Alfred Russel Wallace of the theory of evolution by natural selection, author of *On the Origin of Species*, and hence founder of the mode of evolutionary thinking that pervades biology today.

Darwinism Evolution by natural selection, originally proposed by Charles Darwin. The modern interpretation of the process is called neo-Darwinism; it incorporates all we know about evolution from genetics, ecology, and other disciplines.

debt-for-nature swap The purchasing or forgiving of portions of the debt of poorer countries in exchange for local conservation projects, especially the purchase of land.

deme A population of organisms in which breeding is completely random—an important idealized concept used as a standard to calculate the degrees of inbreeding and genetic drift.

demography The study of birth rates, death rates, age distributions, sex ratios, and size of populations—a fundamental discipline within the larger field of ecology. Also the properties themselves, as in the demography (demographic traits) of a particular population.

density dependence The increasing severity by which factors in the environment slow down growth of a population as the organisms become more numerous and hence densely concentrated. Density-dependent factors include competition, food shortage, disease, predation, and emigration.

DeVries, Philip J. (1952–) Tropical field biologist and author of the widely admired field guide *The Butterflies of Costa Rica* (1987).

Diamond, Jared M. (1937–) Professor at the University of California, Los Angeles, School of Medicine; explorer of New Guinea bird fauna, inventor of the concept of assembly rules in community organization, and an influential scholar of the extinction process.

diploid Having two copies of the same complement of chromosomes in each cell. The diploid condition usually arises from fertilization, during which one set of chromosomes from the male is joined with a second set from the female. Cf. *haploid*.

disharmony In biodiversity studies, the gross overrepresentation of some groups of organisms and underrepresentation or absence of others on an island or continent owing to accidents of dispersal. Example: there are no native woodpeckers or ants on Hawaii but a great variety of honeycreepers and wasps.

diversity See *biodiversity*.

DNA Deoxyribonucleic acid. The fundamental hereditary material of all living organisms; the polymer composing the genes.

dominance In genetics, the expression of one form of a gene over another form of the same gene when both occur on different chromosomes in the same organism; the gene for normal blood clotting, for example, is dominant over the one for hemophilia (failure to clot) in human beings. In ecology, the abundance and ecological influence of one species or group of species over others: pines are dominant plants and beetles are dominant animals. In animal behavior, the control of one individual over another in social groupings.

echinoderm A member of the phylum Echinodermata, such as a starfish or a sea urchin.

ecological economics A new interdisciplinary field devoted to protection of the environment and to attainment of sustained economic production.

ecology The scientific study of the interaction of organisms with their environment, including the physical environment and the other organisms living in it.

ecosystem The organisms living in a particular environment, such as a lake or a forest (or, in increasing scale, an ocean or the whole planet), and the physical part of the environment that impinges on them. The organisms alone are called the community.

ecosystem services The role played by organisms in creating a healthful environment for human beings, from production of oxygen to soil genesis and water detoxification.

ecotourism Tourism focused on attractive and interesting features of the environment, including the fauna and flora.

Ehrlich, Paul R. (1932–) Professor at Stanford University; a leading researcher on population dynamics and the extinction process; in addition his many books and articles with Anne H. Ehrlich have brought environmental problems to the attention of a worldwide audience.

Eisner, Thomas (1929–) Professor at Cornell University; a foremost entomologist and founder of chemical ecology; developed the concept of chemical prospecting.

Eldredge, Niles (1943–) Curator of fossil invertebrates at the American Museum of Natural History; a leading authority on trilobites and originator, with Stephen Jay Gould, of the thesis of punctuated equilibrium.

electrophoresis A method by which substances, especially proteins, are separated from one another on the basis of their electric charge and molecular weights. Used in the study of diversity among both species and organisms belonging to the same species.

endangered Near extinction. Referring to a species or ecosystem so reduced or fragile that it is doomed or at least fatally vulnerable.

endemic A species or race native to a particular place and found only there. If it originated in the same place by evolution, it is also called an autochthon.

environment The surroundings of an organism or a species: the ecosystem in which it lives, including both the physical environment and the other organisms with which it comes in contact.

eon The major division of geological time. The most recent such division, the Phanerozoic eon, spans the past 550 million years.

epiphyll A plant that grows on the leaves of other kinds of plants, hence a specialized form of epiphyte.

epiphyte A plant specialized to grow on other kinds of plants in a neutral or beneficial manner, not as a parasite. Examples: most species of orchids and bromeliads.

epoch The division of geological time just below the period in rank. We live in the Recent epoch, which began 10,000 years ago with the end of the Pleistocene epoch, or Ice Age.

equilibrium See *species equilibrium*.

era A major subdivision of geological time, just below the eon in rank. The Phanerozoic eon, for example, is divided into three eras: the Paleozoic (oldest), Mesozoic, and Cenozoic (most recent).

Erwin, Terry L. (1940–) Curator in entomology at the United States National Museum; a leading expert on beetles, best known for his estimation of the diversity of insects and other arthropods in rain forests.

ethnobotany The study of plant biology as understood by other cultures, especially those of preliterate peoples, and the practical uses of plants made by those cultures.

eukaryote An organism whose DNA is enclosed in nuclear membranes. The vast majority of kinds of organisms are eukaryotic; only bacteria and a few other microscopic forms lack such a nuclear envelope.

evolution In biology, any change in the genetic constitution of a population of organisms. Evolution can vary in degree from small shifts in the frequencies of minor genes to the origins of complexes of new species. Changes of lesser magnitude are called microevolution, and changes at or near the upper extreme are called macroevolution.

evolutionary agent (or force) Any factor in the external environment or in the bodies of the organisms themselves that induces shifts in the frequencies of genes within populations, hence evolution.

evolutionary biology An umbrella term for a broad array of disciplines that have in common their focus on the evolutionary process and hence the creation of biodiversity. Evolutionary biology includes the study of molecular evolution, ecology, population biology, systematics, biogeography, and comparative aspects of anatomy, physiology, and animal behavior.

extinction The termination of any lineage of organisms, from subspecies to species and higher taxonomic categories from genera to phyla. Extinction can be local, in which one or more populations of a species or other unit vanish but others survive elsewhere, or total (global), in which all the populations vanish. When biologists speak of the extinction of a particular species without further qualification, they mean total extinction.

extractive reserve A wild habitat from which timber, latex, and other natural products are taken on a sustained yield basis with minimal environmental damage and, ideally, without the extinction of native species.

family In the hierarchical classification of organisms, a group of species of common descent higher than the genus and lower than the order; hence a group of genera. Examples: Felidae (cats) and Fagaceae (beeches and oaks).

fauna All the animals found in a particular place.

flora All the plants found in a particular place.

food chain Part of the food web of a particular community of organisms, consisting of predators and their prey, predators that feed on the predators, and so on all the way from the photosynthesizing plants to top predators (such as eagles and cats) and decomposers that consume the remains of dead organisms.

food web The complete array of food links in a particular habitat, represented in diagrams by the direction in which energy and nutrients flow from consumed to consumer.

fossil Any remains left by an organism, such as a track or mineralized bone, that has been preserved through geological time, by which is usually meant a timespan of 10,000 years or longer.

gene   The basic unit of heredity.

gene frequency   For the population as a whole, the percentage of genes at a particular locus that are of one form (allele) as opposed to another, such as the allele for sickle-cell hemoglobin that can be distinguished from the allele for normal hemoglobin.

gene pool   All the genes in all the organisms belonging to a population.

genera   The plural of genus; groups of closely related, similar species.

genetic drift   Evolution in the genetic constitution of a population by chance processes alone.

genome   All the genes of a particular organism or species.

genotype   The genetic constitution of an organism, either prescribing a single trait (such as eye color) or a set of traits (eye color, blood type, etc.).

Gentry, Alwyn H. (1945–)   Tropical botanist of the Missouri Botanical Garden; a principal modern explorer of South American plants.

genus   A group of similar species of common descent. Examples: *Canis*, comprising the wolf, domestic dog, and similar species; and *Quercus*, the oaks.

geographic speciation   Also called allopatric speciation. The divergence to species level by populations that originally belonged to the same species but were isolated by a physical barrier such as a sea strait, river valley, or mountain range.

Goksøyr, Jostein (1922–).   Professor of microbiology at the University of Bergen, Norway; a pioneer in estimation techniques of bacterial diversity.

Gould, Stephen Jay (1941–)   Professor of geology and curator of fossil invertebrates at Harvard University; the most influential modern popularizer and commentator on evolutionary biology; originator with Niles Eldredge of the thesis of punctuated equilibrium.

Grant, Peter R. (1936–)   Professor of zoology at Princeton University; vertebrate ecologist and a leader in the study of the ecology and microevolution of Darwin's finches.

Great American Interchange   The migration of North American mammals south and South American mammals north when the Panamanian land bridge came into existence 2.5 million years ago. The process continues to the present day. Attention has been focused on mammals because of their excellent fossil record, but plants and other animals participated as well.

guild   A group of species found in the same place that share the same food resource. Examples: the insects of a Rhode Island field that feed on goldenrod pollen; the hawks of a Bolivian rain forest that prey on songbirds.

habitat An environment of a particular kind, such as lake shores or tall-grass prairie; also a particular environment in one place, such as the mountain forest of Tahiti.

habitat island A patch of habitat separated from other patches of the same habitat, such as a glade separated by forest or a lake separated by dry land. Habitat islands are subject to the same ecological and evolutionary processes as "real" islands.

haploid Possessing a chromosome set composed of only one chromosome of each kind, usually encountered in eggs and sperm and characterizing the haploid generation in the alternation of generations.

hectare Metric unit of area; equal to 2.47 acres.

heterozygous Possessing two gene forms (alleles) at the same chromosome position but on different chromosomes. A person carrying an allele for sickle-cell hemoglobin on one chromosome and an allele for normal hemoglobin on the other chromosome is said to be heterozygous for those traits. Cf. *homozygous*.

homology In biology, a similarity in structure, physiology, or behavior in two species due to inheritance from a common ancestor, whether or not the function is the same. Example: human arms and bat wings. Also two chromosomes of the same type found in the same individual are spoken of as homologous. Cf. *analogy*.

homozygous Possessing the same gene form (allele) on both chromosomes. A person carrying alleles for sickle-cell hemoglobin on both chromosomes is said to be homozygous for that condition. Cf. *heterozygous*.

host race A genetically distinct population of organisms that feeds on one kind of plant and lives in the midst of other populations of the same species feeding on other kinds of plants; thought to be an intermediate stage in the formation of full species.

hot spot A region of the world, such as the island of Madagascar, that is both rich in endemic species and environmentally threatened.

Hubbell, Stephen P. (1942–) Professor of biology at Princeton University; a leading tropical ecologist and originator of the long-term study of tree diversity on Barro Colorado Island, Panama.

hybrid The offspring of parents that are genetically dissimilar, especially of parents that belong to different species.

intrinsic isolating mechanism Any hereditary difference between species that prevents them from interbreeding freely under natural conditions. Examples: different breeding seasons, courtship behavior, or local habitats.

invertebrate Any animal lacking a backbone of bony segments that enclose the central nerve cord. Most animals are invertebrates, from sea anemones to earthworms, spiders, and butterflies.

island biogeographic theory The concepts and mathematical models that account for the number of species of organisms found on islands and fragments of habitats. A central idea in the theory is the equilibrium in the species numbers attained when new species arrive and old residents go extinct at the same rate.

Janzen, Daniel H. (1939–) Professor at the University of Pennsylvania; a leading tropical biologist widely known for his program to regenerate the threatened deciduous forest of Central America.

keystone species A species, such as the sea otter, that affects the survival and abundance of many other species in the community in which it lives. Its removal or addition results in a relatively significant shift in the composition of the community and sometimes even in the physical structure of the environment.

kilometer 1,000 meters; equal to .62 mile.

kingdom The highest category used in classification. Five kingdoms are commonly recognized: Plantae (plants), Animalia (animals), Fungi (mushrooms and other fungi), Protista (or Protoctista, algae and single-celled "animals"), and Monera (bacteria and close relatives).

Knoll, Andrew H. (1951–) Professor of paleobotany at Harvard University; a principal scholar of the history of life, from the earliest Precambrian microorganisms to modern flowering plants.

latitudinal diversity gradient The trend, widespread but not universal among plants and animals, toward greater diversity when passing from polar regions toward the equator.

lichen A compound organism composed of a fungus that harbors either cyanobacteria or single-celled algae. The symbiosis of the two kinds of organisms is mutually beneficial.

life cycle The entire lifespan of an organism from the moment it is conceived (usually at fertilization) to the time it reproduces.

Lovejoy, Thomas E. (1941–) Assistant secretary for external affairs, Smithsonian Institution; expert on South American birds, originator of the giant Forest Fragment;Project in the Brazilian Amazon.

MacArthur, Robert H. (1930–1972) Professor at the University of Pennsylvania and Princeton University; a brilliant theoretician of ecology and originator of the theory of island biogeography.

macroevolution Large-scale evolution, entailing major alterations in anatomy or other biological traits, sometimes accompanied by adaptive radiation. Cf. *microevolution*.

mammal An animal of the class Mammalia, characterized by production of milk in the female mammary gland and possession of a body covering of hair.

marsupial An animal, such as an opossum or a kangaroo, characterized (in most species) by a pouch, the marsupium, containing milk glands and serving as a receptacle for the young.

Martin, Paul S. (1928–) Paleontologist and professor at the University of Arizona; chief architect of the hypothesis of mass extinctions of the megafauna by prehistoric humans.

May, Robert M. (1936–) Professor of ecology at Oxford University; a foremost theoretical population biologist and an important scholar of the natural processes underlying biodiversity.

Mayr, Ernst (1904–) Professor emeritus at Harvard University; doyen of evolutionary biology; architect of neo-Darwinism and the biological-species concept.

megafauna The largest animals, weighing over 10 kilograms, such as deer, big cats, elephants, and ostriches.

meiosis Cell division leading to reduction of the number of chromosomes from two sets to one, which in most kinds of higher organisms leads directly to the production of sex cells. Cf. *mitosis*.

Mesozoic era Age of Reptiles or Age of Dinosaurs, extending from 245 million years to 65 million years ago.

metamorphosis A radical change in body form, physiology, and behavior during the growth and development of an organism.

metapopulation A set of partially isolated populations belonging to the same species. The populations are able to exchange individuals and recolonize sites in which the species has recently become extinct.

meter Basic metric unit of length; equal to 39.37 inches.

microbial mat A thin layer of bacteria and cyanobacteria ("blue-green algae") that forms on bare surfaces, sometimes secreting a carbonate base called a stromatolite; one of the earliest of ecosystems, still persisting in some modern environments such as shallow intertidal waters.

microbiology The scientific study of microscopic organisms, especially bacteria.

microevolution Evolutionary change of minor degree, such as an increase in size or body part, usually controlled by a relatively small number of genes. Cf. *macroevolution*.

millimeter One-thousandth of a meter; equal to 0.04 inch (25 millimeters is approximately 1 inch).

mitosis Cell division in which the chromosomes are exactly duplicated without loss in numbers. Cf. *meiosis*.

mollusk An animal belonging to the phylum Mollusca, such as a snail or clam.

mutation Broadly defined as any genetic change in an organism, either from an alteration of DNA composing individual genes or from a shift in the structure or number of chromosomes. Mutations form the new material for evolution.

mutualism Symbiosis in which both of the partner species benefit.

mycorrhiza A symbiotic association between fungi and plant roots.

Myers, Norman (1934–) British botanist and conservation biologist; identifier of hot spots; a major scholar in biodiversity studies.

natural selection The differential contribution of offspring to the next generation by various genetic types belonging to the same population; the mechanism of evolution proposed by Darwin. Distinguished from artificial selection, the same process but carried out with human guidance.

neo-Darwinism The modern study of the evolutionary process that assigns a central role to natural selection, the idea originally suggested by Darwin and now informed by substantial new knowledge from genetics, ecology, and other modern disciplines of biology.

niche A vague but useful term in ecology, meaning the place occupied by a species in its ecosystem—where it lives, what it eats, its foraging route, the season of its activity, and so on. In a more abstract sense, a niche is a potential place or role within a given ecosystem into which species may or may not have evolved.

nucleus In biology, the dense central body of the cell, surrounded by a double nuclear membrane and containing the chromosomes and genes.

Olson, Storrs L. (1944–) Curator in paleontology at the United States National Museum; authority on fossil birds and pioneer in the study of extinction of island bird faunas, especially those on Hawaii (conducted with Helen F. James).

paleontology The scientific study of fossils and all aspects of extinct life.

period A division of geological time just below the era. The Mesozoic era (Age of Reptiles), for example, is divided into three periods: Triassic, Jurassic, and Cretaceous.

Permian period Last period of the Paleozoic era, extending from 290 million years to 245 million years ago and closing with the greatest extinction spasm of all time.

Phanerozoic eon The major division of geological time during which most biodiversity has evolved and existed, 550 million years ago to the present.

phenotype   The observed traits of an organism, created by an interaction of the organism's genotype (hereditary material) and the environment in which it developed.

phylogeny   The evolutionary history of a particular group of organisms, such as antelopes or orchids, with special reference to the family tree of the species composing the group.

phylum   The highest level of classification below the kingdom. Examples: phylum Mollusca (snails, clams, octopuses) and phylum Pterophyta (ferns).

phytoplankton   The plant part of the plankton, as opposed to the zooplankton, the animal part.

Pimm, Stuart L. (1949–)   A theoretical population biologist and ecologist at the University of Tennessee; important scholar of the extinction process.

placental   Pertaining to a group of mammals characterized by use of a placenta to nurture the unborn young; comprises the great majority of living mammalian species. Cf. *marsupial*.

plankton   Organisms that float passively in the sea and air, comprising mostly microorganisms and small plants and animals.

Pleistocene epoch   The span of geological time preceding the Recent epoch, during which continental glaciers advanced and retreated and the human species evolved. The epoch began about 2.5 million years ago and closed with the end of the Ice Age 10,000 years ago.

polyploidy   The condition in a cell or an organism in which the number of complete sets of chromosomes (per cell) is greater than two. Polyploidy is a common means by which species multiply, especially in plants.

population   In biology, any group of organisms belonging to the same species at the same time and place.

prokaryote   An organism whose DNA is not enclosed in nuclear membranes; hence the cells of a prokaryotic organism do not contain a well-defined nucleus. Most prokaryotes are bacteria. Cf. *eukaryote*.

protistan   A member of the kingdom Protista (or Protoctista), comprising the protozoans, algae, and related forms.

protozoan   A member of a group of single-celled organisms, including amoebas and cilitates, usually placed in the kingdom Protista.

Quaternary period   The second and final period of the Cenozoic era, following the Tertiary period and including the Pleistocene and Recent epochs, thus extending from about 2.5 million years ago to the present.

rain forest   See *tropical rain forest*.

Raup, David M. (1933–) Professor of paleontology at the University of Chicago; a foremost contributor to the analysis of diversification and extinction.

Raven, Peter H. (1936–) Director of the Missouri Botanical Garden, St. Louis; authority on tropical botany, initiator of studies of plant diversity around the world.

restoration ecology The study of the structure and regeneration of plant and animal communities, aimed at the enlargement or restitution of threatened eco-systems.

seed bank A central facility for the storage of seeds representing a diversity of species and genetic strains, especially of domestic plants and their wild relatives.

selection See *natural selection*.

Sepkoski, J. John, Jr. (1948–) Professor of paleontology at the University of Chicago; with his collaborator David Raup a leading scholar of diversification and extinction.

sibling species Species so similar to each other as to be difficult to distinguish, at least by human observers.

sickle cell A hereditary condition prescribed by a change in a single gene that causes a warping of the red blood cells; induces anemia when present in double dose.

Simberloff, Daniel S. (1942–) Professor of ecology at Florida State University; a pioneer of island biogeography.

Soulé, Michael E. (1936–) Professor of environmental studies at the University of California, Santa Cruz; a founder of conservation biology.

source-sink model The hypothesis that species diversity, especially in tropical forests, builds up when restricted localities favorable to certain species allow them to produce a surplus of emigrants, hence to be a source of new individuals dispersing to less favorable sites nearby, the sinks.

Southwood, T. R. E. (1931–) Sir Richard Southwood, vice chancellor of Oxford University; a major contributor to the theory and measurement of diversity.

speciation The process of species formation: the full sequence of events leading to the splitting of one population of organisms into two or more populations reproductively isolated from one another.

species The basic unit of classification, consisting of a population or series of populations of closely related and similar organisms. In sexually reproducing organisms, the species is more narrowly defined by the biological-species concept: a population or series of populations of organisms that freely interbreed with one another in natural conditions but not with members of other species.

species-area curve See *area-species curve*.

species equilibrium The steady-state number of species, or biodiversity, found on an island or isolated patch of habitat due to a balance between the immigration of new species and the extinction of old residents. See also *island biogeographic theory*.

species selection The differential multiplication and extinction of species as a result of differences in certain traits possessed by the organisms belonging to the various species, and causing a spread of the favoring traits through the fauna or flora as a whole.

Stanley, Steven M. (1941–) Professor of paleobiology at Johns Hopkins University; authority on fossil invertebrates and a developer of the theory of species selection.

Steadman, David W. (1951–) Senior scientist in zoology at the New York State Museum; with Storrs Olson a leading researcher on the fossil history and extinction history of island birds, especially in Polynesia.

strip logging Removal of timber in narrow strips along contours, allowing rapid regrowth, sustained yield, and protection of the native fauna and flora.

subspecies Subdivision of a species. Usually defined narrowly as a geographical race: a population or series of populations occupying a discrete range and differing genetically from other geographical races of the same species. Cf. *host race*.

sustainable development The use of land and water to sustain production indefinitely without environmental deterioration, ideally without loss of native biodiversity.

symbiosis The living together of two or more species in a prolonged and intimate ecological relationship, such as the incorporation of algae and cyanobacteria within fungi to create lichens.

sympatric Occurring in the same place, as in the case of two species sharing parts of their geographic ranges.

sympatric speciation The splitting of an ancestral species into two daughter species without an intervening geographic barrier that first breaks the ancestral population into isolated populations.

systematics The scientific study of the diversity of life. Sometimes used synonymously with taxonomy to mean the procedures of pure classification and reconstruction of phylogeny (relationship among species); on other occasions it is used more broadly to cover all aspects of the origins and content of biodiversity.

taxonomy The science (and art) of the classification of organisms. See also *systematics*.

tephra Fragmented rock and ash ejected during a volcanic eruption.

Terborgh, John (1936–) Professor of biology at Duke University; botanist and zoologist, best known for long-term studies of bird and mammal ecology in rain forests.

Tertiary period The first period of the Cenozoic era, beginning with the end of the Mesozoic era (Age of Reptiles) 66 million years ago and closing with the start of the Pleistocene epoch about 2.5 million years ago; succeeded by the Quaternary period (Pleistocene plus Recent epochs).

Thornton, Ian W. B. (1926–) Professor of zoology at La Trobe University, Australia; leader of modern expeditions to Krakatau.

triploid A cell or organism possessing three complete sets of chromosomes.

trophic level A group of organisms that obtain their energy from the same part of the food web of a biological community. Examples: the primary producers, which are mostly plants, and herbivores, the animals that consume plants.

tropical rain forest Also known more technically as *tropical closed moist forest*: a forest with 200 centimeters of annual rainfall spread evenly enough through the year to support broad-leaved evergreen trees, typically arrayed in several irregular canopy layers dense enough to capture over 90 percent of the sunlight before it reaches the ground.

Vavilov center A region containing crop plants in both wild and cultivated states, hence a center of unusual genetic diversity in the species. Named after the Russian botanist Nikolai Vavilov.

vertebrate Any animal that possesses a backbone of bony segments enclosing the central nerve cord. There are five major groups of living vertebrates: fishes, amphibians (frogs, salamanders, and caecilians), reptiles, birds, and mammals.

Vrba, Elisabeth S. Professor of geology at Yale University; authority on the fossil history of African mammals and a leading contributor to species-selection theory.

Webb, S. David (1936–) Curator of vertebrate zoology at the Florida State Museum; major contributor and theorist of mammalian evolution in the New World.

World Continent fauna The dominant fauna that evolved in Africa, Europe, Asia, and North America (the World Continent) during the Cenozoic era (the past 66 million years). The several constituent land masses have been connected closely enough to allow periodic exchanges of species, illustrated especially by the mammals.

zooplankton The animal portion of the plankton, as opposed to the phytoplankton, the plant part.

Acknowledgments

In a sense, the preparation of this book began when I was a student at the University of Alabama in the late 1940s, working my way around red-clay gullies and toxic creeks in search of remnants of the natural environment. Though often discouraged, I always hoped that the world was in better shape elsewhere. My intellectual journey gathered momentum in 1953 during a field trip to Cuba's Sierra Trinidad, as I toiled up muddy roads in search of rain forest, past logging trucks on their way to Cienfuegos with the final fragments of the trees. I repeated that experience many times in other countries in later years. The world, I discovered, was *not* in better shape elsewhere. The book took solid form in my mind in September 1986 during the National Forum on Biodiversity, held in Washington, D.C., under the auspices of the National Academy of Sciences and the Smithsonian Institution. There I joined sixty biologists, economists, agricultural experts, and related professionals to consider, at long last comprehensively and with unaccustomed attention from the media, the full scale of global diversity as a central issue of the environment.

The study of biological diversity as it relates to contemporary human affairs is an eclectic subject just beginning to coalesce. In attempting this synthesis I have benefited from the advice and encouragement of colleagues in a wide range of disciplines. It is a pleasure to list them here, while exonerating them from the errors and omissions that, like hidden tripwires, may survive as the book goes into production (February 1992).

Larry D. Agenbroad (Quaternary megafauna extinction)

Peter S. Ashton (tropical floras)

Richard O. Bierregaard Jr. (rain-forest diversity)

Elizabeth Boo (ecotourism)

Keneth J. Boss (mollusks)

William H. Bossert (area-species modeling)

Bryan C. Clarke (mollusks)

Rita R. Colwell (microbiology)

Simon Conway Morris (Cambrian diversity)

Jared M. Diamond (extinction)

Eric Dinerstein (bioreserve analysis)

Victoria C. Drake (debt-for-nature)

Donald A. Falk (U.S. plants)

Richard T. T. Forman (landscape, · policy analysis)

Charles H. W. Foster (bioregionalism)

David G. Furth (beetle diversity)

Douglas J. Futuyma (evolutionary theory)

Alwyn H. Gentry (tropical floras)

Thomas J. Givnish (butterflies, metapopulations)

Jostein Goksøyr (bacterial diversity)

Jerry Harrison (world natural reserves)

Karsten E. Hartel (ichthyology)

Gary S. Hartshorn (forestry, public policy)

Michael Huben (mites)

Helen F. James (Hawaiian birds, extinction)

David P. Janos (mycorrhizal fungi)

Robert E. Jenkins (biodiversity inventories)

Carl F. Jordan (tropical forestry)

Laurent Keller (entomology)

Andrew H. Knoll (geological history of life)

Russell Lande (genetic diversity)

Robert J. Lavenberg (sharks)

Karel F. Liem (ichthyology)

Hans Löhrl (European birds)

Jane Lubchenco (marine ecosystems)

Ariel E. Lugo (tropical forests, extinction)

Denis H. Lynn (protozoan diversity)

David R. Maddison (genetics, systematics)

Michael A. Mares (extinction)

Ernst Mayr (species formation)

Kenton R. Miller (conservation and public policy)

Russell A. Mittermeier (conservation biology)

Gary Morgan (Cenozoic mammals)

Norman Myers (deforestation, extinction)

Storrs L. Olson (Hawaiian birds, extinction)

Michael O'Neal (plant extinctions)

Raymond A. Paynter Jr. (ornithology)

Tila M. Pérez (mites)

David Pilbeam (human evolution)

Mark J. Plotkin (economic botany, ethnobotany)

James F. Quinn (mammalian extinctions)

Katherine Ralls (cetacean diversity)

David M. Raup (paleontology, extinction)

Peter H. Raven (plant diversity, ethnobotany)

Jamie Resor (economics, foreign aid)

Michael H. Robinson (zoological parks)

Gustavo A. Romero (orchids)

Jose P. O. Rosado (reptiles)

Cristián Samper K. (South American forests)

G. Allan Samuelson (beetles)

J. William Schopf (geological history of life)

Richard E. Schultes (ethnobotany)

Raymond Siever (Cenozoic era)

Daniel S. Simberloff (extinction)

Tom Simkin (Krakatau)

Otto T. Solbrig (plant evolution)

Andrew Spielman (mosquitoes)

Steven M. Stanley (geological history of life, evolutionary theory)

David W. Steadman (Pacific birds, extinction)

Martin H. Steinberg (sickle-cell anemia)

Peter F. Stevens (plant diversity)

Roger D. Stone (conservation and policy analysis)

Nigel E. Stork (arthropod diversity)

Jorgen B. Thomsen (parrots)

Ian W. B. Thornton (Krakatau)

Barry D. Valentine (beetle diversity)

Noel D. Vietmeyer (economic botany)

Elisabeth S. Vrba (evolutionary theory, mammalian evolution)

S. David Webb (mammalian evolution)

T. C. Whitmore (tropical forestry, extinction)

Delbert Wiens (plant evolution)

Irene K. Wilson (editorial process)

As I have for all my books and articles back to 1966, I ackowledge the meticulous and invaluable work of Kathleen M. Horton in bibliographical research and preparation of the manuscript. It has also been a pleasure to work with Sarah Landry, George Ward, and Amy Bartlett Wright as they prepared the illustrations, and with Mark Moffett and Darlyne Murawski in selecting photographs from their fine natural-history collections.

Several of the line drawings were derived from previously published work by other authors. **Chapter 8:** The structure of leaf litter and humus in deciduous forest, together with data on the distribution of arthropods living in these strata, is based loosely on figures and data in Gerhard Eisenbeis and Wilfried Wichard, *Atlas zur Biologie der Bodenarthropoden* (Stuttgart: Gustav Fischer, 1985). The species-scape, representing the amount of species diversity in each group by the size of the representative organism, was introduced by Quentin D. Wheeler, "Insect Diversity and Cladistic Constraints," *Annals of the Entomological Society of America*, 83(6):1031–1047 (1990). The figure used here is an interpretation by Amy Bartlett Wright. **Chapter 9:** The idea of representing ecosystem assembly rules as a jigsaw puzzle originated with James A. Drake, "Communities as Assembled Structures: Do Rules Govern Pattern?," *Trends in Ecology and Evolution*, 5(5):159–164 (1990). **Chapter 9:** The depiction of a driver-ant swarm was prepared by Katherine Brown-Wing and published in my book *Success and Dominance in Ecosystems: The Case of the Social Insects* (Oldendorf/Luhe, Germany: Ecology Institute, 1990). **Chapter 10:** The cross section of a microbial mat is derived from David J. Des Marais, "Microbial Mats and the Early Evolution of Life," *Trends in Ecology and Evolution*, 5(5):140–144 (1990). The figure of the rise of local plant diversity is based on a diagram by Andrew H. Knoll, "Patterns of Change in Plant Communities through Geological Time," in Jared M. Diamond and Ted J. Case, eds., *Community Ecology* (New York: Harper and Row, 1986), pp. 126–141. **Chapter 11:** The map of the California checkerspot metapopulation is modified from a map presented by Susan Harrison, Dennis D. Murphy, and Paul R. Ehrlich, "Distribution of the Bay Checkerspot Butterfly, *Euphydryas editha bayensis*: Evidence for a Metapopulation Model," *American Naturalist*, 132(3):360–382 (1988). **Chapter 12:** The diagram of mass extinction of mammalian megafaunas during the past 100,000 years is modified from one by Paul S. Martin, "Prehistoric Overkill: The Global Model," in P. S. Martin and Richard G. Klein, eds., *Quaternary Extinctions: A Prehistoric Revolution* (Tucson: University of Arizona Press, 1984), pp. 354–403. The hot-spot maps are based on the publications of Norman Myers, as cited in my notes. The historical map of deforestation in Ecuador is derived from an illustration provided by Calaway H. Dodson and Alwyn H. Gentry, "Biological Extinction in Western Ecuador," *Annals of the Missouri Botanical Garden*, 78(2):273–295 (1991). **Chapter 14:** The map used in geographical systems analysis is modified from one presented by J. Michael Scott, as cited in my notes. The diagram of strip logging is modified slightly from a figure by Carl F. Jordan, "Amazon Rain Forests," *American Scientist*, 70:394–401 (1982), used with the permission of the author and the publisher, Sigma Xi.

Index

deer, 120
deforestation (logging, clear-cutting), 203, 232, 244, 246, 247, 254–7, 259, 269; of rain forest, 261, 263, 264–5, 266, 291–2, 293–4, 309–10, 311–12; rates, 262–3. *See also* area/species curve; habitat, destruction; slash-and-burn agriculture
demes. *See* genetic drift; inbreeding
Devonian period, 27, 29, 178
DeVries, Philip J., 190
Diamond, Jared M., 159, 215, 239–41, 267
dinosaurs, 23, 25, 26, 27, 83, 112, 115, 117, 178
Dioum, Baba, 306
diploid phase, 63–4, 65, 66, 70
diseases, 86, 241, 266; genetic, 73; of plants, 289
dispersal ability, 83, 97, 98, 259
distance effect, 211, 212
distribution, geographic, 266, 268
DNA, 43, 70, 85, 305; in fishes, 101, 103; in bacteria, 135–6, 137; base pairs, 135; multiple species and, 142; human genome project, 151; in higher organisms, 174. *See also* chromosomes; genes
Docters van Leeuwen, W. M., 20
dodos, 241
Dodson, Calaway, 206, 231, 232
dogs, 120, 155, 234
dominance, 121–2. *See also* genes, dominant/recessive
Dominican Republic, 322
Down's syndrome, 64, 73
dragonflies, 18, 21, 87
ducks, 234, 235–6
dynastic succession, 5, 111, 121, 180, 204, 328

eagles, 34, 234, 235, 247, 257
East Africa, 104; Great Lakes, 10, 103
echidna (mammal), 113
echinoderms, 328
ecological displacement, 160, 161
ecological economics, 293
ecological release, 98, 160
ecology, 83, 85, 153; community, 153–4, 157, 168; assembly of species and, 158–60; source-sink model of, 191; laws of, 207–8
ecosystems, 35–6, 331–2; Black Forest, 147–9; extinctions, 246; endangered

(list), 248–58; importance to biodiversity, 271, 294–5; failure of, 295; economic potential of, 305–8; preservation of, 319. *See also* community ecosystems; conservation; microbial mats
ecotourism, 293
Ecuador, 206, 252, 253, 322, 330; Centinela ridge, 231–2, 252, 265, 266
Edwards, Hugh, 109
Eisner, Thomas, 290, 306
elasticity vs. competition, 162, 163
Eldredge, Niles, 80
electrons, 41–2
electrophoresis, 150–51
elephant birds, 238–9
elephants, 156
El Niño, 258, 259
embryonic development, 85
Emmel, Thomas, 185
endangered and threatened species (general discussion), 58, 216, 245–6, 247–8, 259, 299, 301; birds, 216–18, 223, 228–30; trees, 219–20; reptiles, 220; hot spots of, 249–58, *250*, *251*, 299, 320, 321
Endangered Species Act, U.S., 61, 320, 321
energy, solar, 34–5, 187–8, 189, 191, 194, 260
Energy-Stability-Area (ESA) theory of biodiversity, 187–8
environment: effects on evolution, 146, 175, 193; oxygen in, 175–6; aerobic, 176, 183; specialization of species and, 190; sink and source areas, 191; extinction and, 206. *See also* area/species curve; climate; habitat
enzymes, 70, 150, 151
Eocene period, 194
epiphylls, 165–6
epiphytes, 10, 192–3, 294–5
Erwin, Terry L., 129, 132, 186
eukaryotes, 174, 176
evergreens, 26
evolution, 58, 61, 62, 67, 69, 76, 146, 174–5, 177; vertical, 47, 48, 49, 53, 56, 328; human, 48–50, 61, 153, 174, 199; subspecies and, 62; of populations, 69; by natural selection, 69–70, 74–5, 76, 81, 174, 175; genetic, 77, 85, 95; phenotypic, 78; rates, 80–81, 82; of traits, 163–4. *See also* macroevolution; microevolution; speciation

READ MORE IN PENGUIN

In every corner of the world, on every subject under the sun, Penguin represents quality and variety – the very best in publishing today.

For complete information about books available from Penguin – including Puffins, Penguin Classics and Arkana – and how to order them, write to us at the appropriate address below. Please note that for copyright reasons the selection of books varies from country to country.

In the United Kingdom: Please write to *Dept. EP, Penguin Books Ltd, Bath Road, Harmondsworth, West Drayton, Middlesex UB7 ODA*

In the United States: Please write to *Consumer Sales, Penguin Putnam Inc., P.O. Box 12289 Dept. B, Newark, New Jersey 07101-5289.* VISA and MasterCard holders call 1-800-788-6262 to order Penguin titles

In Canada: Please write to *Penguin Books Canada Ltd, 10 Alcorn Avenue, Suite 300, Toronto, Ontario M4V 3B2*

In Australia: Please write to *Penguin Books Australia Ltd, P.O. Box 257, Ringwood, Victoria 3134*

In New Zealand: Please write to *Penguin Books (NZ) Ltd, Private Bag 102902, North Shore Mail Centre, Auckland 10*

In India: Please write to *Penguin Books India Pvt Ltd, 11 Community Centre, Panchsheel Park, New Delhi 110017*

In the Netherlands: Please write to *Penguin Books Netherlands bv, Postbus 3507, NL-1001 AH Amsterdam*

In Germany: Please write to *Penguin Books Deutschland GmbH, Metzlerstrasse 26, 60594 Frankfurt am Main*

In Spain: Please write to *Penguin Books S. A., Bravo Murillo 19, 1° B, 28015 Madrid*

In Italy: Please write to *Penguin Italia s.r.l., Via Benedetto Croce 2, 20094 Corsico, Milano*

In France: Please write to *Penguin France, Le Carré Wilson, 62 rue Benjamin Baillaud, 31500 Toulouse*

In Japan: Please write to *Penguin Books Japan Ltd, Kaneko Building, 2-3-25 Koraku, Bunkyo-Ku, Tokyo 112*

In South Africa: Please write to *Penguin Books South Africa (Pty) Ltd, Private Bag X14, Parkview, 2122 Johannesburg*

BY THE SAME AUTHOR

On Human Nature

What is the basis of aggression, sex, altruism and religion?

Human nature is genetically determined but to what extent? And does the acceptance of evolutionary theory as the basis for social behaviour diminish our humanity? In this controversial book, Edward Wilson argues that science offers us true liberation.

'A work of high intellectual daring ... here is an accomplished biologist explaining, in notably clear and unprevaricating language, what he thinks his subject now has to offer to the understanding of man and society' *New Republic*

'His vision is a liberating one and a reader of this splendid book comes away with a sense of the kinship that exists among the people, animals and insects that share the planet' *New Yorker*

In Search of Nature

A collection of essays of 'elegance, lucidity and breadth ... A teeming anthill of beautiful ideas about the links between "wild nature" and "human nature"' *Independent*

'A graceful, eloquent, playful and wise introduction to many of the subjects he has studied during his long and distinguished career in science. It is [also] a book that will surely whet the appetites of those who have not read his longer work' Richard Bernstein, *The New York Times*

'An elegant and brilliant collection of essays ... Although Wilson is a dedicated scientist, he writes with lucid confidence about the ethical and aesthetic dimensions of our lives' Maggie Gee, *Daily Telegraph*

'Excellent ... Those who have not read his ecological masterpiece, *The Diversity of Life*, will be able to sample for the first time, in miniature, the passion and command with which he describes the frail beauty and interconnectedness of living systems. The final essay describes the appalling extent of our impact on the biosphere, but it also offers wise and slender hope' Ian McEwan, *Financial Times*